Fuels, Energy and the Environment

Fuels, Energy and the Environment

Edited by **Kurt Marcel**

New York

Published by Syrawood Publishing House,
750 Third Avenue, 9th Floor,
New York, NY 10017, USA
www.syrawoodpublishinghouse.com

Fuels, Energy and the Environment
Edited by Kurt Marcel

International Standard Book Number: 978-1-68286-027-4 (Hardback)

Printed in the United States of America.

Contents

Preface

This book explores all the important aspects of fuel resources and energy production while considering their environmental impacts. Some of the vital concepts related to this field are discussed in detail such as characteristics of fuels, alternative energy resources, negative impacts of fuel consumption on environment, energy conversion and recycling, global warming, etc. Different approaches, evaluations, methodologies and advanced studies in this field have been included in this book. Scientists and students engaged in this field will find this book full of crucial and unexplored concepts.

The information shared in this book is based on empirical researches made by veterans in this field of study. The elaborative information provided in this book will help the readers further their scope of knowledge leading to advancements in this field.

Finally, I would like to thank my fellow researchers who gave constructive feedback and my family members who supported me at every step of my research.

Editor

1

A comparative study of biogas production using plantain/almond leaves and pig dung, and applications

EZEKOYE, V. A.

Department of Physics and Astronomy, University of Nigeria, Nsukka, Nigeria.

Plantain/almond leaves and pig dung were used as substrates in anaerobic biodigester for producing biogas by batch operation method within the mesophilic temperature range of 20.0 to 31.0°C. The study was carried out to compare biogas production potential from plantain/almond leaves and pig dung wastes. The cumulative biogas produced from the plantain/almond leaves was 220.5 L while the cumulative biogas from the pig dung was 882.5 L. Orsat apparatus was used to analysize the gas produced. The methane component of gas from pig dung was 70.2% while that for plantain/almond leaves with algae was 72.7%. The biogas from the almond/plantain leaves became combustible on sixteenth day while the biogas from the pig dung was combustible on fourteenth day. Results showed that pig dung produced more biogas than the almond/plantain leaves within the same period.

Key words: Mesophilic, anaerobically, almond/plantain, combustible, algae.

INTRODUCTION

Biogas originates from the process of bio-degradation of organic material under anaerobic (without air) conditions. In the absence of oxygen, anaerobic bacteria decompose organic matter and produce a gas mainly composed of methane (60%) and carbon dioxide called biogas. This gas can be compared to natural gas, which is 99% methane. Biogas is a 'sour gas' in that it contains impurities which form acidic combustion products (Boyd, 2000). In 2008, about 19% of global final energy consumption came from traditional biomass, which is mainly used for heating and 3.2% from hydroelectricity (Ross, 1996). In other words, animal and agricultural wastes constitute a high proportion of biomass and their utilization is important for economical and environmental aspects possessing suitable climatic and ecological conditions (Oktay, 2006).

Biogas as a renewable energy source could be a relative means of solving the problems of rising energy prices, waste treatment/management and creating sustainable development. Anaerobic digestion has been recognized as an effective way to partially solve the growing concern of solid waste management by reducing the weight of organic waste, as well as controlling the soil pathogens (Ieshita and Sen, 2013). The production of biogas involves a complex biochemical reaction that takes place under anaerobic conditions in the presence of highly sensitive microbiological catalyst that are mainly bacteria. Biogas technology has in the recent times also been viewed as a very good source of sustainable waste treatment/management, as disposal of wastes has become a major problem especially to the third world countries. The effluent of this process is a residue rich in essential inorganic elements needed for healthy plant growth known as bio-fertilizer which when applied to the soil enriches it with no detrimental effects on the environment (Energy Commission of Nigeria, 1998). The raw materials used in many places for the gas production include agricultural wastes and animal manures which are called Biomass. The greatest potential to increase the use of biomass in energy production seems to lie in forest residues and other biomass resources e.g. agro biomass and fruit biomass (Kramer, 2002; Eija et al., 2007). The plantain/almond leaves, algae from sewage and pig dung used in this research work are readily available in our environment. These biomasses are highly degradable in nature. However the rate of efficiency of digestion of feedstock depends on its physical and chemical form.

Plant materials especially crop residues are more difficult to digest than animal manures. This is because hydrolysis of cellulose materials of crop residues is a slow process and can be a major determining step in anaerobic digestion process. Raw plant materials are bound up in plant cells usually strengthened with cellulose and lignin, which are difficult to digest. In order to let the bacteria reach the more digestible foods, the plant material must be broken down (Kozo et al., 1996; Fulford,1998). Addition of bacterial seed or inoculum accelerates biogas generation and also reduced the lag time (number of days required for biogas production to start). The inoculum is known to enrich the bacterial of the digester which will enhance their action on the substrate and hence on the quantity as well as quality of the biogas generated. Biogas microbes play different roles in the conversion of the organic substances, according to their nutrient requirements (Maishanu and Maishanu, 1998). Maximum methane yield requires adequate and efficient nutrient supply for micro-organisms in the digester (Thomas et al., 2006).

Plantain leaves (*Musa paradisiaca*) are readily available in the tropics, while the almond tree (*Terminalia catappa*) popularly known as fruit or umbrella tree are found around the school premises (University of Nigeria, Nsukka), their wastes constitutes a nuisance in the school environment. However, these wastes can be converted to a renewable energy source. This experiment was carried out to find out the effect of parameters such as pH level, temperature, retention time and the biodegradability of these biomass. When the temperature is high, the activity of bacteria is simply more vigorous, so that the fermentation period becomes shorter. When the temperature is low digestion is slow, and the fermentation period is longer (Goodger, 1980).

The anaerobic digestion processes is carried out by a delicately balanced population of various bacteria. These bacteria can be very sensitive to change in their environment. Temperature is a prime example. It has been determined that 35°C is an ideal temperature for anaerobic digestion (Kucha and Itodo, 1998).

AIMS AND OBJECTIVES

This paper determined the effect of environmental and operational parameters on the fermentation/production rate of biogas from organic waste. It also studied the extent to which plantain leaves/almond leaves; algae from sewage pond and pig dung generate gas, and transformed the organic wastes into high quality fertilizer.

The chemistry of biogas production

Generally, the production of this biogas involves a complex biochemical reaction that takes place under anaerobic conditions in the presence of highly sensitive microbiological catalysts that are mainly bacteria. The major products of this reaction are methane (CH_4) and carbon-dioxide (CO_2) (Hashimoto et al., 1980). The anaerobic biological conversion of organic matter occurs in 3 stages Figure 1).

Factors affecting biogas production

Many factors that are affecting the fermentation process of organic substances under anaerobic condition include:

i. Temperature
ii. Nature of raw material
iii. pH of slurry and alkalinity
iv. Stirring
v. Carbon/nitrogen ratio (C/N)
vi. Nutrients addition
vii. Retention time
viii. Total solid
ix. Volatile solid x. Mixing
xi. Inhibition.

The length of fermentation period is dependent on temperature. Keeping the digestion chamber at nearly constant temperature is important.

Smaller particles would provide large surface area for adsorbing the substrate that would result in increased microbial activity and hence increased gas production. pH is an important parameter affecting the growth of microbes during anaerobic fermentation. Stirring of digester content needs to be done to ensure intimate contact between microorganisms and substrate which ultimately improves digestion process. It is generally found that during anaerobic digestion microorganisms utilize carbon 25 to 30 times faster than nitrogen. Thus to meet this requirement, microbes need a 20 – 30: 1 ratio of C to N with the largest percentage of the carbon being readily degradable (Bardiya, and Gaur, 1997; Malik and Tauro, 1995). Addition of inoculum tends to improve both

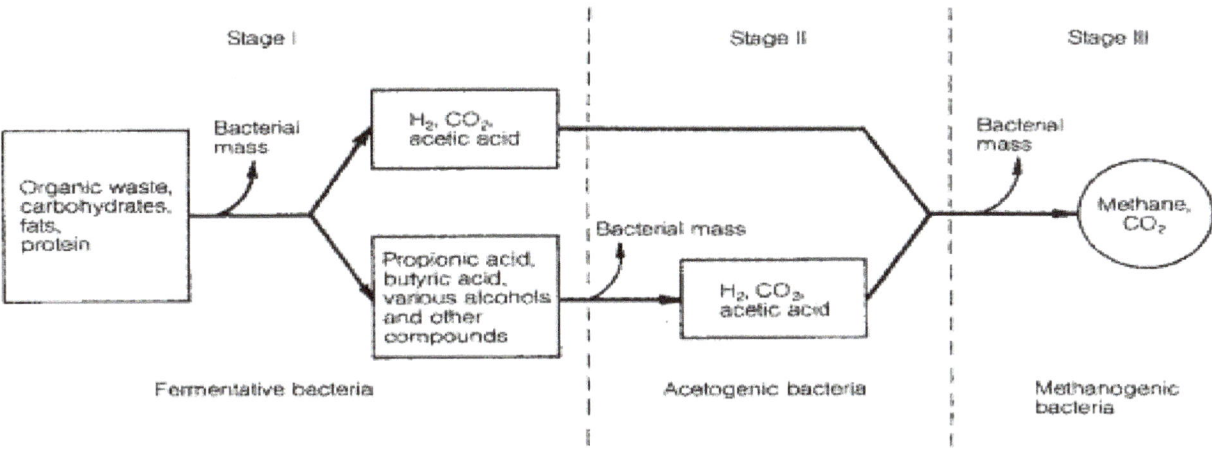

Figure 1. The three-stage anaerobic fermentation of biomass (Hoerz et al., 2008).

Figure 2. Digester used in the experiment.

the gas yield and methane content in biogas. It is possible to increase gas yield and reduce retention period by addition of inoculums (Dangaggo et al., 1996; Kanwar and Guleri, 1995; Kotsyurbenko et al., 1993).Retention time is the average time spent by the input slurry inside the digester before it comes out. Total solid concentration (TS%) is a measure of the dilution ratio of the input material. Some organic materials which contain lignins do not decompose easily. The lignin content even if quite low will decrease the rate of digestion of carbohydrates (cellulose and hemicellulose).

MATERIAL AND METHODS

Apparatus

The biodigester that was used for this study is 0.00147 m^3 in capacity and made of galvanized metal sheet material (Figure 2). Batch operation method was used. Pre-decayed plantain/almond leaves and pig dung were used for this study. The plantain/almond leaves were obtained from University of Nigeria, Nsukka premises. Pig dung was collected from veterinary farm, University of Nigeria, Nsukka.

Characteristics of the wastes used in this study

The slurry of plantain/almond leaves and algae from sewage pond (P/A and A) was obtained by diluting the solid wastes with water in the mass ratio of 1 : 4 (waste : water). This implies that a total of 23 kg of plantain/almond leaves and algae(P/A and A) from sewage pond was mixed with 92 kg of water giving a total mass of 115 kg of slurry. These were measured using a weighing balance of 0 to 50 kg ranges. Both waste and water were thoroughly mixed in a small drum ensuring that no solid (hard) material, which was not decomposable, was present before introducing the mixture into the digester. Due to the high lignin which is non-degradable material, the waste (leaves) was seeded with algae water from sewage (inoculums) which boosted the rate of gas production. The waste

Table 1. Ratio and the temperature of the samples.

Waste	Mixing ratio	Quantity of waste and water (kg)	Ambient temperature range (°C)	Slurry temp range (°C)	Total volume biogas produced (L)
P/A and A	1 : 4	115	27.0 - 57.0	33.0 - 38.5	220.5
Pig dung	1 : 2	120	26.0 - 31.0	31.5 - 38.0	882.5

occupied about 74% by volume of the digester, this is the loading rate. The remaining part was left for gas collection. After introducing the waste, all openings were closed. After 1 day from charging, biogas generation commenced. The gas became combustible from the sixteenth day to the end of digestion. The total volume of gas produced was 96.0 L. The biogas became combustible as from the fourth day to the end of digestion as show in Table 1.

The initially dry pig dung was pulverized and every hard stones were removed. It was dissolved in water in the ratio of 1 : 2. A total of 40 kg of waste was mixed with 80 kg of water in batches in a small drum after which it was introduced into the digester (Figure 2) while keeping gas outlet open to exhaust trapped air. The fermentation and biogas production started after 1 day. The gas became combustible from the fourteenth day to the end of digestion. Batch operation was adopted.

RESULTS

Proximate analysis

The sample were analyzed for Ash, PH, total solid, volatile solid, moisture, phosphorus, fiber contents using the method of Association of Official and Analytical Chemistry (Sharma, 2002). Protein, fat contents were determined using Micro-Kjeldahl method. Carbon content was determined by the method of Walkey and Polpraset (Polpraset and Bitton, 1989 ; Aubart and Farinet, 1983).

The growth and catabolism of microbes need various kinds of nutrients especially elements of carbon, nitrogen and phosphorus. For high quality of methane, carbon is required for building of the cell structure of the methanogenic bacteria. Specific group of bacteria always consume carbon and nitrogen elements in a fixed proportion. From Tables 3 and 4, it was discovered that the value of protein/nitrogen, volatile solid, total solid and carbon in both samples decreased in percentage after digestion. Some of them were used up by the bacteria. The percentage of phosphorous was increased at end of digestion. The biofertilizer was rich with phosphorous and nitrogen which produces better yield when apply on farm land.

Storage of biogas

A gas compressor is a mechanical device that increases the pressure of a gas by reducing its volume. The capacity of the compressor used was 1/5 horse-power. Each cylinder was able to compress biogas of 1.2 bars of pressure. The biogas from pig dung and A/P and A became combustible on the fourteenth and sixteenth days respectively and it burned with blue flame Each sample of biogas produced was analyzed using Orsat apparatus. The measuring principle of Orsat apparatus is the measurement of the reduction which occurs when individual constituents of a gas are removed separately by absorption in liquid reagents (Ezekoye and Okeke, 2006).

The daily ambient temperature and slurry temperature for almond/plantain leaves with algae and pig dung were shown in Figures 3 and 4. Almond/plantain leaves with algae recorded the highest temperature range of 27.0 – 37.0°C and the 2 wastes produced biogas within the mesophilic range of temperature (Itodo and Philips, 2002; Goodger, 1980). Orsat apparatus was used to analysis the gas produced. The methane component of gas from pig dung was 70.2% and for almond/plantain leaves with algae was 72.7%.

The daily volumes of biogas yield of the almond/plantain leaves with algae and pig dung were shown in Figure 5. Careful examination of the curves shows that the pig dung generated the highest gas from the first day to the thirty-seventh day. It was followed by almond/plantain leaves with algae which produced the highest between twelfth and thirteenth days.

On the other hand, the pig dung slurry produced combustible gas on the fourteenth day and almond/plantain leaves with algae (A/P and A) the sixteenth day (Table 2). The cumulative biogas yields of the sample are compared in Figure 6. The pig dung gave the highest yield 882.5 L. The almond/plantain leaves with algae produced 220.5 L.

DISCUSSION

For optimum functioning, the anaerobic micro-organisms require a neutral environment. The optimal pH was found to be 6.00 to 7.5 for pig dung, 6.00 to 7.3 for almond/plantain leaves with algae (Tables 3 and 4). Both acid and methane forming bacteria could not survive the pH values of 4 and 10.

Different wastes were used in feeding the digester to find out which one produced more gas. It was found that organic waste which is easily digestible produced more gas. Material with high lignocellulose produces less amount of gas. Carbon, which constitutes the basic frame of all organic substrates, provides microbes with the energy required for their living activities, and is the source

Table 2. Days of flammability and total biogas produced.

Waste	Flammable time (day)	Retention time (days)	Total biogas produced (L)
P/A&A	16	43	220.5
Pig dung	14	43	882.5

Table 3. Proximate analysis for almond/plantain leaves and algae.

Components	Before digestion	After digestion
Protein	1.62%	1.31%
Fats	2.05%	0.86%
pH	6.50	7.30
Moisture	89.32%	65.97%
Ash	0.50%	2.0%
Fiber	5.75%	Trace
Carbohydrate	24.12%	6.50%
Total solid (TS)	14.83%1	13.00%
Volatile solid (VS)	11.50%	6.67%
Carbon	0.29%	0.19%
Phosphorus	1.38ppm	7.71ppm

PPM; Part per million.

Table 4. Proximate analysis for pig dung.

Components	Before digestion	After digestion
Protein	6.63%	1.06%
Fats	0.78%	1.33%
pH	6.00	7.50
Moisture	85.67%	79.92%
Ash	1.90%	0.44%
Fiber	Trace	Trace
Carbohydrate	10.59%	11.66%
Total solid (TS)	16.00%	12.70%
Volatile solid (VS)	7.70%	5.00%
Carbon	1.09%	0.21%
Phosphorus	3.93ppm	5.76ppm

PPM; Part per million.

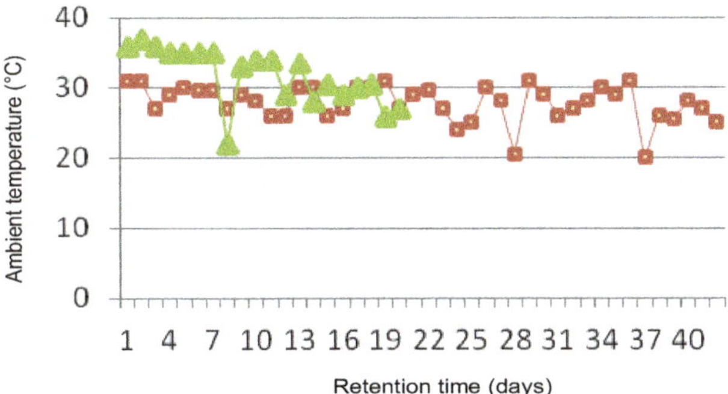

Figure 3. Change in ambient temperature during fermentation.

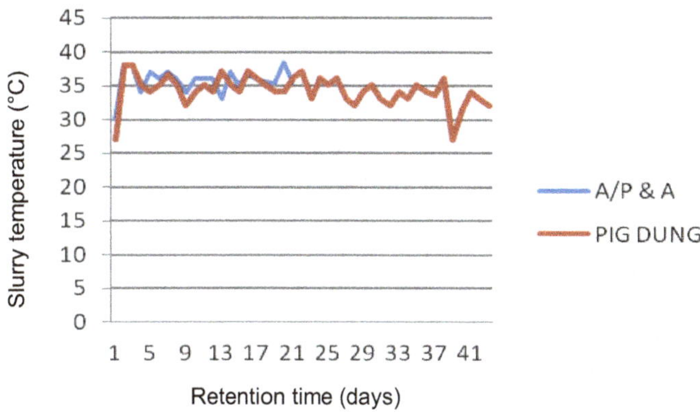

Figure 4. Change in slurry temperature during fermentation.

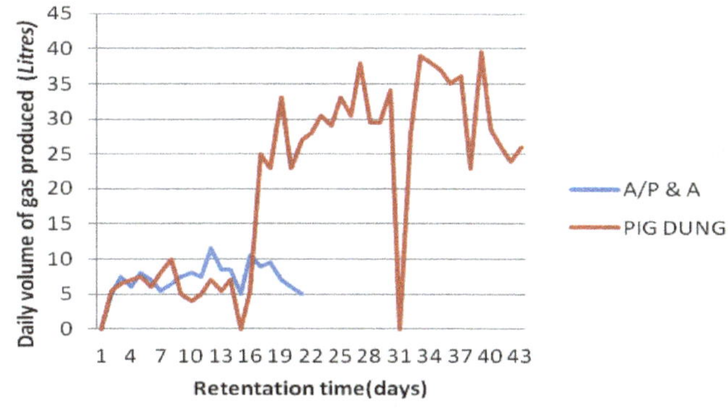

Figure 5. Volume of gas produced by A/P and A and pig dung leaves during fermentation.

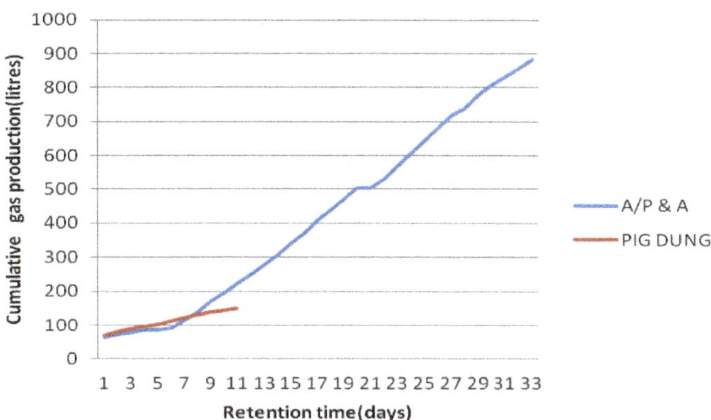

Figure 6. Daily cumulative gas produced by almond/plantain leaves.

Table 5. Percentage of the component of biogas from 2 different wastes using Orsat apparatus.

Waste	Carbon dioxide (CO_2) (%)	Hydrogen sulphide H_2S (%)	Carbon monoxide (CO) (%)	Methane and other components (%)
Pig dung	24.9	1.2	0.5	70.2
A/P&A	17.5	0.8	9.0	72.7

for the formation of biogas. In biogas production, nitrogen provides methanogenic bacteria with ammonia, which is the source of nitrogen for the composition of living matter of new cells. The carbon/nitrogen ratio for pig dung was 8 : 1, for almond/plantain leaves with algae, it was 6 : 1

The slurry should not be too thick nor too dilute. In this experiment the dilution ratios used for pig dung and Almond/Plantain leaves with Algae were 1 : 2 and 1 : 4 respectively. Stirring is necessary for increased gas production. Gas production was found to be low at pH 4 and 9. When the slurry was stirred once in a day, there was an increase in gas production. There was also a drop in gas production, when stirring was completely omitted due to scum formation. The slurry was seeded with bacteria (algae from sewage pond). After the analysis of the slurry, it was discovered that there was an increase in the percentage content of nitrogen, potassium, protein and phosphorus (N P K) after digestion. This shows that the sludge is a better fertilizer to the soil (Tables 3 and 4) (Chemical Land 21, 2006). The importance of inoculum is that it fastened the establishment of anaerobic micro flora and thus eliminates the unnecessary lag phase observed during the start-up.

Table 2 shows that pig dung gives the shortest flammability time of fourteenth (14th) day followed by A/P and A sixteenth (16th) day. This shows that the day the biogas started burning should be in the order: pig dung < A/P and A.

Figure 7 shows the cumulative biogas generation during the fermentation. This connotes that the biogas yield should be in the order: pig dung > A/P and A. This means that pig dung produced the highest biogas. Table 5 shows the composition of biogas produced. Biogas generation has the following applications: production of energy (heat, light, electricity); transformation of organic wastes into high quality fertilizer; reduction of workload, mainly for women, in firewood collection and cooking; and environmental advantages through protection of forests, soil, water and air.

Conclusion

The result of this study showed that almond leaves, plantain leaves, pig dung, and algae from sewage which can cause pollution in the environment can be a useful source of energy by subjecting it to anaerobic digestion for biogas production. The microorganism associated with the fermentation of plantain leaves/almond leaves mainly originated from the inoculum used. The addition of inoculums (algae from sewage pond) to the leaves (almond/plantain) was found to enhance gas production. The pig dung anaerobically digested showed the highest yield for methane production. It is also expected that this will be a source of waste management and pollution control.

ACKNOWLEDGMENT

My sincere appreciation goes to Mr. Emeka of National Center for Energy Research and Development, University of Nigeria, Nsukka who truly helped me during the practical work.

REFERENCES

Aubart C, Farinet JL (1983). Anaerobic digestion of pig and cattle manure in large-scale digesters and power production from biogas. Symp. Pap. Energy Biomass Wastes, pp. 741–766.

Bardiya N, Gaur AC (1997). Effects of carbon and nitrogen ratio on rice straw biomethanation. J. Rural Energy 4:1–16.

Boyd R (2000) .Internallising Environmental Benefits of Anaerobic Digestion of Pig Slurry In Norfolk.University of East Anglia ENV 4:6-10

Chemical Land 21 (2006). NPK, fertilizer E:/NPK (fertilizer) mht.

Dangaggo SM, Aliya M, Atiku AT (1996).The effect of seeding with Bacteria on biogas production rate. Renewable Energy— An Int. J. 9:1045–1048.

Eija A, Antti H, Terhi L , Pirkko V (2007). Biomass Fuel Trade in Europe. Summary report V T T-R-03508-07.

Energy Commission of Nigeria (1998). Rural renewable energy needs and five Supply technologies, pp. 40–42.

Ezekoye VA, Okeke CE (2006). Design, Construction, and Performance Evaluation of Plastic Biodigester and the Storage of Biogas.The Pacific J. Sci.Tech. 7:176-184.

Fulford D (1998). Running Biogas Programing. A handbook. 'How Biogas Works'. Intermediate Technology Publication. 103-05. Southampton Row, London. WCB, UK, pp. 30-34.

Goodger EM (1980). Agricultural Biogas Casebook. Great lakes Regional Biomass Energy Programme Council of Great Lakes Governors.

Hashimoto AG, Chen YR, Vare VI (1980). Theoretical aspects of biogas production.state of art. In proceedings livestock waste: A Renewable Resource 4th International Symposium on livestock wastes. ASAE,pp. 86-91.

Hoerz T,Kramer P, Klingler B, Kellner C, Thomas W, Klopotek FV,Krieg A, Euler H (2008) Biogas Digest Volume i, ii ,iii Biogas Basics. Isat Information and Advisory Service on Appropriate Technology gtz

Itodo IN, Philips TK (2002). Covering Materials for Anaerobic Digesters Producing Biogas.Nig. J. Rene Ene. 10:48-52.

Kanwar SS, Guleri RL (1995). Biogas production from mixture of poultry litter and cattle dung with an acclimatized inoculum. Biogas Forum I. 60:21–23.

Kotsyurbenko OR, Nozhevnikova AN, Kalyuzhnyy S, Zavarzin G (1993). Methanogenic digestion of cattle manure at low temperature.Mikrobiologiya. 62:761–771.

Kozo I ,Hisajima S, Darryl RJ (1996). Utilization of agricultural wastes for biogas Production in Indonesia,: In traditional technology for environmental conservation and sustainable development in Asia Pacific Region. 9th Ed, pp. 137-138.

Kramer JM (2002). Agricultural Diogas Casebook, Resource Strategies,INC,Great Lakes Regional Biomass Energy Program Council Of Great Lakes Governors. pp. 41-42.

Kucha EI , Itodo IN (1998). An empirical relationship for predicting biogas yield from Poultry waste slurry. Niger. J. Rene. Ene. 2:31–37.

Ieshita P, Sen SK (2013). Microbial and physico-chemical analysis of composting process of wheat straw. Indian J. Biotech. 12:120-128.

Maishanu SM , Maishanu HM (1998). Influence of inoculum age on Biogas generation from cowdung. Niger. J. Rene. Energy 6:21-26.

Malik RK, Tauro P (1995). Effect of predigestion and effluent slurry recycling On biogas production. Indian J. Microbiol. 35:205–209.

Oktay Z (2006). Olive Cake As A Biomass Fuel For Energy Production,Energy Sources, Part A: Recovery Utilization And Environmental Effects. 28:329-339.

Polpraset C, Bitton G (1989). Organic Waste Recycling In Waste Water Microbiology ,John Wiley and Sons, New York; U.S.A.

Ross C (1996). Handbook on Biogas Utilization, 2nd Edition Muscle shoals, all Southeastern Regional Biomass Energy Program, Tennessee Valley Authority.

Sharma DK (2002). Studies on availability and utilization of onion storage waste in a rural habitat. PhD dissertation, Centre for Rural Development and Technology, Indian Institute of Technology, Delhi, India.

Thomas A, Barbara A, Vitaliy K, Werner .Z, Karl M, Leon hard G (2007). Biogas Production from maize and dairy cattle manure-Influence Of Biomass Composition On The Methane Yield. Science Direct. Agriculture Ecosystem And Environment 118:173-182.

Adjust of energy with compactly supported orthogonal wavelet for steganographic algorithms using the scaling function $1/\sqrt{2^j}$

Blanca E. Carvajal-Gámez*, Francisco J. Gallegos-Funes, Alberto J. Rosales-Silva and José L. López-Bonilla

National Polytechnic Institute, Professional Unit Engineering and Advanced Technologies, Av. IPN 2580, Electronica, Barrio La Laguna Ticoman, 07740, Mexico D. F., Mexico.

When RGB images are processed for the implementation of steganographic algorithms, it is important to study the quality of the cover and retrieved images, since it typically used digital filters, reaching visibly deformed images. The discrete wavelet transform is useful for adjusting the stego- image quality, because one of their principal qualities which are able to analyze different frequency levels (multi-resolution analysis). By decomposing the image using the discrete wavelet transform, we obtain sub-images representing different levels of frequencies, in each of these sub-bands can expand or contract the obtained adjusting sub image well as possible to a desired quality. There by avoiding one of the main problems of steganography: distortion of stego-image. When a steganographic algorithm is employed, numerical calculations performed by the computer cause errors and alterations in the processed images, so we apply a scaling factor depending on the number of bits of the image to adjust these errors.

Key words: Steganographic algorithms, RGB images, scaling factor, wavelet.

INTRODUCTION

A classification of information hiding techniques is described in Petticolas and Katzenbeisser (2000). According to this classification, the information hiding techniques include convert communication, steganography, and digital watermarking. Steganography is the science that involves the secret communications of data in an appropriate multimedia carrier, such as, audio, image, and video files (Cheddad et al., 2010). It comes under the assumption that if the secret data is visible, the point of attack is evident (Cheddad et al., 2010; Petticolas and Katzenbeisser, 2000), thus, the goal here is to conceal the existence of the embedded data. In the case of image files, a steganographic method employs innocent-looking media called host or cover image to imperceptibly carry hidden data to an intended recipient (Cheddad et al., 2010; Petitcolas et al., 1999; Petticolas and Katzenbeisser, 2000). The camouflage of secret data in stego-images to pass sensitive data between two parties can be regarded as secret communication, but it is different from cryptography. Cryptography encrypts the content of secret data into cipher texts and transmits the data in cipher to a receiver. However, using cryptography techniques merely for secret communication may fail since a cipher text is in meaningless form and thus, easily arouses the curiosity of malicious people who desire to recover or destroy data. Unlike cryptography, steganography conceals the fact that there is secret communication going on. A stego-image has meaning and the hidden data are camouflaged very well so that no one suspects that embedded data exist (Yuan et al., 2007).

For assured security, secret data used in steganography are usually encrypted using cryptography techniques before they are embedded into a host image. It is well known that the main concerns with steganographic methods are the hiding capacity (HC) of the host image and the quality of the stego-image. It is different from the technique of watermarking. The

*Corresponding author. E-mail: becarvajalg@gmail.com.

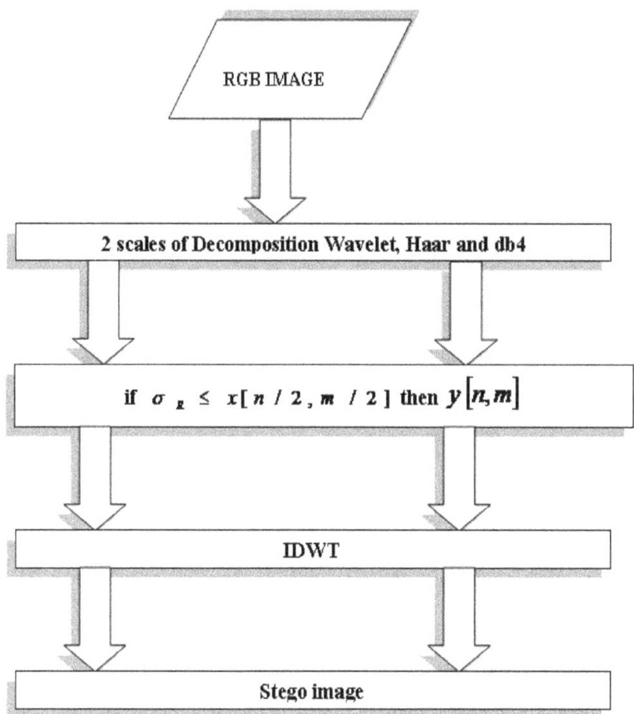

Figure 1. Block diagram of the steganographic algorithm.

purpose of watermarking is to protect the copyright of an image. The image embedded with the hidden data (that is, secret data, copyright notice, serial number) is called the stego-image and it looks as a normal image. Unintended recipients of a stego-image are unaware of the existence of the hidden data. A stego-key is used to control the hiding process so as to restrict detection and/or recovery of the embedded data (Petitcolas et al., 1999). A steganographic technique is usually evaluated in terms of the visual quality and the embedding capacity; in other words, an ideal steganographic scheme should have a large embedding capacity and excellent stego-object visual quality. Unfortunately, the fact is that the visual quality degradation is in proportion to the embedding capacity, and to push one to the limit really means to totally sacrifice the other. The more reasonable way to deal with this trade-off situation is probably to strike a balance between the two (Wu and Hwang, 2007). Judging by whether the human vision sensitivity is considered in the design of the embedding algorithm, we can categorize the schemes into three types: (1) high embedding capacity schemes with acceptable image quality (Chan and Cheng, 2004; Lin and Tsai, 2004; Thien and Lin, 2003; Wang, 2005; Wu et al., 2005), (2) high image quality schemes with a moderate embedding capacity (Chang, 2004; Wang et al., 2008; Wu and Tsai, 2003; Yang et al., 2008; Zhang and Wang, 2005), and (3) high embedding efficiency schemes with a slight distortion (Mielikainen, 2006; Zhang and Wang, 2006a, b). In the first type of schemes, the mechanism of capacity estimation in the embedding procedure functions on a pixel-by-pixel basis without taking the local texture into consideration. The most common embedding algorithm the first type has is the least significant bit (LSB) substitution technique, where the secret messages are hidden into the pixel LSB to create a stego-image. The second type is the adaptive steganographic schemes (Chang, 2004; Wang et al., 2008; Wu and Tsai, 2003; Yang et al., 2008; Zhang and Wang, 2005), where the embedding capacity estimation of a pixel depends on the variation among the immediate neighbor pixels. Finally, the third type is the high embedding efficiency schemes. They focus on how to minimize the image distortion when embedding relatively small amounts of messages, normally less than or equal to two bits per pixel. Zhang and Wang (2006b) introduced the embedding efficiency as the ratio between the number of embedded bits and the distortion energy caused by data embedding.

PROPOSED METHOD

For the method proposed in this study, the scheme chosen number two previously mentioned in the work is justified as a choice in the use of the sensitivity of the human visual system. Usually, this scheme is common to take the use of the frequency domain (Carvajal-Gámez, 2010). Within this domain can work discrete Fourier transform (DFT), discrete cosine transform (DCT) and the discrete wavelet transforms (DWT). Frequency domain methods hide messages in significant areas of the cover image which make them more robust to attacks, such as compression, cropping, and some image processing than the spatial domain methods. However, while they are robust to various kinds of signal processing attacks, they remain imperceptible to the human visual system (Johnson and Jajodia, 1998). The algorithm proposed here for the implementation of steganography is based on the DWT. In the optimization and evaluation of compression algorithms in digital images, the Peak Signal-to-Noise ratio (PSNR) is the most frequently used to evaluate the quality of the images (Huanga et al., 2005).

However, the use of qualitative measures of the image are based on the properties of HVS, then the models usually are embedded into HVS sensitivity to light and sensitivity to spatial frequency (Ginesu et al., 2006). If we apply the DWT, the image resolution is divided into 4 sub-matrices called approximations (a), horizontal (*h*), vertical (v), and diagonal (d), each sub-matrix is a copy of the original one in different levels of the frequency which provides a certain amount of energy (Moon et al., 2007; Walker, 2003). In the subdivision of images, the steganography algorithm is applied only in the sub-matrix *h* and the other sub-matrices are discarded. The proposed steganographic algorithm based on wavelets is shown in Figure 1 (Reddy and Chatterji, 2005). From the third block of Figure 1, σ_R (red channel, from RGB image) that defines the standard deviation of the cover image $x[n,m]$, with it we apply the following criterion of insertion: $\sigma_R \leq x[n/2, m/2]$ then $y[n,m]$, where $x[n,m]$ is the cover image and $y[n,m]$ is the hidden image (Figure 1).

DWT is closely linked to the analysis of multi-resolution (MRA), that is, see the signal at different frequencies (Vetterli and Kočević, 1995), which allows to have a broader knowledge of the signal and facilitates the rapid calculation when the wavelet family is orthogonal (Petrosian and Meyers, 2002; Debnath, 2002;

Sheng, 2002; Bogges, 2009).

It can be obtained wavelets Ψ such that the family moved for *j* and dilated for *n*,

$$\left\{\psi_{j,n}(t)=\frac{1}{\sqrt{2^j}}\psi\left(\frac{t-2^j n}{2^j}\right)\right\}_{j,n\in Z^2} \quad (1)$$

It is an orthonormal base of $L^2(IR)$. These orthogonal wavelets transport information about the changes of the signal to the resolution 2^{-j}. Then, the MRA appears: an image will be modeled with orthogonal projections on vector space of different resolution, $P_{V_j}f, V_j \subset L^2(IR)$. The quantity of information in every projection will depend on the size of V_j. For search orthogonal wavelets it will be necessary to work with approaches of MRA (Petrosian and Meyers, 2002; Debnath, 2002; Sheng, 2002; Daubechies, 1988; Mallat, 1989). For a function $f \in L^2(IR)$, the partial sum of the coefficients wavelet $\sum_{n=-\infty}^{\infty} < f,\psi_{j,n}$ can be interpreted as the difference between two approaches of *f* to the resolutions 2^{-1+j} and 2^{-j}. The MRA approaches calculate the approach of signals to different resolutions with orthogonal projections in spaces $\{V_j\}_{j\in Z}$. Also, the MRA approaches are characterized completely by a particular discrete filter that controls the loss of information along the different resolutions. With these, discrete filters can be designed as orthogonal wavelets bases of simple form (Petrosian and Meyers, 2002; Debnath, 2002; Sheng, 2002; Daubechies, 1988; Mallat, 1989). The approach of a function *f* with a resolution 2^{-j} comes specified by a discrete sampling grid, which provides local averages of *f* in a neighborhood of proportional size to 2^j.

Therefore, a MRA approach consists of embedded networks approach. This means that the approach of a function to a resolution 2^{-j} is defined as an orthogonal projection in a space $V_j \subset L^2(IR)$. The space V_j regroups all the possible approaches to the resolution 2^{-j}. The orthogonal projection of *f* is the function $f_j \in V_j$ that it minimizes $\|f-f_j\|$. The orthonormal wavelets carry the necessary details to increase the resolution of the approach of the signal. The approaches of *f* to the scales 2^j and 2^{j-1}, are respectively equal, to its orthogonal projections in V_j and V_{j-1}. We already know that $V_j \subset V_{j-1}$. Be W_j the orthogonal complement of V_j in V_{j-1}. The orthogonal projection of *f* in V_{j-1} can be written as the sum of orthogonal projections P in V_j and W_j. Then, $P_{V_{j-1}}f = P_{V_j} + P_{W_j}$. The function $f(t)$ can be reconstructed from the discrete wavelets coefficients $W_f(j,n)$ in the following way (Petrosian and Meyers, 2002; Debnath, 2002; Sheng, 2002; Daubechies, 1988; Mallat, 1989),

$$f(t) = A\sum_j\sum_n W_f(j,n)\psi_{j,n}(t) \quad (2)$$

Where *j* is the scale factor and *n* is the movement factor. The wavelets $\psi_{j,n}(t)$ generated of the same wavelet mother function $\psi(t)$ have different scale *j* and place *n*, but they have the same form. Scale factor *j* > 0 is always used. The wavelet is dilated when the scale *j* > 1, and it is contracted when *j* < 1. This way, changing the value of *j* the different range from frequencies is covered. Big values of the parameter *j* correspond to frequencies of minor range, or a big scale of $\psi_{j,n}(t)$. Small values of *j* correspond to frequencies of minor range or a very small scale of $\psi_{j,n}(t)$ (Sheng, 2002; Bogges, 2009; Petrosian and Meyers, 2002; Debnath, 2002; Sheng, 2002; Daubechies, 1988; Mallat, 1989). The continuous wavelet functions with discrete factors of scale and movement are named discrete wavelets. Finally, the signal *f*(*t*) can be compressed or expand in the time. This will have a few certain after effects in the plane of frequencies,

$$f(t)\,compression\ by\ a\ factor\ 2^j(s)f_s(t)=\frac{1}{\sqrt{2^j}}f\left(\frac{t}{s}\right) \quad (3)$$

$$\hat{f}(w)\,compression\ by\ a\ factor\ \frac{1}{2^j}\hat{f}_{2^j}(w)=\frac{1}{\sqrt{2^j}}2^j\hat{f}(2^j w)=\sqrt{2^j}\hat{f}(2^j w)$$

The coefficient of the decomposition of a function *f* in an orthogonal base of wavelets is calculated by a subsequent algorithm of discrete convolution with h and g, and realizes a sampling as follows,

$$x_{low}[k] = \sum_n x[n]h[2k-n] \quad (4)$$

$$x_{high}[k] = \sum_n x[n]g[2k-n] \quad (5)$$

Where $x_{low}[k]$ and $x_{high}[k]$ are outputs of the low pass filter (LPF) and high pass filter (HPF), respectively; $g[2k-n]$ and $h[2k-n]$ represent the impulse response of HPF and LPF, respectively, sub-sampled by a factor of 2 as expressed in Equations 4 and 5 (Walker, 2003; Vetterli and Kočević, 1995; Daubechies, 1988; Mallat, 1989).

These coefficients are calculated by cascades of discrete filters through convolution and sampling.

Then the host image $x[n,m]$ must pass through a series of mirror filters banks in quadrature (Moon et al., 2007; Vetterli and Kočević, 1995; Walker, 2003), these filters are LPF and HPF. The signal from each filter is decimated by a factor of 2. The filter *h* removes de high frequencies of the $x[n,m]$, while g removes the lower frequencies. For the reconstruction, an interpolation is realized, inserting zeros and expanding $x_{high,high}$, $x_{high,low}$, $x_{low,high}$, $x_{low,low}$. This filtering and decimation process in a continuous way is known as sub-band coding. The Figure 2 depicts the decomposition of the discrete wavelet for an RGB image and is interpreted as the decomposition of the sub-matrix R represented by X_R [n,m], for the first decomposition shown in Figure 2 which applies to itself step of the LPF through the rows and columns to obtain sub-matrix *a* previously mentioned of the image, in the second decomposition is applied to LPF by filtering the rows and columns is applied to HPF obtaining the sub-matrix *h*, the third decomposition is similar to *h* but reverses the first filter is the HPF and later the LPF and get the sub-matrix *v* and finally for sub-matrix *d* filtering is applied in rows and columns with the HPF. This operation is interpreted as follows (Vetterli and Kočević, 1995; Walker, 2003; Daubechies, 1988; Mallat, 1989).

The DWT decomposes a discrete signal into two sub-signals of

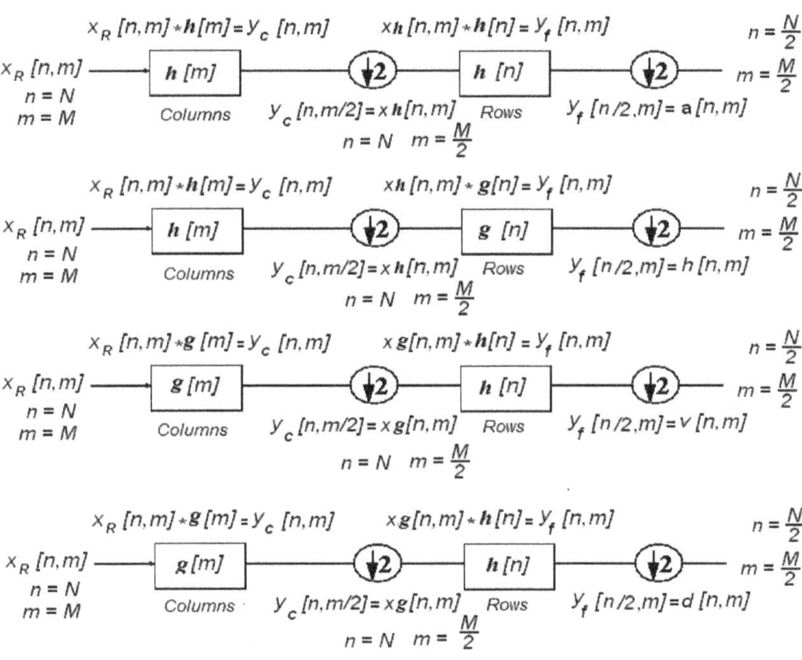

Figure 2. Filters bank for encoding sub-bands where it represents discrete wavelet decomposition of an image.

half of the original length. This sub-signal is known as the approaches and the other one is known as the details (Walker, 2003). The first sub-signal $a^1=(a_1,a_2,\cdots,a_{m/2})$, for the signal *x* is obtained making the average of the signal as follows: the first value a_1 is calculated by taking the first set of values vector $x[m]:(x_1+x_2)/2$ and multiplying it by $\sqrt{2}$, that is, $a_1=(x_1+x_2)/\sqrt{2}$, similarly $a_2=(x_3+x_4)/\sqrt{2}$, etc. It is given by Walker (2003) as follows:

$$a_{m/2}=\frac{x_{2m-1}+x_{2m}}{\sqrt{2}} \tag{6}$$

Where *m* is the total number of vectorial elements. The other sub-signal is also known as the first fluctuation of the signal *x* denoted as: $d^1=(d_1,d_2,\cdots,d_{m/2})$ and is calculated by taking the difference between the first pair of values of *x*, $(x_1-x_2)/2$ and then multiplied and divided by $\sqrt{2}$, and so on. The expression can be written in the following way,

$$d_{m/2}=\left(\frac{x_{2m-1}-x_{2m}}{\sqrt{2}}\right) \tag{7}$$

Where *m* is the vector size. After applying, the DWT got two vectors, which are approximations and details obtained, with a length of half the original vector. Finally, continuing the recovery of the vector,

$$f[n]=\left\{\frac{a_1+d_1}{\sqrt{2}},\frac{a_1-d_1}{\sqrt{2}},\ldots,\frac{a_{n/2}+d_{n/2}}{\sqrt{2}},\frac{a_{n/2}-d_{n/2}}{\sqrt{2}}\right\} \tag{8}$$

We note that the terms a_1 and d_1 in Equations 6 and 7 can be interpreted as follows

$$\varepsilon_{(a^1|d^1)}=a_1+\cdots+a_{n/2}+d_1+\cdots+d_{n/2},$$

$$a_1+d_1=\left[\frac{f_1+f_2}{\sqrt{2}}\right]^2+\left[\frac{f_1-f_2}{\sqrt{2}}\right]^2=\frac{f_1^2+2f_1f_2+f_2^2}{2}+\frac{f_1^2-2f_1f_2+f_2^2}{2}=f_1^2+f_2^2 \tag{9}$$

and similarly for each set of vectorial approaches and details (Walker, 2003). So, the conservation of energy in wavelets is mentioned, the factor $1/\sqrt{2}$ is mentioned too (Carvajal-Gámez, 2010). By applying the steganographic algorithm to the *h* sub-matrix, it is necessary to use the scaling factor, but as the work is with an 8-bit RGB image, this scaling factor is closely related to energy conservation applied to wavelet theory for grayscale images as shown in most applications. However, for RGB images we propose the following scaling factor,

$$1/\sqrt{2^j} \tag{10}$$

Where *j* is directly dependent on the number of bits that integrate the image (Carvajal-Gámez, 2010).

EXPERIMENTAL RESULTS

Many distortion or quality measures used in the images during the visual information processing belong to the group of measures of difference in distortion (Kutter and Petitcolas et al., 1999; Carvajal-Gámez, 2010), which are based on the difference between the original and modified images. The most common distortion measure is the peak signal to noise relation PSNR,

$$\mathrm{PSNR} = 10 \cdot \log\left[\frac{(255)^2}{\mathrm{MSE}}\right], \mathrm{dB} \tag{11}$$

Where $MSE = \frac{1}{M_1 M_2} \sum_{n=1}^{M_1} \sum_{m=1}^{M_2} \left\| y(n,m) - x(n,m) \right\|_{L_2}^{2}$ is the mean square error, M_1, M_2 are the image dimensions, $y(n,m)$ is the 3D vector value of the pixel (n,m) of the stego-image, $x(n,m)$ is the corresponding pixel in the original cover image, and $\|\cdot\|_{L_1}, \|\cdot\|_{L_2}$ are the L1- and L2-vector norms, respectively; The normalized color deviation (NCD) is used for the quantification of the color perceptual error,

$$NCD = \frac{\sum_{n=1}^{M_1} \sum_{m=1}^{M_2} \left\| \Delta E_{Luv}(n,m) \right\|_{L_2}}{\sum_{n=1}^{M_1} \sum_{m=1}^{M_2} \left\| E^{*}_{Luv}(n,m) \right\|_{L_2}} \tag{12}$$

Here, $\left\| \Delta E_{Luv}(n,m) \right\|_{L_2} = \left((\Delta L^{*}(n,m))^2 + (\Delta u^{*})^2 + (\Delta v^{*})^2 \right)^{1/2}$ is the norm of color error; ΔL^{*}, Δu^{*}, and Δv^{*} are the difference in the L^{*}, u^{*}, and v^{*} components, respectively, between the two color vectors that present the stego and cover images for each a pixel (*n*,*m*) of an image, and $\left\| E^{*}_{Luv}(i,j) \right\|_{L_2} = \left[(L^{*})^2 + (u^{*})^2 + (v^{*})^2 \right]^{1/2}$ is the norm or magnitude of the host image pixel vector in the $L^{*}u^{*}v^{*}$ space.

The quality index (Q) is provided to demonstrate the quality of the stego-images (Carvajal-Gámez, 2010),

$$Q = \frac{4\sigma_{xy} xy}{(\sigma_x^2 + \sigma_y^2)(x^2 + y^2)} \tag{13}$$

Where x and y are the mean values of host and stego-images, respectively, σ_x^2 and σ_y^2 are the variances of host and stego-images, respectively, and $\sigma_{xy} = \frac{1}{N-1} \sum_{i}^{N} (x_i - x)(y_i - y)$.

Some tests were conducted with 8-bit RGB images to show our mentioned scaling factor in this paper. As mentioned previously, a filter can distort the images, as will be shown in subsequent tests applied filters to distort the host image with the DWT. However, applying the proposed scaling scheme can be seen as an improvement in visual images. The proposed scheme has been evaluated, and its performance is compared with the LSB (Chan and Cheng, 2004), LSB optimized (LSBO) (Yuan et al., 2007) steganographic methods. These methods were implemented to compare them with the proposed scheme approach. The reason for selecting these methods of comparison is that their performances have been compared with various well known methods, and they were used accordingly as the reference ones. From these results we present a discussion in terms of the objective image quality, the HC, the color spaces, and the subjective visual results, and we provide recommendations to use the proposed methods under different circumstances. Table 1 shows the performance results in terms of PSNR, mean absolute error (MAE), COI (Correlation), Q, NCD, and HC. In the case of different *j* values in the scaling factor by using the 320 × 320 color images "Mandrill" (Vitterbi, 2012) as cover image and "Lena" (Vitterbi, 2012) as hide image, from Table 1, one can see when the *j* value increases the performance results increase too. Figure 3 depicts the processed images for stego-image Mandrill (Figure 3a, b, and c) and retrieved secret image Lena (Figure 3d, e, and f) according with Table 1. Table 1 shows that as the value of *j* is increased, measures considerably improve visual quality. It can be seen that when *j* = 0, the values of stego-imagen "Mandrill" in PSNR is equal to 31.5084%db with a COI equal to 78.51%, similarly to the recovered image "Lena" which is a test that more affected by the distortion has a PSNR equal to 16.5327 db and a COI equal to 38.8%. Also, one can observe that when *j* is increased to 5, it obtained PSNR value of 36.1233 db for stego-image and the recovered image with a PSNR of 27.2474 db. We observe from Figure 3c and f that the best results are obtained when *j* = 9, where *j* represents the number of bits resolution of the image to hide. It is observed that when *j* is gradually increased, the quality image is significantly improved. From Figure 3d, e, and f one can see that when the value of proposed scaling factor increases the subjective quality of images increases too, it is observed that Figures 3a, b, and c have in the upper part a certain lineal distortion, which can be interpreted as external information inserted to the cover image. As *j* takes the value of 10 or 24 of this lineal distortion in the top of the stego-image not easily identified visually. We also present the wrong images. Table 2 shows the results using comparative methods, LSB, LSBO and WLSB.

Test were performed with *j* = 10, in Table 1 and shows the performance results in the case of use in the scaling factor. More tests were performed to increase the value of *j* to *j* = 24, amount determined by the image type manipulated. The significant changes of stego-image and the recovered image were observed. Figure 4 presents the visual results according to Table 3. The proposed scaling factor $1/\sqrt{2^j}$ for each test presents a different result as can be seen in the previous tests, the scaling

Table 1. Performance results for different values of *j* with cover image "Mandrill" and hidden image "Lena".

j		PSNR db	COI %	NCD	Q	MAE
j = 0	Cover image	31.5084	78.52	-	0.7836	10.4878
	Recovered image	16.5327	38.80	0.4748	0.3586	4.9403
j = 2	Cover image	31.4999	80.46	-	0.8040	9.9490
	Recovered image	16.9537	38.89	0.4749	0.3587	4.9471
j = 5	Cover image	36.1233	97.81	0.0020	0.9792	3.2086
	Recovered image	27.2474	99.85	0.0020	0.9962	2.7714
j = 9	Cover image	36.1233	99.08	8.015e-4	0.9913	2.0309
	Recovered image	31.0781	99.80	0.0020	0.9962	2.7714
j = 10	Cover image	36.1233	99.34	6.0486e-4	0.9934	1.7022
	Recovered image	32.5167	99.55	0.0020	0.9962	2.7714

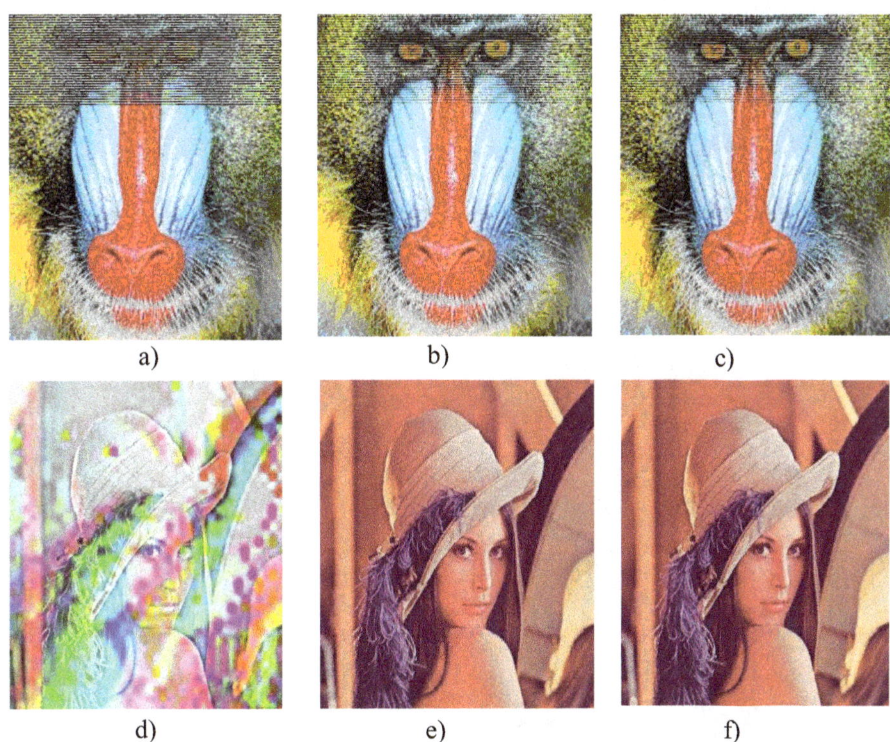

Figure 3. Visual results for different values of *j* in the scaling factor, a) and d) column with *j* = 2, b) and e) column with *j* = 5, and c) and f) *j* = 9.

factor does not affect the steganographic algorithm and preserving the energy of images. It can be seen in the Lena error image of Figure 4d, the difference in values between the host image and the recovered image is approximately zero. In general, we can say that in the case of the cover images with uniform regions (Table 4, 'Splash' image as cover image, Figure 5), although conserved, acceptable results quality image, compared with LSB, LSBO and WLSB methods there is no significant difference of values. Significantly for uniform color images insertion is usually more complicated because of the few regions of abrupt changes. In the opposite case, the best scenario is when the cover images contain several details (Table 3, 'Mandrill' image as cover image) where we can observe that the PSNR, COI, NCD, MAE and Q of the proposed method is higher than other methods used as comparative providing the best image quality in favor of the proposed methods. Then, the hidden information is almost imperceptible. The results shown in Table 3 using the scaling factor *j* =24

Table 2. Performance results for different steganographic methods with cover image "Mandrill" and hidden image "Lena".

Parameter		PSNR db	COI %	NCD	Q	MAE
LSB	Cover image	34.5602	99.78	7.3506e-4	0.9977	3.5758
	Recovered image	36.1233	99.54	0.0020	0.9962	2.7714
LSBO	Cover image	26.9908	99.76	0.0022	0.9948	10.6396
	Recovered image	29.2421	99.54	0.0033	0.9944	8.0316
WLSB	Cover image	28.8347	99.77	0.0053	0.9963	8.8469
	Recovered image	31.2219	99.54	0.0033	0.9954	5.0316

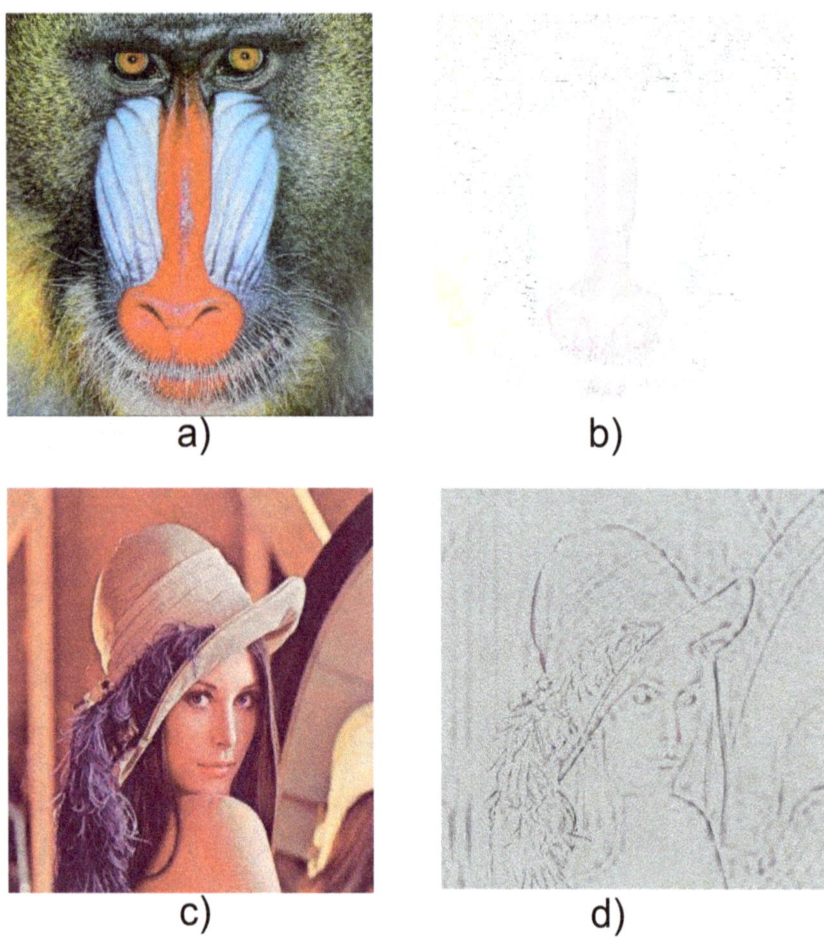

Figure 4. Visual results in the case of j = 10, a) host image "Mandrill, b) error image "Mandrill, c) hide image "Lena", d) error image "Lena".

are superior to those shown in Table 2. Being the best method, LSB with PSNR equal to 34.5602 db is exceeded by the second PSNR of 36.7973 db with j = 24. Tables 4 and 5 show the results for the stego-image "Splash" (image to hide "Tiffany") and the recovered image "Tiffany", respectively. Figure 5 presents the results obtained from subjective visual Tables 4 and 5.

Conclusions

The RGB images are altered in their energy contribution in each sub-matrix of wavelet decomposition when the steganographic algorithm is applied. It is known that the value of $1/\sqrt{2}$ is the key factor in the adjustment of the wavelets energy; this adjustment value has been applied

Table 3. Performance results in the case of j = 24 in the scaling factor with cover image "Mandrill" and hidden image "Lena".

Cover image "Mandrill"	Hide image "Lena"
Q = 0.9968	Q = 0.9962
PSNR = 36.7973 dB	PSNR = 36.1232 dB
COI = 99.67%	COI = 99.56%
NCD = 5.7204 e-4	NCD = 19.735e-4
MAE = 1.4721	MAE = 2.7714

Table 4. Performance results for different steganographic methods with cover image "Splash" and hidden image "Tiffany".

Test	Steganographic algorithms			
	LSB	LSBO	WLSB	Proposed method j = 24
PSNR dB	30.1606	25.8026	26.0296	34.5830
MAE	6.1696	11.3249	11.4953	3.5761
COI %	99.10	99.08	99.18	99.60
Q	0.9871	0.9828	0.9881	0.9919
NCD	0.0065	0.0101	0.0086	0.0028

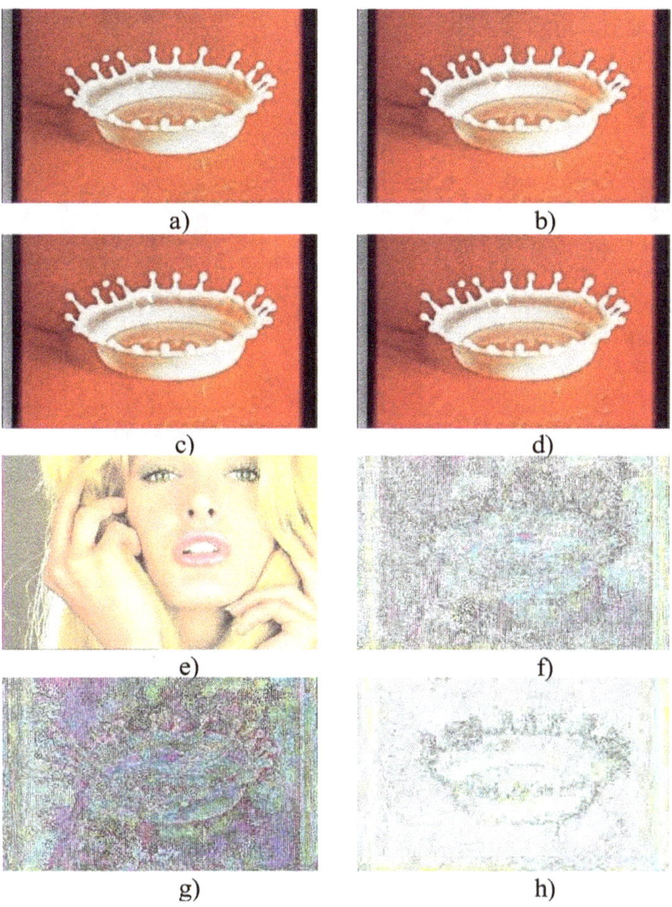

Figure 5. a) Cover image "Splash", Stego-images; b) LSB; c) LSBO; d) proposed method with j = 24; e) Image Original "Tiffany", error stego-images; f), 4LSB; g), 4LSBO; h), proposed method.

Table 5. Comparative performances of the recovered image "Tiffany".

Test	Steganographic algorithms			
	LSB	**LSBO**	**WLSB**	**Proposed method j = 24**
PSNR dB	34.3755	29.7652	32.2852	34.3755
MAE	3.4600	6.9246	4.4938	3.4600
COI %	98.37	98.06	97.99	98.37
Q	0.9713	0.9650	0.9788	0.9713
NCD	0.00065	0.0012	0.0028	0.00065

only in gray scale images tests. However, the energy conservation factor mentioned above is not valid for true color images, such as the one formed by 3 sub-matrices (RGB) which separately provides some level of energy, must have a value close to the total energy calculated up to the image, because if it is not right or there are energy gap in the components, the images may have a poor display as distortion. By applying the proposed scaling factor, $1/\sqrt{2^j}$, where j represents the number of bits that integrates each image, there is an adjustment factor for the energy input in each sub-matrix. It is also noted that when it is changing the value of j, adjust the sharpness and image clarity providing a visible improvement of the visual image. The results shown here with varying scale factor were showing improvement in each of the visual quality measures, comparing j = 10 exceeded the values obtained as the LSB, LSBO and WLSB steganographic methods. Agree with the experimental results we observed that there are some variations in the PSNR, MAE, NCD, COI and Q results between different versions of the steganography methods but in general, we recommend the application of scaling factor because they guarantee good results in comparison with other methods used as comparative, and sometimes these provide a higher performance than other variants of the steganography methods. By other way, the scaling factor, provide the best results in almost all the tests by balancing the performance of the image quality among different types of images (with uniform regions or several details).

ACKNOWLEDGMENTS

This work is supported by National Polytechnic Institute of Mexico and National Council on Science and Technology (CONACYT) Mexico.

REFERENCES

Bogges A, Narcovich F (2009). A first course in wavelets with Fourier Analysis, 2th Ed., Wiley pp. 1-305.

Carvajal-Gámez BE, Gallegos Funes FJ, Lopez Bonilla JL (2010). Computational Techniques and algorithms for Image Processing. Ed. Lambert Academic Publishing Chapter 13.

Chan C, Cheng (2004). Hiding data in images by simple LSB substitution. Pattern Recogn. 37:469-474.

Cheddad A, Condell J, Curran K, Mc Kevitt P (2010). Digital image steganography: Survey and analysis of current methods. Signal Process. 90:727-752.

Daubechies I (1988). Orthornormal bases of compactly supported wavelets. Commun. Pure Appl. Math. 41:909-996.

Debnath L (2002). Wavelets and Signal Processing. Birkhauser, Berlin P. 106.

Ginesu G, Massidda F, Giusto DD (2006). A multi-factors approach for image quality assessment based on a human visual system model. Signal Process. Image Commun. pp. 316-333.

Huanga KQ, Wub ZY, Fungc GSK, Chan HY (2005). Color image denoising with wavelet thresholding based on human visual system model. Signal Process. Image Commun. pp. 115-127.

Johnson NF, Jajodia S (1998). Exploring Steganography: Seeing the Unseen. Comput. Pract. IEEE pp. 26-34.

Johnson NF, Jajodia S (1998). Steganalysis of Images Created Using Current Steganography Software. Lecture notes in computer science. Springer Verlag (1525):273-289.

Kutter M, Petitcolas F (1999). A fair benchmark for image watermarking systems. Electronic Imaging '99. Security and Watermarking of Multimedia Contents. Int. Soc. Opt. Eng. 3657:1-14.

Lin C, Tsai W (2004). Secret image sharing with steganography and authentication. J. Syst. Softw. 73:405-414.

Mallat S (1989). A theory for multiresolution signal decomposition: the wavelet representation. IEEE Trans. Pattern Anal. Mach. Intell. 11(7):674-694.

Mielikainen J (2006). LSB matching revisited. IEEE Signal Process. Lett. 13:285-287.

Moon H, You T, Sohn M, Kim H, Jang D (2007). Expert system for low frequency adaptive image watermarking: Using psychological experiments on human image perception. Expert Syst. Appl. pp. 674-686.

Petitcolas F, Anderson RJ, Kuhn MG (1999). Information hiding a survey. Proceed. IEEE 87(7):1062-1078.

Petrosian A, Meyers F (2001). Wavelets in signal and image analysis. Comput. imaging vision Springer. P. 19.

Petticolas F, Katzenbeisser S (2000). Information hiding techniques for steganography and digital watermarking. Artech. House. Boston. pp. 15-30.

Reddy A, Chatterji B (2005). A new wavelet based logo-watermarking scheme. Pattern Recog. Lett. pp. 1019-1027.

Sheng Y (2002). The transforms and applications handbook. Ed. 2th, CRC Press EEUU.

Thien C, Lin J (2003). A simple and high-hiding capacity method for hiding digit by digit data in images based on modulus function. Pattern Recog. 36:2875-2881.

Vetterli M, Kočević J (1995). Wavelets and subband coding. Prentice-Hall, New Jersey. pp. 1-9, 97-202, 209-298, 414-438.

VITTERBI (2012). Data Base Images of the University of Southern of California, Signal and image processing institute. http://sipi.usc.edu/database.

Walker J (2003). A primer on wavelets and their scientific applications, Chapman and Hall/CRC, London. pp. 5-26, 41-53.

Wang C, Wu N, Tsai C, Hwang M (2008). A high quality steganographic

method with pixel value differencing and modulus function. J. Syst. Softw. 81:50-58.
Wang J (2005). Steganography of capacity required using modulo operator for embedding secret image. Appl. Math. Comput. 164:99-116.
Wu D, Tsai W (2003). A steganographic method for images by pixel value differencing. Pattern Recogn. Lett. 24:1613-1626.
Wu H, Wu N, Tsai C, Hwang M (2005). Image steganographic scheme based on pixel-value differencing and LSB replacement methods. IEEE Proceed. Vision Image Signal Process. 152:611-615.
Wu N, Hwang M (2007). Data hiding: Current status and key issues. Int. J. Network Secur. 4:1-9.
Yang C, Weng C, Wang S, Sun H (2008). Adaptive data hiding in edge areas of images with spatial LSB domain systems. IEEE Trans. Inform. Forensics Secur. 3:488-497.
Yuan HY, Chin CC, Iuon CL (2007). A new steganographic method for color and grayscale image hiding. Comput. Vision Image Unders. 107:183-194.
Zhang X, Wang S (2005). Steganography using multiple-base notational system and human vision sensitivity. IEEE Signal Process. Lett. 12:67-70.
Zhang X, Wang S (2006a). Dynamical running coding in digital steganography. IEEE Signal Process. Lett. 13:165-168.
Zhang X, Wang S (2006b). Efficient steganographic embedding by exploiting modification direction. IEEE Commun. Lett. 10(11):781-783.

The effects of cow dung inoculum and palm head ash-solution treatment on biogas yield of bagasse

Ofomatah A. C.[1]* and Okoye C. O. B.[2]

[1]National Centre for Energy Research and Development, University of Nigeria, Nsukka, Enugu State, Nigeria.
[2]Department of Pure and Industrial Chemistry, University of Nigeria, Nsukka. Enugu State, Nigeria.

Blends of ash solution and water-soaked bagasse with cow dung inoculum were studied for biogas production. Two sets of five biodigesters each of 50 kg capacity were used. The blends were respectively charged into the 50 kg metal prototype biodigesters in the ratio of 3:1 (water:waste). Proximate analyses as well as total solids, volatile solids, carbon content, and calorific value were conducted on the wastes while microbial level, pH and temperature were determined on the slurry. The wastes were subjected to anaerobic digestion for 35 days at mesophilic temperature range of 29 to 31°C. Relative humidity, ambient temperature, pH, slurry temperature, and volume of gas were monitored and recorded on daily basis. Cumulative biogas yields of all the ash solution treated bagasse blends were higher than those of the water-soaked blends by up to 400%. However, the blending of bagasse with cow dung did not improve the biogas yield of bagasse; instead, a steady state was established due to mutual inhibitions. Onset of gas flammability was observed on the 4th day for bagasse and its blends while for cow dung alone, it was observed on the 5th day.

Key words: Bagasse, cow dung inoculum, ash solution treatment, anaerobic digestion, mutual inhibition.

INTRODUCTION

Dwindling energy resources has become a global problem. Now there is an urgent need for alternative energy sources. In Nigeria, rural communities who rely on wood and kerosene for domestic energy are not finding it easy due to scarcity and high cost of these two products. This has made it imperative to search for new sources of domestic energy. The quest for wood as a source of domestic energy has led to deforestation and erosion in the southern parts and near desertification in the northern parts of the country (Ilochi and Nwachukwu, 1989). Since both wood and kerosene are costly and often scarce, biogas, which is renewable and sustainable, offers the best alternative as rural energy supply source (Ilochi and Nwachukwu, 1989). Biogas had been generated from various biomass wastes. It is a mixture of methane (50 to 70%), CO_2 (30 to 40%) and traces of other gases such as carbon (II) oxide (CO), nitrogen (II) oxide (NO), water vapour, ammonia (NH_3), and hydrogen sulphide (H_2S) (Garba et al., 1996).

Sugar cane is a total utility crop (Smith, 1993) widely grown in tropical and semi-tropical countries of South and Central America and the Caribbean Islands as well as in Africa and Asia (Wayman and Parekh, 1990). In Nigeria, it is grown mainly in the northern parts of the country. Bagasse is the fibrous residue remaining after extraction of sugar from sugar cane. In various parts of Nigeria, eaters of sugar cane easily litter our towns especially the market places with bagasse, dirtying the environment. Heaps of bagasse at market places in Nigeria are often burnt leading to air pollution (Busari, 2004). Cancado et al. (2006) reported that air pollution from bagasse burning led to increased hospitalization of children and the elderly due to respiratory problems.

Biogas production from lignocelluloses materials requires an initial treatment to break down some of their

*Corresponding author. E-mail: ofomatony@yahoo.co.uk.

Table 1. Ratios of waste blends.

Waste (Kg)	A	B	C	D	E
BG	9.4	6.6	4.7	2.8	0.0
CD	0.0	2.8	4.7	6.6	9.4
BG:CD (%)	100	70:30	50:50	30:70	100

lignocelluloses in order to increase their volatile solids content and create a favourable environment for micro organisms to thrive. Lignocellulose materials can be treated with 50% KOH (Ofoefule et al., 2011), Steam at 130 to 170°C (Bougrier et al., 2006), 0.1 g Ozone per gram total solids (Yeom et al., 2002; Weemaes et al., 2000) and 0.7% solution of sodium hydroxide (NaOH), Potassium hydroxide (KOH), Magnesium hydroxide [$Mg(OH)_2$] or Calcium hydroxide [$Ca(OH)_2$] (Kim et al., 2003). Bagasse had been treated with 0.5% NaOH, 5% sodium chlorite (NaCl) and steam at 120°C (Ishihara et al., 1988). Treatment with lime (Saska and Gray, 2006), various blends of (NaoH) containing urea (Suksombat, 2004), as well as $Ca(OH)_2$ (Firdos et al., 1989) have also been reported. In this study, the effects of treatment with solution of spent oil palm head ash and co-digestion with cow dung for improved biogas production were investigated.

EXPERIMENTAL

Bagasse was collected from dumps around Ose market at Onitsha, Anambra State, while cow dung was collected from the abattoir in Nsukka market in Enugu State, Nigeria. The bagasse was sorted and thoroughly washed with tap water, dried and later ground in a mill. Some of the bagasse were soaked in tap water in the ratio of 1:2 (bagasse:water) while some were soaked in spent oil palm head ash solution for two weeks. The soaked bagasse were transferred into a sieve placed on a container for the liquid to drain off for 24 h. The effluent was kept to be further used to prepare the slurry for anaerobic digestion.

Analyses of waste blends

All chemicals used for various analyses were analytical grade procured from BDH Chemicals Ltd, Poole England. Waste blends were prepared by thoroughly mixing pulverized bagasse portions with cow dung portions in the ratios shown in Table 1. Proximate analyses were carried out as follows: Nitrogen/crude protein was found by the micro-Kjedahl method (Pearson, 1976). Crude fibre, moisture and ash were determined by standard methods (AOAC, 1990). Energy content was determined according to AOAC (1975) using a bomb calorimeter; model XRY-1A, Changji, China. Carbon content was determined by the method of Walkley and Black (1934). Total and volatile solids were determined by the method described by Meynell (1982) while microbial analyses were done by the method of Miles and Misra (1938).

Digestion of wastes

Two sets of five 50 L capacity biodigesters of fixed dome prototype constructed in the Mechanical Engineering unit of Energy Research Centre, University of Nigeria, Nsukka, were used (Figure 1). Each was charged up to 75% of its capacity. Blends of the ash solution (alkaline)-soaked bagasse with cow dung were charged into one set, while blends of the water-soaked bagasse and cow dung were charged into the other set of 5 biodigesters. For each set, the effluent was shared equally into five and made up with tap water to 28 L which were added to each blend in a 1:3 waste:water ratio and properly stirred to give a homogenous slurry. The digesters were made air-tight to ensure anaerobic environment. The liquid slurry was analysed for microbial levels, pH, and slurry temperature. Daily gas productions were measured using the water displacement method (Itodo et al., 1995), and stirring was carried out daily using the in-built stirrer throughout the period of 35 days when biogas production was no more substantial. Pure bagasse (A) and pure cow dung (E) served as controls.

RESULTS AND DISCUSSION

Table 2 shows the proximate composition of the water-soaked bagasse and cow dung blends, while Table 3 shows those of the alkaline-soaked bagasse blends. Fat content was similar in both the water treated waste (0.44-1.88%; Table 2) and in the ash solution treated waste (0.44 to 1.64%; Table 3). The bagasse only wastes (A) showed the highest content of fat while cow dung alone (E) showed the lowest. Ash was highest in pure bagasse, (A) than in the blends (B-D) and cow dung (E). The range was 3.42 to 4.55% in the water-soaked wastes and 3.37 to 8.04% in the alkaline-soaked wastes. Moisture was quite low (14.38%) in bagasse compared to cow dung (78.20%). Protein contents were similar in bagasse (A) and cow dung (E) and in the blended wastes while fibre and carbohydrate were much higher in bagasse than in cow dung with ranges of 8.61 to 42.01% and 3.70 to 31.94%, respectively. With respect to the alkaline-treated bagasse and its blends (Table 3), only insignificant changes occurred in the parameters considered in Table 2 except in ash content. In other words, alkaline treatment using palm head ash solution does not significantly affect the fat, protein, fibre, and carbohydrate contents of the waste. The increase in ash content was due to the soaking of the bagasse in palm head ash solution which is rich in sodium, potassium, and calcium, these being major ions in plant materials (Okoye, 2001).

The physico-chemical properties of the water-soaked waste blends were presented in Table 4, while those of the ash solution soaked blends were presented in Table 5.

Comparing Tables 4 and 5, total solids, carbon content and calorific value decreased, while volatile solid

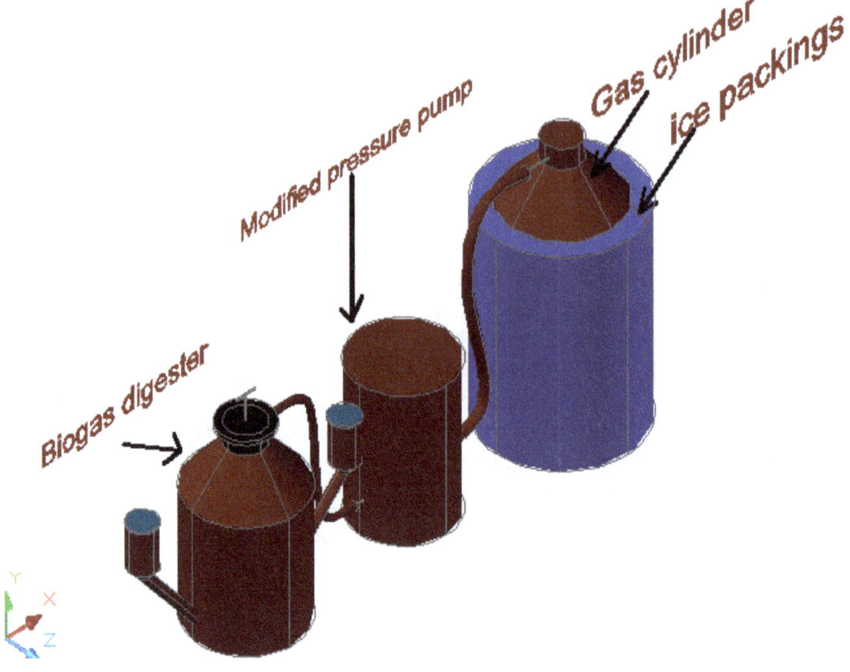

Figure 1. A fixed-dome digester.

Table 2. Proximate composition of the water- soaked waste blends.

Blends	Fat (%)	Ash (%)	Moisture (%)	Protein (%)	Fibre (%)	Carbohydrate (%)
A'(BG only)	1.88	4.55	14.38	5.24	42.01	31.94
B'(70:30) BG:CD	1.51	4.17	33.53	5.30	31.96	23.53
C'(50:50) BG:CD	1.23	4.02	46.10	5.36	25.29	18.00
D'(30:70) BG:CD	0.86	3.76	58.95	5.50	18.61	12.32
E'(CD only)	0.44	3.42	78.20	5.63	8.61	3.70

Table 3. Proximate composition of the ash solution treated waste blends.

Blends	Fat (%)	Ash (%)	Moisture (%)	Protein (%)	Fibre (%)	Carbohydrate (%)
A(BG only)	1.64	8.04	15.24	4.50	40.36	30.22
B(70:30) BG:CD	1.30	7.10	34.06	4.82	30.42	22.30
C(50:50) BG:CD	1.05	6.50	46.80	5.15	21.80	18.70
D(30:70) BG:CD	0.79	5.90	59.45	5.41	15.52	12.93
E(CD only)	0.44	3.37	78.55	5.76	8.14	3.74

increased in the alkaline treated wastes. These are indications of some initial degradation of the alkaline-treated bagasse and its blends. Tables 2 to 5 show that the wastes contained suitable nutrients needed for anaerobic digestion (Bryant, 1974). The increase in volatile solids due to the alkaline hydrolysis is evidence of an initial breakdown of the lignocelluloses.

The micro organisms identified in the wastes include *Escherichia coli*, *Pseudomonas* spp; *Bacillus subtilis*, Yeast, *Salmonella*, *Staphylococcus*, Proteus *spp*; *Aerobacter aerogens* etc. Among these are pathogens like *E. coli*, *Salmonella*, *Staphylococcus* and *Pseudomonas* spp. which can cause numerous diseases including skin infections, urinary tract infections, wound infections and food poisoning in human beings and animals (Green-wood, 1984). The daily mean

Table 4. Physico-chemical properties of the water-soaked waste blends.

Blends	T.S (%)	V.S (%)	Carbon content (%)	Cal.Value (Kj/kg)	N (%)	C:N
A'(BG only)	85.62	60.34	37.36	19,314.69	0.84	44.48
B'(70:30) BG:CD	66.47	46.79	31.92	17,208.98	0.85	37.55
C'(50:50) BG:CD	53.90	37.50	28.61	16,312.83	0.86	33.27
D'(30:70) BG:CD	41.05	28.28	25.39	15,831.46	0.88	28.85
E'(CD only)	21.80	14.77	20.30	14,734.50	0.90	22.56

Table 5. Physico-chemical properties of the ash solution treated waste blends.

Blends	T.S (%)	V.S (%)	Carbon content (%)	Cal.Value (Kj/kg)	N (%)	C:N
A(BG only)	84.76	65.20	30.51	16,921.92	0.72	42.38
B(70:30) BG:CD	65.94	50.32	26.60	16,204.44	0.77	34.55
C(50:50) BG:CD	53.20	40.52	24.84	15,693.89	0.82	30.29
D(30:70) BG:CD	40.55	31.14	23.23	15,449.89	0.87	26.70
E(CD only)	21.45	16.20	19.66	13,956.32	0.92	21.37

Table 6. Cumulative biogas yield (litres) for the water-soaked and ash-solution treated wastes.

Sample	Water-treated	Ash-treated	Percentage increase
A(BG only)	47.9	198.7	314.8
B(70:30) BG:CD	55.2	203.5	268.7
C(50:50) BG:CD	96.5	220.2	128.2
D(30:70) BG:CD	107.5	221.4	106.0
E(CD only)	297.3	292.8	-

$$\% \text{ Increase} = \frac{\text{Yield (ash-treated)} - \text{Yield (water-treated)}}{\text{Yield (water-treated)}} \times 100$$

temperatures ranged from 29 to 31°C throughout the period of digestion. The pH of the water treated blends ranged from 4.8 to 5.8, largely acidic, while those of the ash solution treated wastes ranged from 7.2 to 7.8. Speece and Mcarthy (1964) reported that, biogas production is favoured by pH in the range of 6.6 to 7.6 with optimum range at 7.0 to 7.2. The alkaline metal salts in the ash-solution increased the pH. Microbial population was in the order of 10^4 to 10^5 in blends of water treated bagasse while in blends of ash solution treated bagasse, the order was 10^6 to 10^7, an increase of two orders of magnitude due to a more favourable pH.

The carbon (C):Nitrogen (N) ratio for optimum biogas yield was reported to be between 20:1 to 30:1. This ratio is higher in the pure bagasse wastes (A), 44.5:1 and 42.4:1 (Tables 4 and 5) Also for the pure cow dung, (E) C:N ratio was approximately 22.56:1 and 21.37:1, which are within the optimum range. For the blends, C:N was in the range of 22.85 to 37.55:1. The blending of bagasse with cow dung helped to bring the C:N ratio closer to the range of 20:1 to 30:1, however, that did not lead to higher biogas yield by blended bagasse wastes as evidenced in Table 6, where for pure bagasse (A), there was 314.8% increase in biogas yield of ash solution treated bagasse wastes over that of the water-soaked bagasse. However, for the blended wastes, the increases were 268.7 to 106% becoming smaller as cow dung increases. Thus, instead of cow dung inoculum enhancing biogas production by bagasse, the reverse is the case.

To further investigate likely inhibition, theoretical gas yields were calculated for the ash solution treated wastes based on the yields of pure ash-treated bagasse, A (100%) and pure inoculum, E (100%). The calculated yields were compared with the experimental yields and are presented in Table 7. In all cases, the experimental yield was lower than the calculated yield with ratios or yield efficiencies of 83.2 to 89.4%. In other words, the blending of bagasse with cow dung has not improved the biogas yield of bagasse; instead, a steady state was established due to mutual inhibition, leading to 11 to 17%

Table 7. Gas yield efficiency of ash-solution treated wastes.

Biodigester	Blend	Experimental yield	Calculated yield	Yield efficiency (%)
A	100% BG	198.7	-	-
B	70:30	203.5	227.59	89.41
C	50:50	220.2	246.85	89.20
D	30:70	221.4	266.11	83.20
E	100 % CD	292.8	-	-

Calculated yield = % BG in waste × 100 % (Pure) BG yield + % CD in waste × 100 % (Pure) CD yield.

reduction in biogas yield.

Conclusion

The mixing of different wastes or co-digestion is meant to improve biogas production. Mixtures such as cattle + pig, cattle + poultry, cattle + sewage, poultry + sewage and sewage + weeds had been reported (Srinivasan et al., 1997). Blending of wastes in this way can lead to improved digestion and enhancement of biogas production through synergistic effects. These benefits could be possible as a result of the supply of the required nutrients or reduction of substances that can inhibit bioconversion. However, the blending of bagasse with cow dung did not seem to enhance biogas production by bagasse, instead, it led to a steady state and inhibition as well as introduction of pathogens.

The treatment of bagasse with the palm head ash helped in raising the pH of the bagasse from 4.8 to 7.2 which was favourable for microbial growth. As a result, more gas was produced by the alkaline-treated wastes compared with the water-treated wastes since more microbes were able to thrive in the alkaline-treated waste because of the favourable pH. The treatment of bagasse with palm head ash solution helped in breaking down some of the lignocelluloses present in the bagasse. This resulted in improved volatile solids content and hence more gas production since the volatile solids represent the nutrients that are digestible by the microbes. The cumulative volume of biogas produced from ash-treated bagasse was over four times higher than the one produced from water treated bagasse. This was far higher than 1.8 and 2.1 reported by Ishihara et al. (1988) when bagasse was treated with NaoH and sodium chlorite solutions, respectively. Alkaline solution from palm head ash could be milder to microbes than solution of pure alkali salts, thus leading to higher biogas production by the ash solution treated bagasse.

This study has shown that, alkaline treatment of bagasse using palm head ash enhanced its biogas yield by over 400% compared with 180 and 210% reported by Ishihara et al. (1988) using NaoH and sodium chlorite respectively. The ash provided cheap source of alkali and better condition for anaerobic digestion. The volume of biogas produced by bagasse was not enhanced by adding adjunct like cow dung; instead, a steady state was established as a result of mutual inhibition.

ACKNOWLEDGEMENT

The authors will like to express their profound gratitude to the staff of Energy Research Centre Laboratory for their support during the research work.

REFERENCES

AOAC (1975). Official Methods of Analysis. Association of Analytical Chemists. In: Onwuka GI (2005). Food Analysis and Instrumentation (1st edition) Nigeria. Naphthali prints. pp. 5867.

AOAC (1990). Official Methods of Analysis. Association of Analytical Chemists (14th edition). Arlington, Virginia. P. 222.

Bougrier C, Delgene JP, Carr're H (2006). Combination of thermal treatments and anaerobic digestion to reduce sewage sludge quantity and improve biogas yield. Process Saf. Environ. Prot. 84(B4):280284.

Bryant MP (1974). Nutritional features and ecology of predominant anaerobic bacteria of the intestinal tract. Am. J. Clin. Nutr. 27:131.

Busari AD (2004). Sugarcane and sugar Industry in Nigeria. The bittersweet lesson (first edition). Spectra Books Limited, Ibadan. P. 286.

Cancado JE, Saldiva PH, Pereira LA (2006). The impact of sugar cane burning emissions on the respiratory system of children and the elderly. Environ. Health Perspect. 114(5):2023.

Firdos T, Khan AD, Shah FH (1989). Improvement in the digestibility of bagasse pith by chemical treatment. J. Islamic Acad. Sci. 2(2):8992.

Garba B, Zuru A, Sambo AS (1996). Effect of slurry concentration on biogas production from cattle dung. Niger. J. Renew. Energy 4(2):3843.

Greenwood BM (1984). Scientific basis of infection. In: Oxford Textbook of Medicine. DJ Ledinghem and Warrel L (eds) Oxford University Press London. 5:175.

Ilochi EE and Nwachukwu A (1989). Proc. Int. Symposium Biotechnology for Energy. Dec 1621 Faisabad.

Ishihara M, Toyama S, Yonaha K (1988). Biogas production from methane fermentation of sugarcane bagasse. The Science Bulletin. Faculty of Agriculture, University of Ryukyu. 35:4551.

Itodo IO, Onuh CE, Ogar BB (1995). Effect of various total solid concentration of cattle waste on biogas yield. Niger. J. Energy 13:3639.

Kim J, Park C, Kim TH, Lee M, Kim S, Kim SW, Lee J (2003). Effects of various pretreatments for enhanced anaerobic digestion with waste activated sludge. J. Biosci. Bioeng. 95(3):271275.

Meynell PJ (1982). Planning a digester. Prison Press, Stable Court, Chalmington, Dorset. pp. 1015.

Miles and Misra (1938). Surface Viable Count Method. A Standard Laboratory Technique in Pharmaceutics and Pharmaceutical

Microbiology. P. 2450.
Ofoefule AU, Onyeoziri MC, Uzodinma EO (2011). Comparative study of biogas production from chemicallytreated powdered and unpowdered rice husks. J. Environ. Chem. Ecotoxicol. 3(4):7579.
Okoye COB (2001). Trace Metal Concentrations in Nigerian Fruits and Vegetables. Int. J. Environ. Stud. 58:501509.
Pearson D (1976). The Chemical Analysis of Foods. New York: Churchill Livingston. pp. 120148.
Saska M, Gray M (2006). Pretreatment of Sugarcane leaves and bagasse pith with lime impregnation and steam explosion for enzymatic conversion to fermentable sugars. 28th Symposium on Biotechnology for Fuels and Chemicals. Nashville, T.N. April 30–May 3, 2006.
Smith LL (1993). The case for cane In: CERES –The F. A. O Review on Development. July to August 25(143):4144.
Speece RE, McCarthy PL (1964). Nutrient requirements and biological solid accumulation in anaerobic digestion In: International Conference on Water Pollution Research, New York. Permagon Press. P. 305.
Srinivasan SV, Jayanthis S, Sundarajan R (1997). Synergistic effect of kitchen wastes. P. 522.
Suksombat W (2004). Comparison of different alkaline treatment of bagasse and rice straw. J. Anim. Sci. 17(10):14301433.
Walkley A, Black IA (1934). A Standard Analytical Laboratory Technique in the Department of Soil Science, University of Nigeria, Nsukka, Enugu State, Nigeria. P. 26.
Wayman M, Parekh SR (1990). Biotechnology of Biomass Conversion: Fuel and Chemicals from Renewable Resources (1st edition). Open University Press, Milton Keynes. P. 235.
Weemaes M, Grootaerd H, Simoens F, Verstraete W (2000). Anaerobic digestion of ozonized biosolids. Water Res. 34(8):23302336.
Yeom IT, Lee KR, Lee YH, Ahn KH, Lee SH (2002). Effects of ozone treatment on the biodegradability of sludge from municipal waste water treatment plants. Water Sci. Technol. 46(45):421425.

4

Twenty-five years of wind data in south-eastern locations of Nigeria: Modeling and prediction of wind potential

F. C. Odo[1,2]*, D. O. Ugbor[2] and P. E. Ugwuoke[1]

[1]National Centre for Energy Research and Development, University of Nigeria, Nsukka, Nigeria.
[2]Department of Physics and Astronomy, University of Nigeria, Nsukka, Nigeria.

Weibull distribution is often invoked to interpret and predict wind characteristics needed for effective design of wind power systems for different locations. In this paper, daily average wind data for Enugu (6.4°N; 7.5°E), Onitsha (6.8°N; 6.1°E) and Owerri (5.5°N; 7.0°E) over a 25-year period is modeled in terms of the Weibull distribution in order to accurately predict wind potentials for the locations. The monthly and annual wind speed probability density distributions at 10 m meteorological height were analyzed and the Weibull shape and scale factors were empirically determined for the locations. The predicted and measured wind speed probability density distributions of the locations are compared and the accuracy of the model determined for each location using Pearson product moment correlation coefficient (*r*) and root-mean-square error (*ξ*). We find *r* and *ξ* to be 0.64, 1.40, 0.67, 1.17 and 0.93, 1.55, respectively, for Enugu, Onitsha and Owerri. The results suggest that the model can be used, with acceptable accuracy, for predicting wind energy output needed for preliminary design assessment of wind machines for the locations.

Key words: Renewable energy-general, wind, Weibull distribution.

INTRODUCTION

Energy plays a central role in economic development and industrialization of any nation. Fossil fuels have been the major resources that supply the world energy demand. However, fossil fuel reserves are limited and usage of fossil fuels to generate energy has negative environmental effects. The world energy demand is continuously increasing with increasing population such that the present fossil fuel reserves cannot meet this demand (Kamau et al., 2010). As a result, energy policies of many nations are geared towards ensuring a supply of reliable, economical and environmentally friendly energy resources in a form that supports the targets for growth and social development (Ucar and Balo, 2009). Wind energy applications have been recently (Weisser and Garcia, 2006) described as an economic and environmentally friendly solution to the urgent energy problems of many countries.

Wind is an effect caused as a result of pressure differences over regions and heights in the atmosphere resulting in bulk motion of air masses. The force carried by the moving air mass (wind) can be harnessed for useful purposes such as grinding grain (in windmills) and generating electricity (in wind turbine generators). It is estimated that between 1.5 to 2.5% of the global solar radiation received on the surface of the earth is converted to wind (Vosburgh, 1983). Hence, wind energy, which contributes very little pollution and few greenhouse gases to the environment, is a valuable alternative to the non-renewable and environmentally hazardous fossil fuels (Taylor, 1983). Thus, the utilization of wind energy has been increasing around the world at an accelerating pace.

*Corresponding author. E-mail: finboc@yahoo.com.

The extent to which wind can be exploited as a source of energy depends on the probability density of occurrence of different speeds at the site, which is essentially, site-specific. However, the development of new wind projects continues to be hampered by the lack of reliable and accurate wind resource data in many parts of the developing world. To optimize the design of a wind energy conversion device, data on speed range over which the device must operate to maximize energy extraction is required, which requires the knowledge of the frequency distribution of the wind speed. Among the probability density functions that have been proposed for wind speed frequency distributions of most locations, the Weibull distribution has been the most acceptable and forms the basis for commercial wind energy applications and software (Seyit and Ali, 2009). Some of the wind energy software based on the Weibull distribution includes the Wind Atlas Analysis and Application Program (WAsP) and the recently developed Nigerian Wind Energy Information System (WIS).

In previous papers (Enibe, 1987; Ugwuoke et al., 2008; Odo et al., 2010), the theoretical potentials of wind at various heights above the ground, based on annual average values of wind speed, have been assessed for many Nigerian locations. These analyses were carried out using measured data over various periods ranging from 1 to 10 years. In these analyses, little or no attention was given to the frequency distribution patterns of wind speed over the studied periods for the locations. In this paper, the frequency distribution of daily averages of wind speed for three locations in south-eastern Nigeria, namely; Enugu, Onitsha and Owerri, over a longer period of up to 25 years are examined. The observed data for these three locations are modeled in terms of the Weibull distribution, to enable an accurate prediction of the wind potentials of the locations and the results are compared. The results of this analysis are expected to be very useful to designers of wind turbines, for various wind energy applications, for the locations.

MATERIALS AND METHODS

In this study, we use 25 years (1978 to 2003) daily averages of wind speed data at 10 m meteorological height, for Enugu (6.4°N; 7.5°E), Onitsha (6.8°N; 6.1°E) and Owerri (5.5°N; 7.0°E) obtained from the data bank of Nigerian Meteorological Agency (NIMET). The data gives information on the daily average wind speed distributions of the locations over the study period, from which the monthly and yearly average data were calculated for the current analysis.

Weibull probability density function

The Weibull probability density distribution is a two-parameter function characterized by a dimensionless shape (k) parameter and scale (c) parameter (in unit of speed). It is a mathematical idealization of the distribution of wind speed over time for most locations. The function gives the probability of wind speed being in a range of 1 m/s about a particular speed (v), taking into account all variations for the period covered by the statistics. The Weibull distribution is a statistical function given (Walker and Jenkins, 1997; Gipe, 2004) by:

$$f_{(v)} = \frac{k}{c}\left(\frac{v}{c}\right)^{k-1} \exp\left(-\frac{v}{c}\right)^{k}, \tag{1}$$

Where $f_{(v)}$ is the probability density defined as the frequency of occurrence of wind speed (v), c is the scale parameter (in unit of m/s), which is closely related to the wind speed for the location, and k is the dimensionless shape parameter, which describes the width of the distribution and measures the probability of extraction of wind energy at a given characteristic wind speed. The Weibull distribution is therefore characterized for any location by the two parameters c and k. The cumulative form of the Weibull distribution $F_{(v)}$ which gives the probability of the wind speed exceeding the value v is expressed (Justus et al., 1978; Walker and Jenkins, 1997) as:

$$F_{(v)} = \exp\left(-\frac{v}{c}\right)^{k} \tag{2}$$

On the other hand, the power derivable from the wind is a cubic function of the wind speed such that, in the Weibull distribution, the power density (P_A) of the wind at any speed is given (Seyit and Ali, 2009) by:

$$P_{(A)} = \frac{1}{2}\rho\int_0^{\infty} v^3 f_{(v)} dv, \tag{3}$$

Where ρ is the density of air. However, the power derivable from the wind scales with the height (h) above the ground $\left(h \propto \frac{1}{\rho}\right)$ according to the Hellman's exponential law given (Walker and Jenkins, 1997; Gipe, 2004) by:

$$\frac{v}{v_0} = \left(\frac{h}{h_0}\right)^{\alpha} \tag{4}$$

Where h_0 is any reference height and v_0 is the wind speed at h_0, while α is the Hellman's constant which varies from one location to another. Equation 4 suggests that the derivable power increases with increasing height only if the change in density of air is negligible. It has been shown (Walker and Jenkins, 1997) that within the troposphere ($h \leq 10$ km) the density of air varies very little for any location.

Analysis of Equation 3 using Equation 1 shows that the power density could be expressed as a gamma function (Γ) defined in general x-variable (Dass, 1998) as:

$$\Gamma x = \int_0^{\infty} x^{n-1} e^{-x} dx. \tag{5}$$

Using Equations 5 and 1 in Equation 3, the average wind power

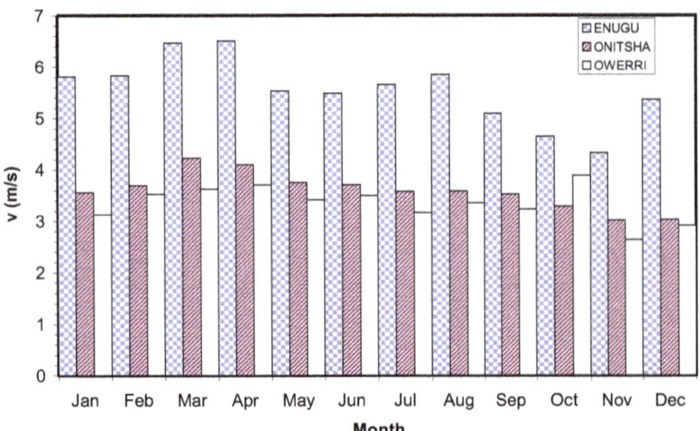

Figure 1. Time series distributions of wind speed for the locations.

Table 1. Wind speed distributions parameters for Enugu, Onitsha and Owerri.

Location	Latitude	Longitude	v_{mean} (m/s)	k	c (m/s)
Enugu	6.4°N	7.5°E	5.5 ± 0.6	2.0	6.4
Onitsha	6.8°N	6.1°E	3.6 ± 0.4	1.5	5.6
Owerri	5.5°N	7.0°E	3.3 ± 0.3	1.9	5.1

density (P_{av}) based on Weibull distribution can be expressed (Ucar and Balo, 2009) in the form:

$$P_{(av)} = \frac{1}{2}\rho\omega^3 \frac{\Gamma\left(1+\frac{3}{k}\right)}{\left[\Gamma\left(1+\frac{1}{k}\right)\right]^3} \tag{6}$$

Where ω is the characteristic wind speed of the location. However, meteorologists have characterized the distributions of wind speeds for many of the world's wind regimes in terms of the speed distribution patterns. For example, in temperate climate (mid latitudes), a typical shape parameter $k \approx 2$ offers a good approximation (Gipe, 2004). For $k = 2$, Equation 1 or 2 reduces to Rayleigh wind speed distribution. Thus, the Rayleigh distribution is a special case of the Weibull distribution developed for estimation of wind potential in temperate climate locations. Wind characteristics are essentially location specific and performance of real wind conversion devices which are designed based on the Rayleigh distribution may greatly differ if actual wind conditions at the location differ from those standard speed distributions.

A method has been suggested (Iheonu et al., 2002; Gipe, 2004) for estimating the shape (k) factor of a set of wind speed data using the mean wind speed (v) and standard deviation (σ) in a simple relation of the form:

$$k = \left(\frac{\sigma}{v}\right)^{-1.086} \tag{7}$$

Equation 7 therefore suggests that the probability of capturing the wind by a turbine at a mean wind speed is small if the shape factor is high for that location, since k could be used as a measure of dispersion (Pallabazzer, 2003) in a distribution. This shows that knowledge of the exact value of k provides preliminary information on the wind speed regime for which wind turbines should be designed for optimum performance in any given location. Similarly, wind speed is a real valued random variable and most locations show wide dispersion (Justus et al., 1978) so that the use of mean values as the characteristic speed for designs may not be very reliable (Jaramillo and Borja, 2004).

In this paper, we use analytic method in which $F_{(v)}$ is plotted against v on double logarithm scales and apply a one-dimensional regression on the plots to obtain values for k and c for the locations.

RESULTS AND DISCUSSION

We calculated the monthly and annual average wind speed distributions at 10 m meteorological height for the three locations over the studied period. The time series distributions of the monthly average values for the locations are shown in Figure 1, while the annual average values are shown in Table 1. The distributions give annual mean values of 5.5 ± 0.6 m/s, 3.6 ± 0.4 m/s and 3.3 ± 0.3 m/s, respectively, for Enugu, Onitsha and Owerri. It could easily be observed from Figure 1 that the distributions of the monthly average wind speed for the three locations are fairly similar, peaking in the month of March and having minimum values in November. However, wind speed is highest in Enugu and lowest in Owerri. Perhaps, the result is as expected since difference in wind speed distributions may be related to the difference in altitude between the locations. To model the data in terms of the Weibull distribution, we took twice logarithm of Equation 2 to obtain

$$\ln(-\ln F_v) = k\ln v - k\ln c\,. \tag{8}$$

The plots of ln (-ln F_V) as a function of ln v for the three locations, on the same scale, are shown in Figure 2. Linear regression of the plots gives

$$\ln(-\ln F_v) = 2.0\ln v - 3.7\,,$$

$$\ln(-\ln F_v) = 1.5\ln v - 2.6$$

and

$$\ln(-\ln F_v) = 1.9\ln v - 3.1$$

respectively, for Enugu, Onitsha and Owerri. By comparing each of the equations with Equation 8, the values of k and c were deduced for each location. The summary of the results is shown in Table 1.

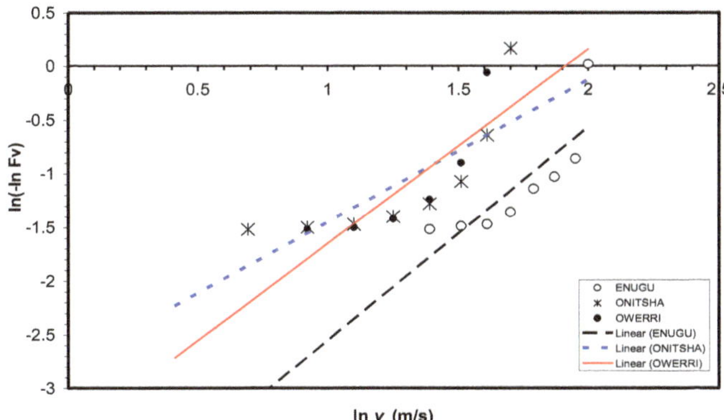

Figure 2. Plot of ln(-ln F_v) against lnv for the locations.

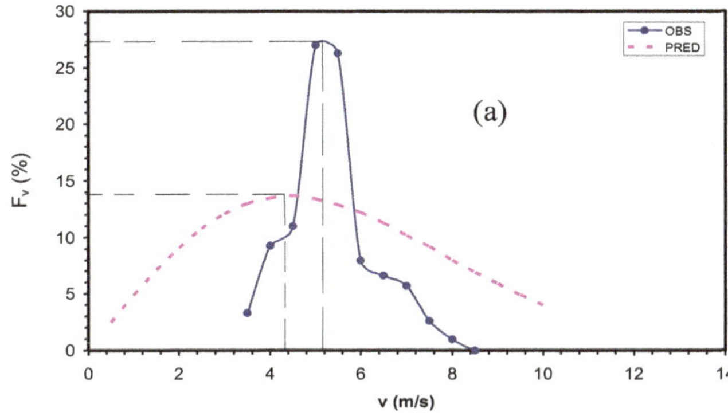

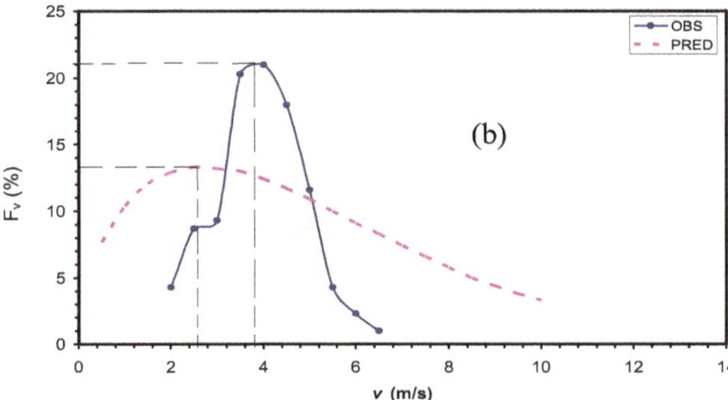

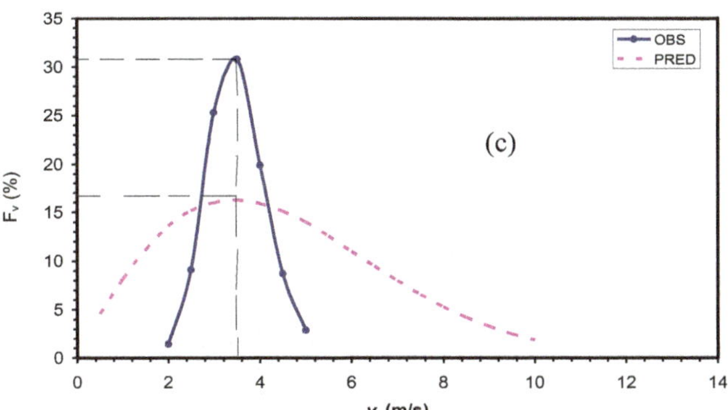

Figure 3. Observed and predicted probability density for: (a) Enugu, (b) Onitsha, (c) Owerri.

Furthermore, using the results for k and c obtained from the regression analysis for each location in Equation 1, the probability densities (%) of occurrence of different wind speeds were determined. The different speeds used for this analysis were chosen from the range $0 \leq v \leq 10$ m/s as covered by observational data in eight (8) comparable speed bins, for effective overlap. The results of the prediction were compared with those calculated from observed data. The measured and predicted probability density distributions for different wind speed bins for the three locations are shown in Figure 3, while the summary of the comparison is displayed in Table 2. Similarly, the suitability of the model in predicting wind potential for each location was determined using two non-parametric statistics, namely, the Pearson product moment correlation coefficient (r) and root-mean-square error (ξ). The Pearson correlation coefficient is defined (Aalen, 1978) as:

$$r = \left\{ 1 - \frac{\sum_{i=1}^{N} (y_i - x_i)^2}{\sum_{i=1}^{N} (y_i - \bar{y})^2} \right\}^{1/2}, \tag{9}$$

Where N is the number of observations in each data set, y and x, respectively are the measured and predicted probability density, while $\bar{y}$ is the mean of the measured values. This statistic varies from 0 (for a null association) to ±1 (for a perfect association). A correlation is statistically significant at a set level of significance if $r \geq 0.5$, otherwise, it is not significant. The present analyses give correlation coefficients $r \sim 0.6$, 0.7 and 0.9, respectively, for Enugu, Onitsha and Owerri at 5% level of significance. Thus, all the correlations are statistically significant at 5% level.

On the other hand, the root-mean-square error is a statistic that determines the degree of departure of two data sets from a supposed association and is defined (Joanes and Gills, 1998) as:

$$\xi = \left[\frac{1}{N} \sum (y_i - x_i)^2 \right]^{1/2} \tag{10}$$

The results give $\xi \approx 1.40$, 1.17 and 1.55, respectively, for Enugu, Onitsha and Owerri. Results of all these analyses are summarized in Table 2.

Modeling and prediction of wind characteristics are major design inputs in the development of wind power systems for any location. However, the wind speed distribution for many of the world's wind regimes have been characterized and wind power systems are

Table 2. Comparison of measured and predicted probability density.

Location	v_{peak} (m/s)		$F_{(v)\ peak}$ (%)		r	ξ
	Measured	Predicted	Measured	Predicted		
Enugu	5.2	4.4	22.0	14.0	0.64	1.40
Onitsha	3.8	2.6	21.0	13.5	0.67	1.17
Owerri	3.5	3.5	31.0	17.0	0.93	1.55

optimized based on these standard distributions. In fact, the Rayleigh distribution (k = 2) is often employed by most wind power system developers (Vosburgh, 1983). The result is that many wind power systems perform poorly in many different locations because the actual wind conditions at the locations differ largely from those standard distributions (Ramachandra et al., 2005; Al-Mohamad and Karmeh, 2003).

It could easily be observed from the time series distributions of the monthly average wind speed that the three locations studied in south-eastern Nigeria show similarity in climatology of wind. Perhaps, this observation could be attributed to their proximity in geographical extent. The three locations lie within a latitudinal stretch of about 1.3° and longitudinal stretch of 1.4°. Similarly, the distributions do not show any latitudinal or longitudinal dependence either. Thus, wind speed in south-eastern locations of Nigeria is essentially location specific, which may be driven by environmental factors, rather than geographical dependence.

We have also shown in the results that the Weibull shape factor is 2.0, 1.5 and 1.9, respectively for Enugu, Onitsha and Owerri, while the corresponding scale factor is 6.4, 5.6 and 5.1 m/s. The values of the shape factor presented in this paper suggest that while the data for Enugu and Owerri are in close agreement with Rayleigh distribution, the data for Onitsha departs significantly from the standard Rayleigh distribution. The results further suggest that the wind speed distributions in the studied locations are widely dispersed, with those of Enugu and Owerri being much wider than that of Onitsha. The implication of these results is that any wind turbine which is optimized based on Rayleigh distribution may be suitable for Enugu and Owerri, but not for Onitsha. It therefore becomes necessary that wind turbines for utility generation in these locations be designed locally rather than relying on importation of already designed systems.

A more comprehensive wind speed evaluation and energy assessment is achieved by the use of real life frequency distribution. The frequency distribution obviously indicates the percentage of the time of occurrence of the various wind bins/spectra and provides information on when a particular rated turbine in the location is expected to yield power (Amonye and Hassan, 2010). Thus, the most frequently occurring wind spectra defines the characteristic wind speed for the location, which is a more reliable value for the design of wind turbines for any location than the average value, which is affected by the skewness of the distribution (Iheonu et al., 2002; Ucar and Balo, 2009). A major outcome of our result is the statistical accuracy of the Weibull distribution model in predicting wind potentials of the locations as determined by the correlation coefficient. In fact, for Owerri, the model gives almost a perfect prediction of the characteristic wind speed. However, for all the locations, the correlation coefficients are statistically significant. These results show that the predicted and measured data are acceptably related by the Weibull distribution. Hence, within the region of overlap, the Weibull distribution can be used, with acceptable accuracy, to predict wind potentials for these locations.

On the other hand, it is obvious in Figure 3 that there is a wide departure of the theoretical predictions from the real life wind speed frequency distribution patterns of the studied locations. In fact, for each of the locations, within the region of overlap, the theoretical model appears to underestimate the probability density of every speed bin. This is further supported statistically by the large values of root-mean-square error calculated for all the locations. Perhaps, this departure is attributable to the coarse approximations arising from daily averages which fail to account for the short time-scale variations of wind characteristics of the locations. Wind speed is a real valued random variable and observations over smaller time-scales, such as hourly averages, may help to improve the results.

REFERENCES

Aalen OO (1978). An Introduction to Categorical Data Analysis. Ann. Stat. 6:701.

Al-Mohamad A, Karmeh H (2003). Wind Energy Potential in Syria. Renew. Energy 28:1039.

Amonye M, Hassan B (2010). Basic Steps in Wind Turbine Development, TradeNews, Nigeria. P. 33.

Dass HK (1998). Advanced Engineering mathematics, S. Chand & company Ltd, New Delhi. P. 1158.

Enibe SO (1987). A Method of Assessing Wind Energy Potential in a Nigerian Location. Nig. J. Solar Energy 6:14.

Gipe P (2004). Wind Power: Renewable Energy for home Farm and Business, Chelsea Green, USA. P. 17.

Iheonu EE, Akingbade FO, Ocholi M (2002). Wind Resource Variation over Selected Sites in West Africa. Nig. J. Renew. Energy 10:43.

Jaramillo OA, Borja MA (2004). Wind Speed Analysis in La Ventosa, Mexico: A Bimodal probability distribution case. J. Renew. Energy 29:1613.

Joanes DN, Gill CA (1998). Comparing measures of sample Skewness and Kurtosis. J. Royal Stat. Soc. 47:183.

Justus CG, Hargraves WR, Mikhail A, Graber D (1978). Methods for Estimating Wind Speed Frequency distribution. J. Appl. Meteorol. 17:350.

Kamau JN, Kinyua R, Gathua J (2010). Six years of wind data for Marsabit: An Analysis of wind energy potentials. J. Renew. Energy 35:1298.

Odo FC, Igwebuike MN, Ezugwu CI (2010). Preliminary Assessment of Enugu Location for Setting up a 1kW Wind turbine for Electricity Generation. Nig. J. Solar Energy 21:182.

Pallabazzer R (2003). Parametric analysis of wind sitting efficiency. J. Wind Eng. 91:1329.

Ramachandra TV, Rajeev KJ, Vansee K, Shruthi B (2005). Wind Energy Potential Assessment Spatial Decision Support System. Energy Educ. Sci. Technol. 14:61.

Seyit AA, Ali D (2009). A New method to estimate Weibull parameters for Wind Energy applications. Energy Convers. Manag. 50:1761.

Taylor RH (1983). Alternative Energy Sources for the Centralised Generation of Electricity, Adam Hilger, Bristol. P. 56.

Ucar A, Balo F (2009). Investigation of wind characteristics and generation Potentiality in Uludag-Bursa, Turkey. J. Appl. Energy 86:333.

Ugwuoke PE, Bala EJ, Sambo AS, Argungu GM (2008). Essessment of Wind Energy Potentials for Electricity Generation in the Nigerian Rural Setting. Nig. J. Solar Energy 19:98.

Vosburgh PN (1983). Commercial Application of Wind Power, Van Nostrand Rainhold, New York. P. 12.

Walker JF, Jenkins S (1997). Wind Energy Technology, John Wiley & Sons, Chinchester. P. 24.

Weisser D, Garcia RS (2006). Instantaneous wind Energy Penetration in Isolated Electricity Grids. J. Renew. Energy 30:1299.

5

Development of two-dimensional solver code for hybrid model of energy absorbing system

Z. Ahmad[1] **and J. Campbell**[2]

[1]Faculty of Mechanical Engineering, Universiti Teknologi Malaysia, 81300, Johor Bahru, Malaysia.
[2]School of Engineering, Cranfield University, Cranfield, Bedfordshire MK43 0AL, United Kingdom.

This paper treats the dynamic and energy absorption responses of nonlinear spring-mass system using a developed two dimensional solver code. The influence of spring stiffness and various mass configurations on the impact response was investigated using validated models. Results are quantified in terms of important impact response and indicate that an assembly of spring-mass system can be employed to examine the impact response of energy absorption system prior to the detailed numerical analysis and experiment. The developed solver code will be useful for educational purposes to preliminarily understand the behavior and impact response of energy absorbing system without learning a piece of complicated nonlinear commercial finite element codes namely LS-DYNA. The developed code could facilitate the early stage of evaluation for impact applications.

Key words: Spring-mass, dynamic, solver, energy absorption, finite element, impact.

INTRODUCTION

For several decades, an increased emphasis has been placed on crashworthiness as a structural design requirement for occupant-carrying vehicles. Structural systems in vehicles are continuously subjected to increasing safety requirement since vehicles are exposed to multidirectional impact conditions namely frontal, rear, side and roll over impacts. Although, destructive testing provides the most reliable check of the structural performance in an impact event, trial and error approach is impractical as it is relatively expensive and time consuming. The most important decision in the development of a new crashworthy structure is made in the early stage of design so that significant economic gains can be achieved by using a reasonably accurate theoretical method to predict the relevant collapse properties (Deb et al., 2004; Sheh et al., 1992; Pifko and Winter, 1981; Deb and Ali, 2004; Wu and Yu, 2001). In order to meet crashworthiness criteria, it is indispensible that an adequate crashworthiness evaluation method may be used as early as possible in the design process. In general, various types of reliable crash analysis using experimental, theoretical, hybrid, finite element and analytical models can be conducted to examine the performance of energy absorption system in terms of energy absorption capacity, deceleration effect, dynamic force experienced by occupants and so forth (Jones, 1989; Jones and Wierzbicki, 1983). There have been numerous studies on the crash analysis of vehicular system available particularly in using experimental and finite element models as a primary approach (Benson, 1992; Wu and Cheng, 1997). However, most of the investigation emphasize on the detailed analysis rather than on the preliminary design concept in the early stage of design. It has been established that a hybrid analysis can promote a simple way of analyzing the effectiveness of energy absorption system with minimum effort and computational time (Wu and Yu, 2001; Ruan and Yu, 2003).

Hybrid analysis integrates a component testing and a simplified finite-element-based analysis. This integration of experimental and analytical approach significantly reduces development time and it also permits the study of a much wider variety of designs compared with conducting experimental testing alone (Gouddreau and Hallquist, 1982). In addition, this method requires less computer time than finite element methods leading to shorten design cycles and reduce development costs as well as providing useful preliminary design information relatively quickly or gross estimates of structural response. Moreover, by using this method, design errors can be easier to rectify before producing a full-scale model. Consequently, simplified hybrid models enable a quick parametric study to identify an optimum distribution of crush components, thus setting priority design targets. Hybrid method is also known as combined experimental and numerical methods in which a structure is divided into a number of masses which is connected each other using subassemblies that are treated as beams or nonlinear spring elements (Pifko and Winter, 1981; Hofmeister, 1978). A hybrid approach relies on experimental data gained separately from the load-deflection curves of full scale tests of each structural component in conjunction with a theoretical model (Liaw et al., 1988; Kecman, 1997). The use of a relatively simple theoretical model aids better understanding of involved collapse phenomenon. If the load-deflection response (stiffness characteristic) of components (beam or nonlinear spring elements) is known, a single or multi degree of freedom mass-spring systems can be possibly developed to simplify a model of the structure without concern of the material used (Liaw et al., 1988). The load carrying capacity of a structure is assumed to be insensitive to the deformation rate (Kirkpatrick et al., 2000). This approach, however, may give a limited accuracy of the results. In a pioneer research on hybrid model, Herridge and Mitchell (1978) have modeled a vehicle collision with two dimensional crash model. It appears that the model may anticipate the behavior of the structures in a brief manner. Sheh et al. (1992) have developed a simple hybrid model, namely a lumped parameter (LP) model in order to simply perform a frontal vehicle crash simulation. The model mainly consists of lumped masses and non-linear springs for a system with a single degree of freedom and the experimental data were used to assist the development of models. It is evident that by using this model, acceptable levels of safety in passenger automobile can be initially analyzed. In another study, Liaw et al. (1988) have used three dimensional hybrid models consist of lumped masses, nonlinear spring and beam using computer KRASH program. The nonlinear properties of the corresponding components were obtained from the static crush test and then as an input for the simulation using the computer program KRASH.

The present study treats the dynamic and energy absorption responses of crashworthy system under two dimensional impact loading. Dynamic response and feasibility of component assembly in energy absorbing system has been examined by varying component masses, crush characteristic of individual components and impact velocity. A solver code developed by using MATLAB programming (Thomson, 1988; Venkataraman, 2002), validated using existing commercial finite element code LS-DYNA was employed to investigate the relative effect of varied parameter on the dynamic response of the energy absorption system. Overall the results demonstrate the advantages of using the developed code as an analysis tool aids for academic and industrial usages. The primary outcome of this study is a simple solver code for educational purposes to preliminarily understand the behavior and impact response of energy absorbing system without learning a piece of complicated nonlinear commercial finite element codes namely LS-DYNA. The developed code could facilitate the early stage of design evaluation for impact applications specifically in automotive and aviation industries.

DEVELOPMENT OF HYBRID SOLVER AND COMPUTATIONAL MODELLING

Lagrangian computational method

The main contribution of this paper is the development of solver code using MATLAB algorithm. This programming solver was developed in accordance with the explicit Lagrangian computational method (Benson, 1992; Gouddreau and Hallquist, 1982). The solutions are advanced in time using an explicit integration scheme and they must be resolved accurately in both space and time domains (Cook et al., 1989; Jonsén et al., 2009). Recently the development of Lagrangian hydrocode is continued for a wide range of application namely automotive and aviation industries. Fundamentally, Lagrangian hydrocode applies Lagrange formulation for spatial discretization (Hallquist, 2006). Spatial discretisation can be carried out in either the Lagrangian or Eulerian framework. This hybrid solver code is a simplified hydrocode thereby becoming easier and straightforward. In particular, the time derivatives are material and thus simple partials of Lagrangian frame where coordinates are assigned to material points. As iterated in Benson (1992), a general iteration of Lagrangian code is outlined as follows.

(i) Stress, pressure, hourglass forces at t^n in each element, calculate the force at the node and followed by acceleration calculation,
(ii) Integration of acceleration to get velocity at $t^{n+1/2}$,
(iii) Integration of velocity to get displacement at t^{n+1},

(iv) Integration of constitutive model for the strength from t^n to t^{n+1},
(v) Calculation of shock viscosity and hourglass viscosity from $u^{n+1/2}$,
(vi) Updating internal energy based on work done between t^n and t^{n+1},
(vii) Pressure is calculated from the equation of state,
(viii) New time step is calculated,
(ix) Advance the time and return to initial step.

where t^n and t^{n+1} are full time intervals while $t^{n+1/2}$ and $u^{n+1/2}$ are half time intervals.

From the above iteration, it should be noted that the present study has only calculated the force, thus ignoring fifth and seventh steps in consideration of the type of elements used and complicated calculation for pressure, shock and hourglass viscosity. Second, third, fourth, sixth, eighth and ninth steps were accordingly devoted in this solver code. For the fourth step, the evaluation of force deflection curve was performed rather than the integration of constitutive model for the strength. For clarity of the developed solver code, the subroutines used are outlined as follows.

(i) Set-up problem: Input variables, material models, material properties, define curves,
(ii) Initialize problem: Calculate stiffness, critical timestep, state initialization, initial length, element direction vector
(iii) Solution: Solver,
(iv) Plotting results: Displacement, velocity and acceleration.

The equations used in the solver are based on Lagrangian formulation, equation of motion and fundamental spring-mass system.

Central difference time integration

Computational method using Lagrange formulation can solve the position of the mesh at discrete points in time (Cook et al., 1989). The solution is advanced from t^n to time t^{n+1} without any iterations and the time step $\Delta t^{n+1/2}$. In order to minimize the storage required, the solution is stored for only one time t^n within the program and the initial solution of the time step is overwritten by the solution at the end of the step. The central different method was used to advance the position of the mesh in time and it is based on second order accurate central difference approximation. From the Taylor series derivation, the integration rule can be carried out. The integration rule for the Lagrange formulation is expressed by Equations (1) and (2). In the present study, discrete element was used and do not involve any element meshes.

$$x^{n+1} = x^n + u^{n+1/2} \Delta t^{n+1/2} \tag{1}$$

$$u^{n+1/2} = u^{n-1/2} + \dot{u}^n \Delta t^n = u^{n-1/2} + (F/m)\Delta t^n \tag{2}$$

Herein, x, u and $\dot{u}$ are the displacement, velocity and acceleration respectively. In Equation (2), F and m are denoted to force and mass, respectively.

Central difference method is prevalently used in explicit direct integration methods to integrate the equations of motion in time (Benson, 1992; Gouddreau and Hallquist, 1982; Cook et al., 1989). Equation of motion is evaluated at the central time. In this present solver, the damping matrix was ignored as the system is an undamped system. Thereby, the equations of motion can be expressed by Equations (3) and (4) for the linear problem and nonlinear problem, respectively. In those equations, M, K and f^{ext} are a group of matrices for mass, stiffness and external force, respectively.

$$[M]\{\dot{u}\} + [K]\{x\} = \{f^{ext}\} \tag{3}$$

$$[M]\{\dot{u}\} + [K(x)]\{x\} = \{f^{ext}\} \tag{4}$$

For linear problems, the stiffness matrix is constant in time while for the nonlinear problem the stiffness matrix is a function of displacement as shown in Equuation (4). The central difference algorithm calculates the acceleration and velocity at the full time intervals, t^{n-1}, t^n, and t^{n+1} and the velocity at the half time interval, $t^{n-1/2}$ and $t^{n+1/2}$. In addition, this algorithm has a finite stable time step (Hallquist, 2006; Livermore Software Technology Corporation, LS-DYNA Keyword User's Manual, 2003). The critical time step must be determined to keep the time step for stability. Theoretically, central difference method is a second order accuracy with a single evaluation. Therefore, it can save the computational cost compared to second order Runge-Kutta. It is evident that the chosen method is problem dependences. In general, direct integration method is more expedient for a complicated nonlinearity. However, central difference method is more preferable for nonlinear dynamic problems (Stronge and Yu, 1993).

Specifications of solver code

Geometry

In this study, discrete element and nodal mass were employed for the spring and component masses, respectively. This type of element can represent a large system when subjected to a gross motion. In addition, discrete element and masses provide a capability for modeling a simple spring-mass system and can be used for more complicated mechanism. It should be noted that the discrete element is deemed to be massless and

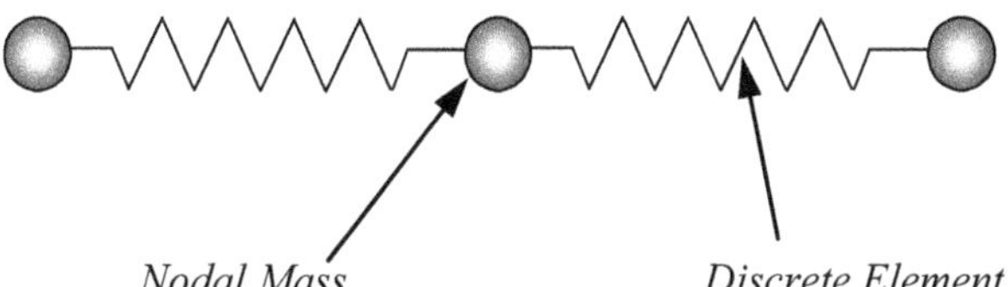

Figure 1. The discrete element and nodal masses for a spring-mass system.

consists of two nodes for each element. In particular, the nodal masses are specified at this unconstraint node or joining node to separate the element. Figure 1 shows the discrete elements and nodal masses for a spring-mass system consisting of two springs and three masses.

Force-displacement curve is an element property that can individually be included to represent the stiffness of the discrete element. By using discrete elements, various stiffness properties can accordingly be assigned to each element. The stiffness property is represented by stiffness matrix for a linear problem while load curve property is employed for a nonlinear problem. The nodal masses represent the crash component such as engine, bumper, sub floor, landing gear and other crashworthy components. Furthermore, element and nodal forces can easily be determined by using discrete element and nodal masses. A force of the local element coordinates can be expressed by the following equation.

$$F_{node} = \sum_{n} F_{element} \tag{5}$$

n = Elements attached to node.

Thereafter, kinematics responses namely acceleration, velocity and displacement for each nodal point mass can be computed directly with the time increment.

Material model

For the spring mass system, basic material models were defined excluding the strain rate effect and complicated equation of state. This programming solver code offers independent strain rate material models for the discrete elements such as linear elastic, nonlinear elastic and general nonlinear. Linear elastic discrete element is a typical force-displacement relations which is similar to Hooke's law as expressed in Equation (6) where k is the element's stiffness and Δl is the change of element's length.

$$\hat{F} = k\Delta l \tag{6}$$

Thus, the linear elastic describes the system in which the force increases linearly with the displacement. The spring

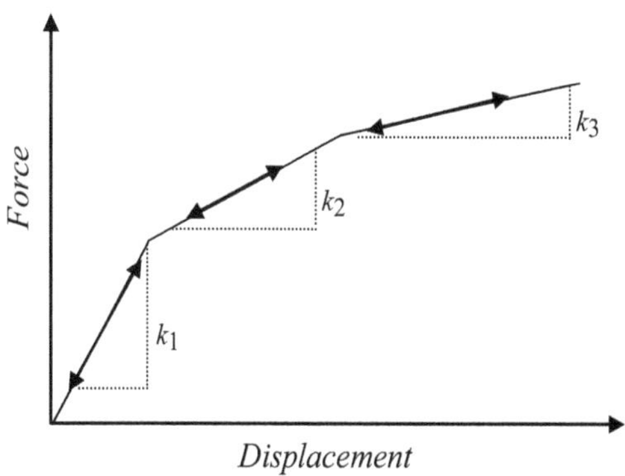

Figure 2. Material model of nonlinear elastic.

element remains in a linear elastic region without yielding when subjected to loading. Nonlinear elastic discrete element exhibits a nonlinear behaviour with varying stiffness properties. Nonlinear elastic model expresses a tabulated force-displacement relation as in Equation (7) where $k\ (\Delta l)$ is the tabulated stiffness relation depending on the total change of element's length.

$$\hat{F} = [k(\Delta l)](\Delta l) \tag{7}$$

In particular, when a nonlinear elastic element starts to be loaded, it follows the loading path with varying stiffness properties. It remains in loading paths without yielding. For an unloading case, a nonlinear element is unloaded back along the loading paths. Figure 2 depicts the behaviour of the nonlinear elastic material model under loading and unloading cases.

General nonlinear material model is a primary material model of this study. This material model can offer better solution for the crash analysis. In many real situations, structure always responses under a nonlinear behaviour and involves in numerous nonlinear problems. In general, a nonlinear problem is inherently more complex to be analyzed than linear problems. It is worth noting that the model can simulate the response of structural components with the inclusion of independent and nonsymmetrical linear loading and unloading paths including the initial yield forces. It is noted that a general nonlinear element yields in either tension or compression. For instance, if the element is subjected to tension, the discrete element loads and unloads along the loading or unloading path as long as the load does not reach a yield point. Upon exceeding yield point and at all subsequent time, the element follows the specified loading and unloading curves. In addition, the displacement origin of the unloading curve is arbitrary and it is shifted to the present displacement where the element starts to unload. Then, the element reloads along the unloading path until it reaches the new updated yield point and follows the loading curve subsequently as demonstrated in Figure 3.

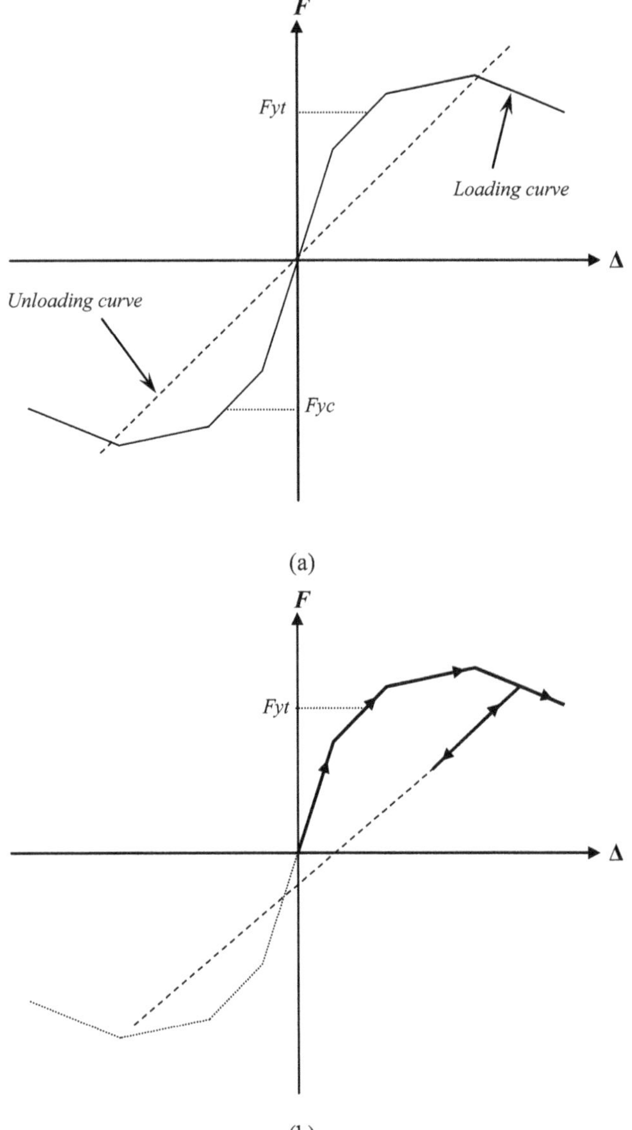

Figure 3. Material model of general nonlinear (a) Loading and unloading curve (b) Element yields in tension.

Initial, boundary and loading condition

In the programmed solver code, initial conditions are necessary in order to determine the kinematic behaviour and internal forces. The initial conditions required are initial velocity and initial coordinate. An initial velocity is needed to evaluate the motion (acceleration or deceleration) of the nodal masses. Alternatively, an initial acceleration can also be specified for the same purpose. Moreover, an initial coordinate is defined in the solver for capturing the translational motion of the element due to the change of the coordinate.

The stationary nodal masses were fully constrained in translation. The present solver also includes the prescribe motion boundary condition. This boundary condition can be devoted to a prescribe motion for nodal motion such as displacement, velocity and acceleration imposed as a prescribed function of time. This is convenient when individual or a set of nodal masses have a required velocity or acceleration time history.

There are only two types of loading conditions that have been specified in the solver namely, a concentrated load and a load with the inclusion of time variation (defined load curve). Any specified load curve has a tabulated force time relation.

Solver output

The present solver code is capable of providing beneficial design information which is pertinent to the structure crash analysis. A visualized output namely time-history plotting can interactively be presented to give a basic understanding of structure behaviour. Time history plot comprises of coordinate-time, displacement-time, velocity-time, acceleration-time and force-time. From those plots, the outputs are integrated to produce force-displacement curve. It is well known that force-displacement curve is dispensable in measuring the crush and energy absorption capacity of the system. In addition, this solver can also produce the kinetic and internal energy outputs. Therefore, the amount of dissipated energy can be evaluated for various spring-mass assemblies. A nodal mass contributes a kinetic energy while a spring discrete element contributes an internal energy.

Description of programming solver

The spring-mass system represents an assembly of crash component in structural analysis. The system consists of nodal masses, linear spring, nonlinear spring and general nonlinear spring. In developing a spring-mass system for crashworthiness analysis, it involves a topology study on the number of masses and springs to be connected and how are they connected. Mass distribution also needs to be justified in order to get an acceptable representation of the physical system.

Initially, a single spring with a pair of masses was used to understand the behaviour of the model and to validate linear, nonlinear and general nonlinear spring elements. For validation, the mass and stiffness property values were chosen for the single spring validation. Figure 4 shows a single spring-mass system in the preliminary model. Subsequently, two springs were duly used to evaluate the spring element behaviour due to the combination of linear elastic, nonlinear elastic or nonlinear elastic and general nonlinear material model.

To ascertain whether the solver code was sufficiently accurate and feasible for analyzing more complicated

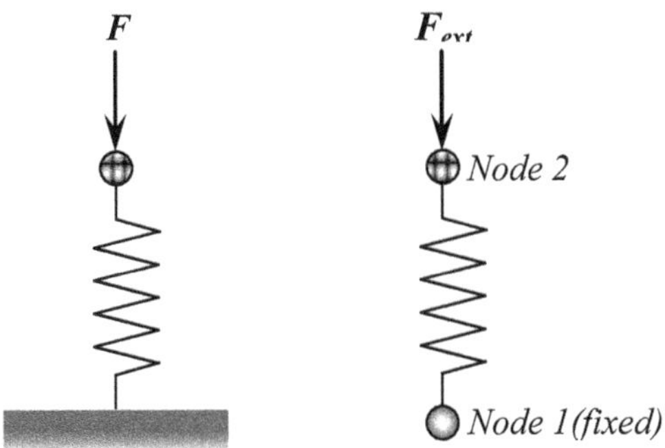

Figure 4. A spring-mass system in the preliminary model.

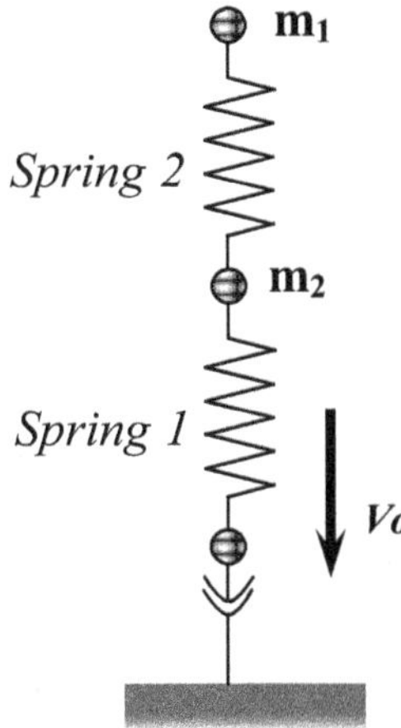

Figure 5. A spring-mass system of an idealized helicopter sub floor vertical impact.

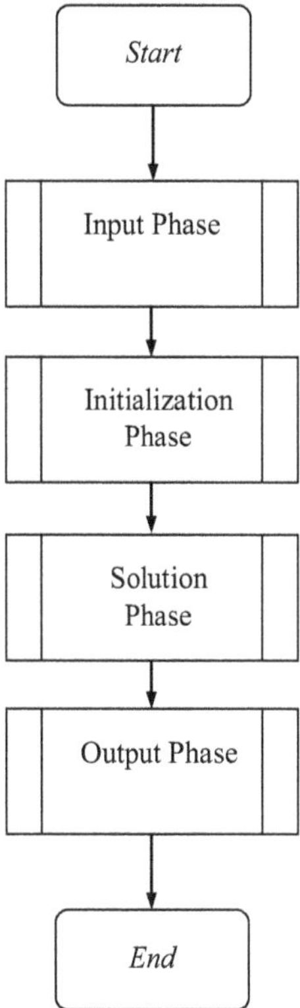

Figure 6. Programming flow for the solver code.

crash problems which consists of multi spring-masses, the code was programmed to perform analysis for multi spring-mass system representing a helicopter sub-floor structure as depicted in Figure 5. From Figure 5, the spring elements 1 and 2 represent a helicopter sub-floor and helicopter cabin structures, respectively. For the nodal masses, mass 1 represents a helicopter transmission and mass 2 represents a troop compartment.

In general, programming flow comprises several phases namely input, initialization, solutions and output phases. The detailed programming flow of the entire solver code can be illustrated in Figure 6.

VALIDATION AND DEMONSTRATION OF THE SOLVER CODE

The solver code was corroborated directly against the numerical results obtained from LS-DYNA solver. Here, the validation is categorized for the single spring element and multi-spring element leading to the practical application of the structural crash.

Single spring element

Single spring element was initially simulated to duly comprehend the spring element behaviour under pertinent loading condition. For the single spring element, LS-DYNA was not employed to validate the response. For validation, force-time, displacement-time and load-displacement curves are mainly of interest as those curves are sufficient to show the validity of the material model for a single spring element. Most importantly, the single spring element can present a reasonable trend prior to the subsequent validation of the multi-spring element.

Linear elastic

Figures 7 and 8 show the validation results for the single spring element. Obviously, the simulated curve trend shows curve response as expected when employing the

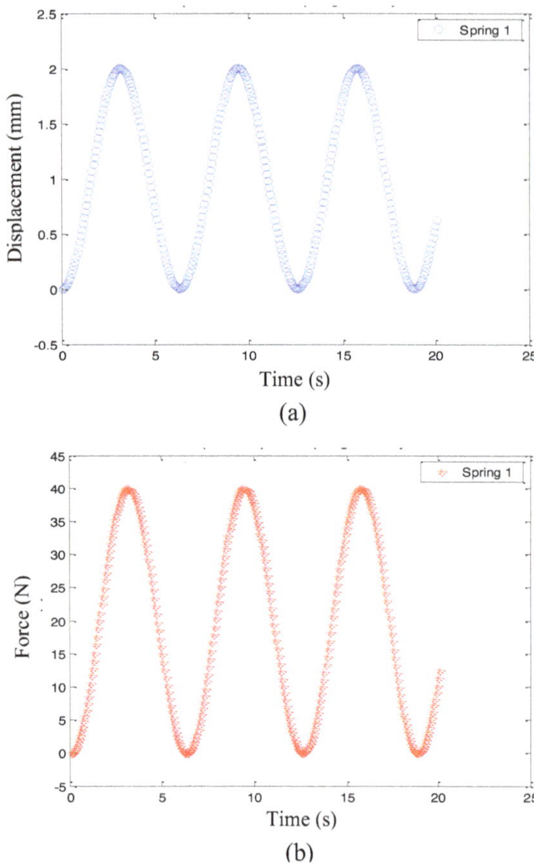

Figure 7. (a) Displacement-time (b) Force-time for a linear elastic spring element.

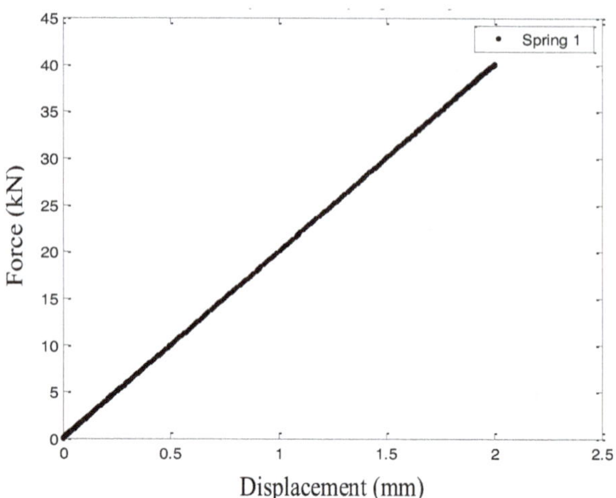

Figure 8. Force-displacement curve for a linear elastic spring element.

Table 1. Input data of force-displacement for nonlinear elastic material model.

Force-displacement curve	
Displacement (X coordinate) (mm)	Force (Y coordinate) (kN)
0.0	0.0
0.4	20.0
1.5	26.0

fundamental theory in which force and displacement oscillate accordingly to the corresponding solution time, thus resulting in proportional trend of a force-displacement curve. It is evident that the slope of the force-displacement curve is 20 N/mm which is similar to the defined spring stiffness, *k* in the solver code.

Nonlinear elastic

Nonlinear elastic material model can be defined by using the solver code for solving a simple nonlinear problem. Table 1 tabulates the input data of the force-displacement points. Figures 9 and 10 show accurate results plotted by the solver code for a force and displacement time history plot and a force-displacement curve respectively. Three points are indicated in Figures 9 and 10 in order to show loading and unloading paths within the solution time. In Fig. 9, the spring element is loaded rapidly to Point 1 and unloaded to Point 2. From this point onward, it is reloaded until it reaches Point 3. It is evident from Figure 10 that the material model of the solver agrees well with the postulated behavior of nonlinear elastic model shown in Figure 9. The results obviously give an accurate correlation between force and displacement time plot (Figure 9) with the force-displacement curve (Figure 10).

General nonlinear

The most complicated material model is a general nonlinear element. For this type of model, validation has been divided into two parts: Tension and compression loadings.

Tension loading: To ascertain whether the solver was sufficiently accurate for more complicated material model, it was virtually validated using force-displacement curve. Force-displacement curve for general nonlinear behavior is specified for both curves: Tension loading and unloading phases as indicated in Tables 2 and 3, respectively. In tension loading, States 0, 1 and 2 are involved in this solver code as demonstrated in Figure 11. Each state condition is as follows:

(i) State 0: Force has not reach tension and compression yield,
(ii) State 1: Force exceeded tension yield,
(iii) State 2: On unloading curve,

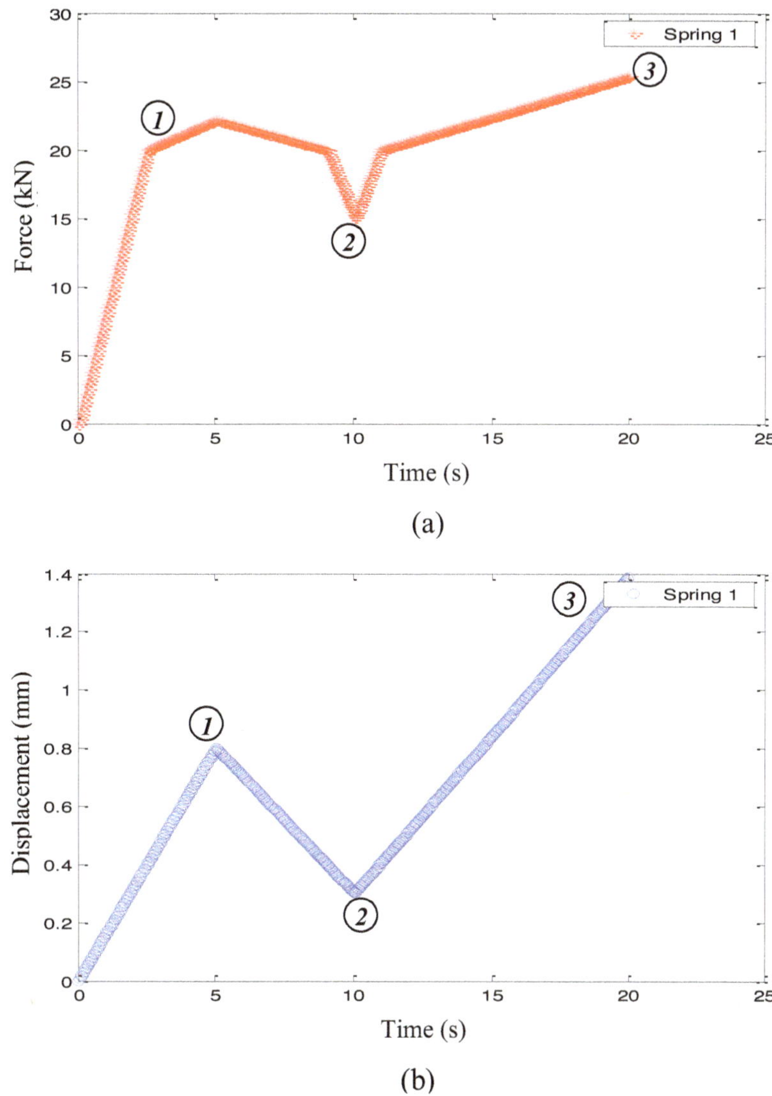

Figure 9. (a) Force-time (b) Displacement-time for a nonlinear elastic spring element.

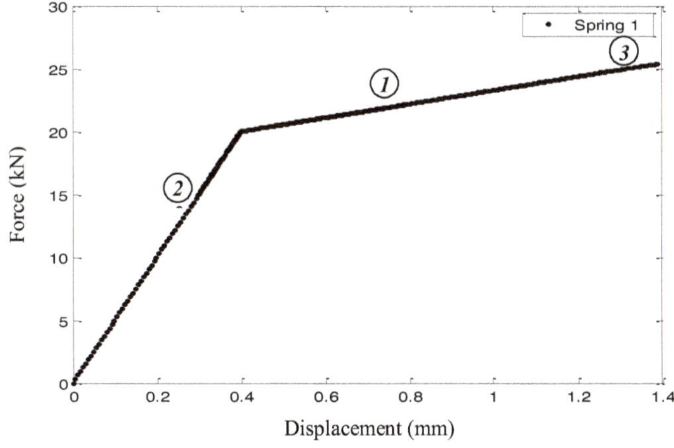

Figure 10. Force-displacement curve for a nonlinear elastic spring element.

Table 2. Input data of force-displacement curve for unloading phase in general nonlinear behaviour.

Force-displacement (Unloading curve)	
Displacement (X coordinate) (mm)	**Force (Y coordinate) (kN)**
0.0	0.0
0.1	1.0
7.0	70.0

(iv) State 3: Force exceeded compression yield.

Obviously, a reliable result is obtained from the direct comparison of the displacement-time plot and

Table 3. Input data of force-displacement curve for loading phase in general nonlinear behaviour.

Force-displacement (Loading curve)	
Displacement (X coordinate) (mm)	**Force (Y coordinate) (kN)**
0.0	0.0
0.23	20.0
0.60	23.0
0.90	18.0
1.25	20.0
1.50	35.0

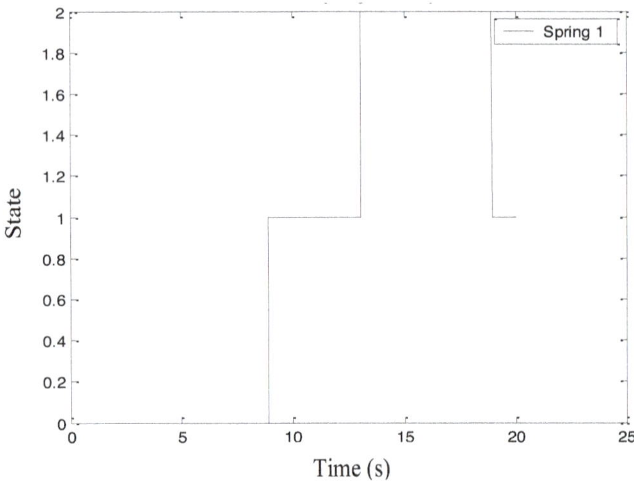

Figure 11. State-time plot for general nonlinear behavior under tension loading.

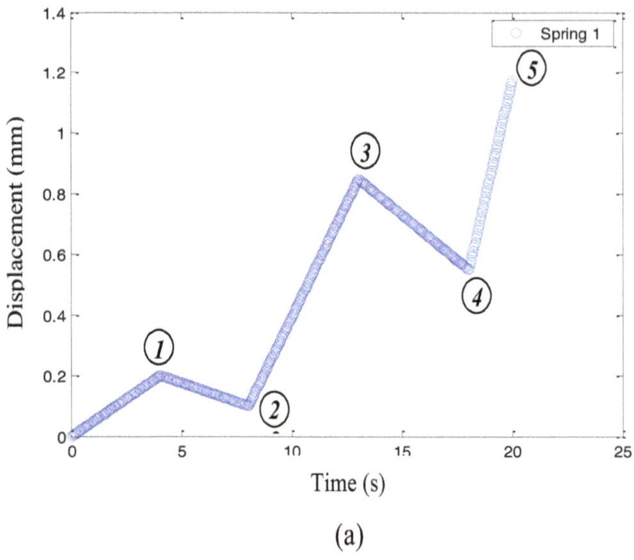

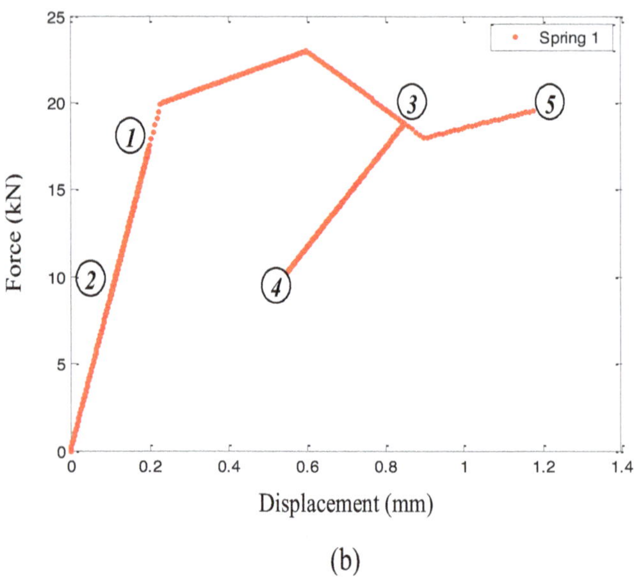

Figure 12. (a) Displacement–time curve; (b) Force-displacement curve for general nonlinear behavior under tension loading.

force-displacement curve (Figure 12). It should be noted that the element is subjected to general loading and unloading curves. In particular, from Point 1 to 2, the element is unloaded on the loading path as the force does not reach an initial yield force, 20 N/mm. However, the element is unloaded on the defined unloading curve upon a yield point as indicated at Points 3 to 4 in Fig.ure 12. From Points 4 to 5, the element is reloaded along the defined unloading curve as the displacement increases. Overall, the solver code shows a good capability of presenting a general nonlinear behavior of system. As such, this solver can perform the task for general nonlinear spring element when subjected to tension loading.

Compressive loading: Most of the crash structure is often exposed to compressive loading. Owing to this, this solver code was also validated under such loading condition. Table 4 tabulated input data of loading and unloading curve for the general nonlinear material model.

Figure 13 shows displacement-time and force-displacement curves for general nonlinear spring element when subjected to compression loading. It is evident that the curves show a good correlation between the loading and unloading conditions.

In particular, when compression load exceeds the initial yield compression, -20 N, it is unloaded on the unloading path rather than on the loading path as expected by using this solver code. This validation result is sufficient enough to show that the solver code is duly feasible and reliable for the element under compressive load. Hitherto, the solver code can subsequently be employed to a real two dimensional crash structure to show its usefulness and

Table 4. Force-displacement input data of loading and unloading curves for Spring 1 (cabin).

Displacement (X coordinate) (m)	Force (Y coordinate) (kN)
Force-displacement (Unloading curve)	
-0.006	-170000
0.0	0.0
0.006	170000
Force-displacement (Loading curve)	
-0.300	-1000000
-0.270	-500000
-0.225	-150000
-0.010	-150000
0.0	0.0

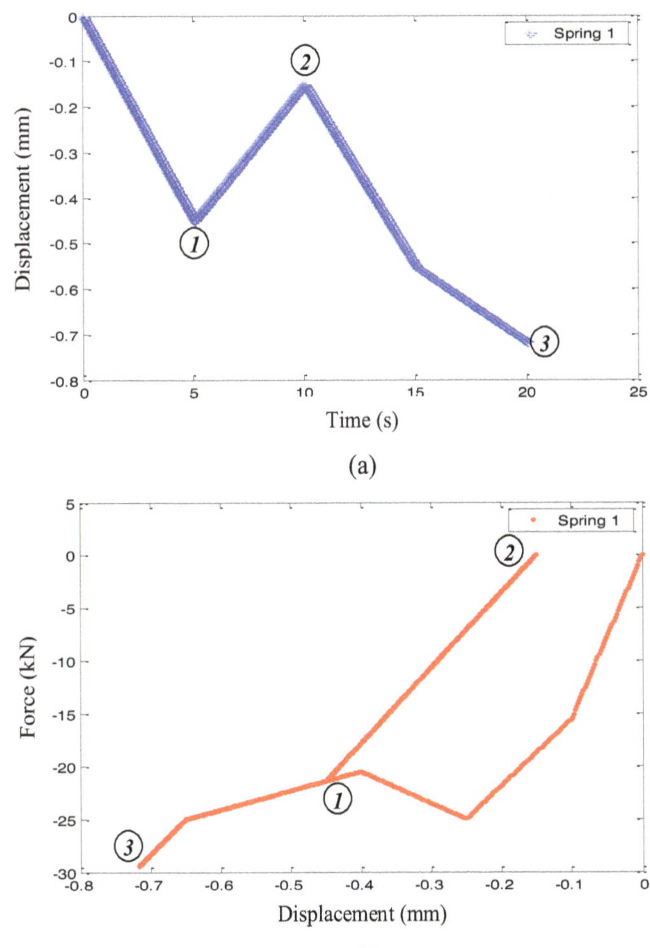

Figure 13. (a) Displacement–time curve (b) Force-displacement curve for general nonlinear behavior under compressive loading.

benefit in structural crash analysis.

Multi-spring element – two-dimensional crash solver code

For the multi-spring element, direct comparisons between the coordinate-time, velocity-time, acceleration-time and force-time curves obtained from existing finite element nonlinear code LS-DYNA and present codes are presented to show the accuracy of this solver code. For instance, a helicopter subfloor structure and frontal automotive structure crashes are demonstrated in this paper. This structural crash scenario can numerically be performed to comprehend the fundamental understanding of crashworthiness analysis since it just consists of two spring elements.

Helicopter sub-floor structure under vertical crash

The assembly of the multi-spring element shown in Figure 5 is used here. Helicopter sub-floor structures are divided into two main crash parts namely a helicopter cabin and an under-floor structure represented by spring element 1 and 2 respectively. The purpose of this crash analysis is to examine the dynamic behaviour of the structure under vertical impact load. Established force-displacement curve for the cabin and under-floor were included in the code prior to the analysis. Tables 4 and 5 show the force-displacement input data for the Spring 1 (cabin) and Spring 2 (under-floor) of helicopter sub-floor crash analysis (Liaw et al., 1988).

From the initial calculation, the undercarriage could absorb sufficient impact energy to reduce the ground impact velocity to 8.2 m/s. As such, this velocity value was defined as the initial velocity. Crash analysis has been performed by using LS-DYNA commercial code and the developed solver code. Figures 14 and 15 demonstrate the accuracy of time history plots obtained

Table 5. Force-displacement data of loading and unloading curves for Spring 2 (underfloor).

Displacement (X coordinate)(m)	Force (Y coordinate)(kN)
Force-displacement (Unloading curve)	
-0.009	-40000
0.0	0.0
0.009	40000
Force-displacement (Loading curve)	
-0.50	-76000
-0.223	-57200
-0.045	-70000
-0.009	-40000
0.0	0.0
10.0	5000

from the solver code compared with the results from commercial software LS-DYNA. In addition to this, a force-time curve is also directly compared and shown in Figure 16.

It is evident that the results obtained from the developed code, indicate good agreement between the two sets of results. The discrepancy becomes obvious upon reaching the solution time. The reason for this is due to the specified boundary condition. In LS-DYNA, rigid wall was defined as replacing translational constraint whereas in the solver code, the boundary condition was fixed. Thus, LS-DYNA code permits nodal mass 1 to bounce up and avoids any tension force occurred during the analysis. If this rigid wall is removed, the result would be better since the solver code does not take into account the contact algorithm. In order to overcome this problem, boundary conditions defined in LS-DYNA can be modified to suit with the non-contact solver code. Initially, the discrepancy has been anticipated due to the contact algorithm. As a recommendation for future research, this solver code can be improved by adding a contact boundary condition subroutine. Nevertheless, the time history plots (displacement, velocity and acceleration) present good agreement with LS-DYNA plot.

Automotive frontal structure crash

Frontal vehicle structure has a more complex spring-mass system. Typical results that produced using this solver code are presented to treat the capability of this solver code. The spring-mass model consists of 8 spring elements and 9 nodal masses as shown in Figure 17.

The important spring properties are accordingly defined to demonstrate the acceptable results of the automotive crash. It is noteworthy that force displacement curves applied on the structure were applied arbitrary on the structure to represent the structure impacting rigid wall. The list of structural component included in this analysis is as follows.

(i) M1: Barrier mass,
(ii) M2: Bumper mass,
(iii) M3: Radiator interface mass,
(iv) M4: Engine mass,
(v) M5: Engine mounting component mass,
(vi) M6: Frontal wheel mass,
(vii) M7: Transmission component mass,
(viii) M8: Aft frame mass,
(ix) M9: Survival room mass.

In this crash analysis, the engine nodal mass and survival compartment mass have been chosen in order to evaluate dynamic behavior of the energy absorbing system. Figure 18 shows the coordinate-time, velocity-time and acceleration-time plots obtained from the solver code. More importantly, the results presented have an acceptable and reasonable finding as presented in Herridge and Mitchell (1978). It is evident that this developed solver code could successfully demonstrate the crash behavior and dynamic response of the multi complicated system in solving one dimensional crash analysis.

Overall, the present solver code has demonstrated its capability in simulating crash analysis involving multi-spring element: helicopter sub-floor structure. Although the code is not as accurate as LS-DYNA code, it can initially be employed to predict the impact response of the structure and to understand the behaviour of structure under impact loading condition.

Conclusion

The purpose of this study was to demonstrate the capability of solver code in performing two dimensional crash analysis of crashworthy system. It is evident that the solver code shows good correlation with the commercial finite element code. This satisfactory validation provides adequate confidence for promoting this code. Hence, this developed code may be employed as an educational tool. Most importantly, the code does not require learning complex pieces of nonlinear explicit code such as LS-DYNA in order to perform crash analysis. Moreover, no user manual is needed to be referred. Output response was quantified with respect to variations in the kinematic parameters and force-displacement thereby representing the energy absorption capacity of the system. The demonstrated multi-spring system shows that the dynamic and energy absorption responses of crashworthy system is significantly influenced by impact time, feasible component assembly

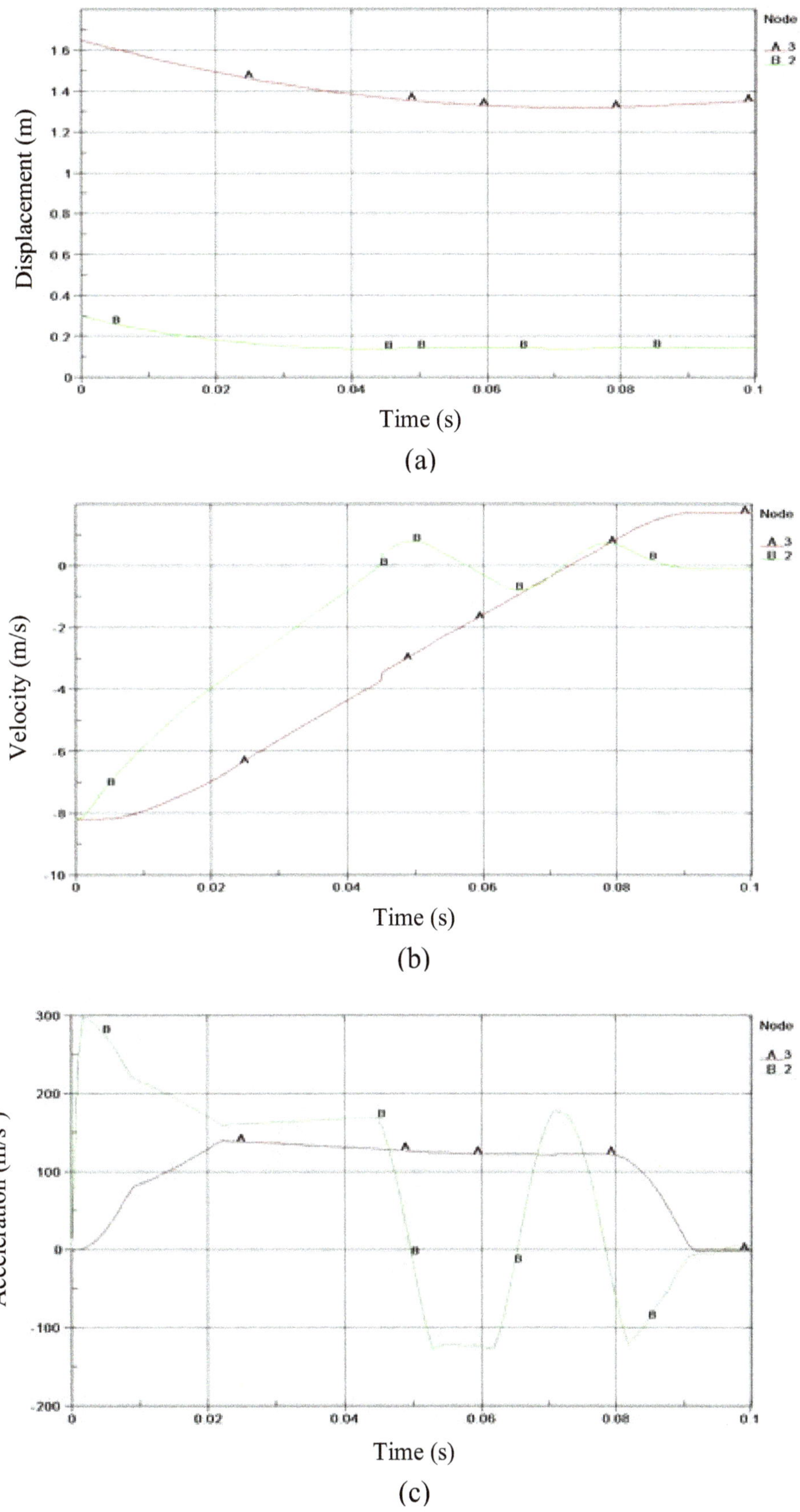

Figure 14. (a) Displacement-time (b) Velocity-time (c) Acceleration-time of helicopter sub-floor crash (LS-DYNA).

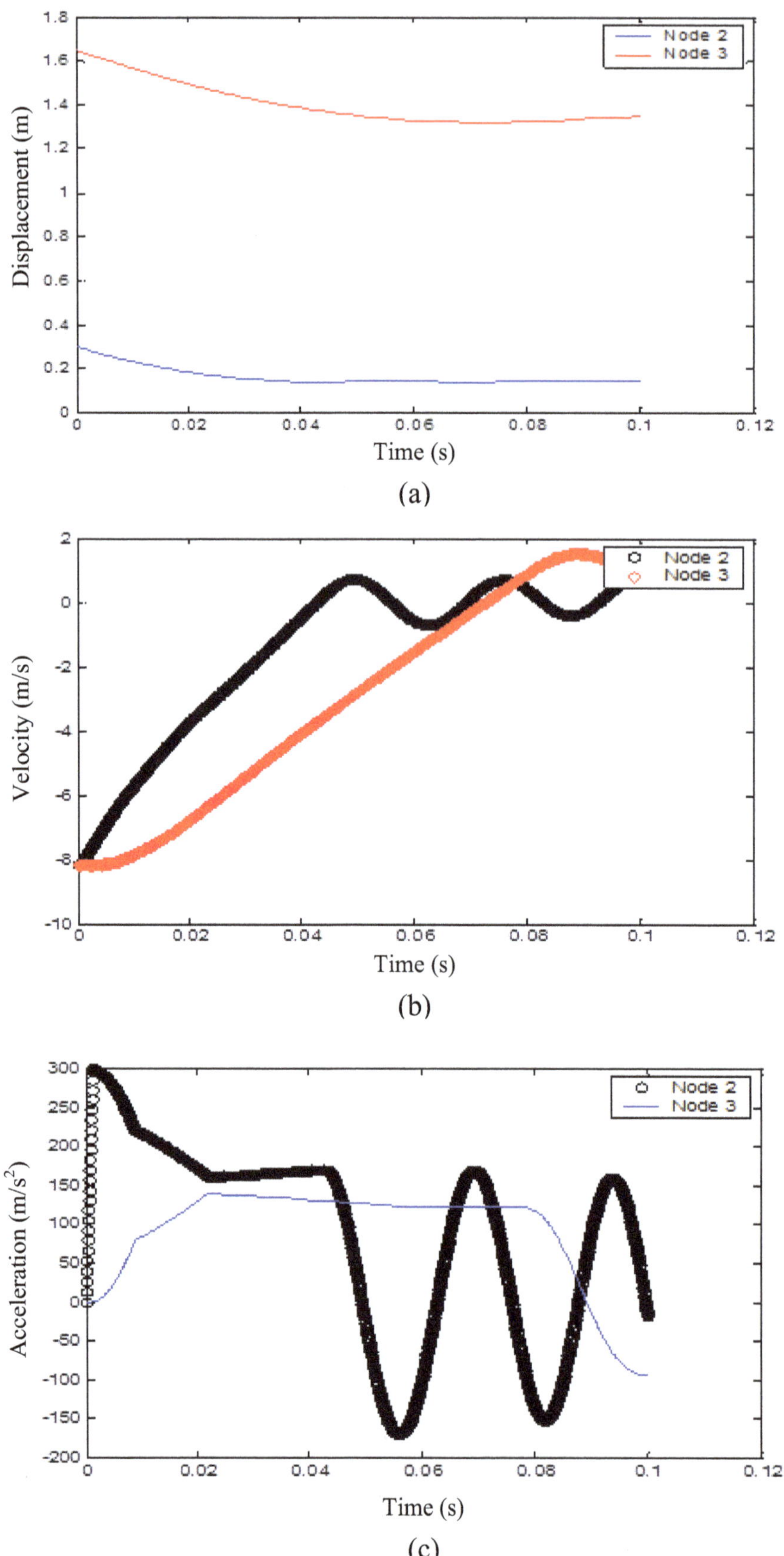

Figure 15. (a) Displacement-time (b) Velocity-time (c) Acceleration-time of helicopter sub-floor crash (solver code).

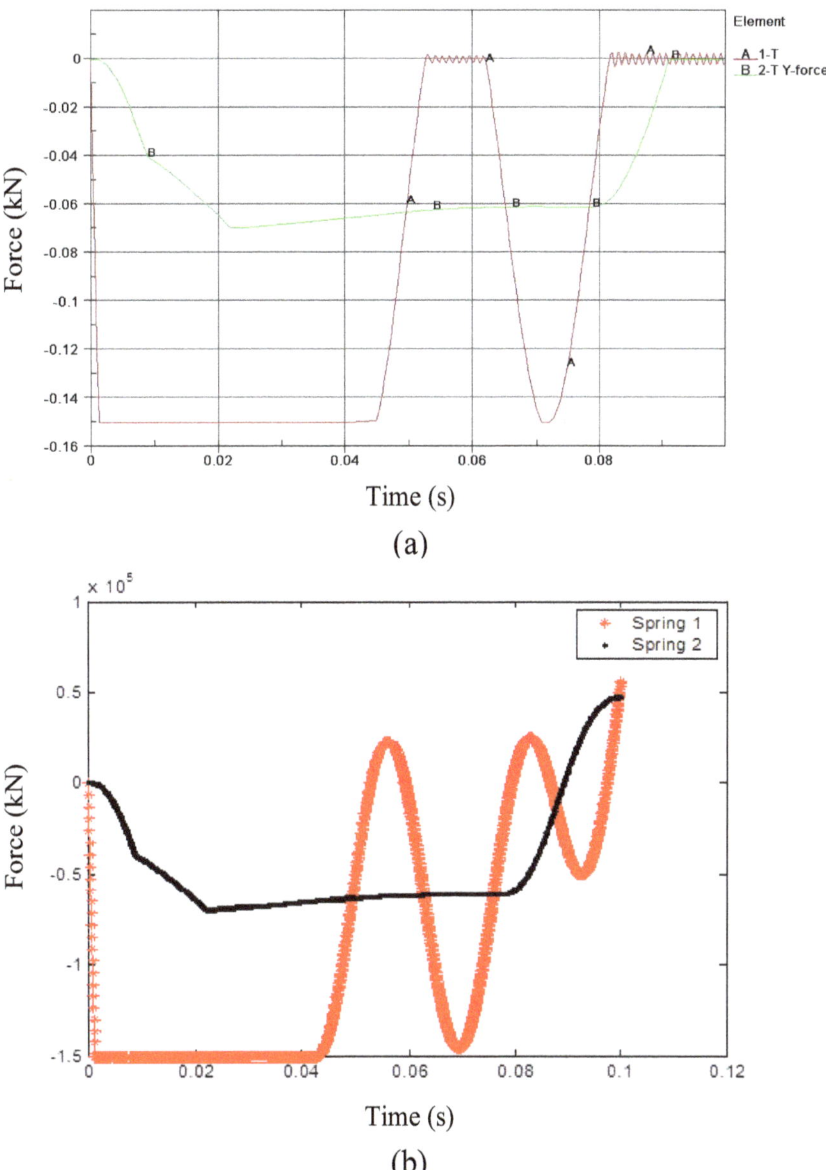

Figure 16. (a) Force-time curve (LS-DYNA) (b) Force-time curve (solver).

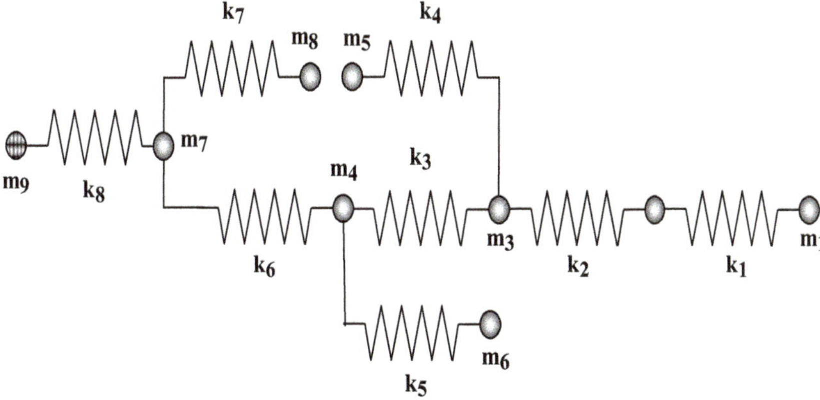

Figure 17. The spring-mass discrete elements for automotive frontal crash analysis.

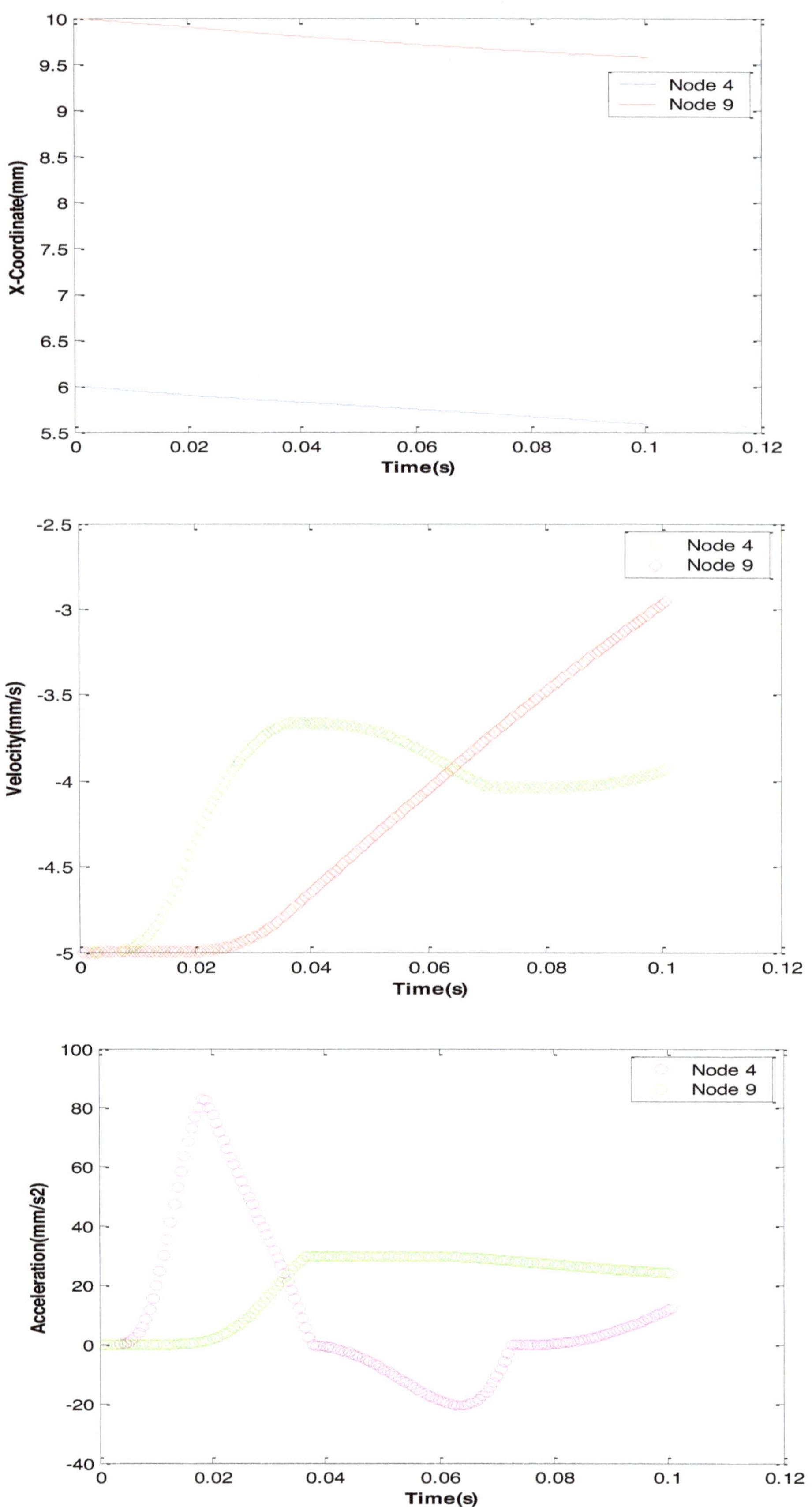

Figure 18. (a) Coordinate-time (b) velocity-time (c) acceleration-time of frontal vehicle crash.

and the stiffness of the component. It should be noted that this two dimensional solver is an innovative code to be used for the high educational level.

REFERENCES

Benson DJ (1992). Computational methods in Lagrangian and Eulerian hydrocodes. Comput. Methods Appl. Mech. Eng. 99:235-394.

Cook RD, Malkus DS, Plesha ME (1989). Concepts and applications of finite element analysis. Wiley Publication. New York.

Deb A, Ali T (2004). A lumped parameter-based approach for simulation of automotive headform impact with countermeasures. Int. J. Impact Eng. 30:521-539.

Deb A, Mahendrakumar MS, Chavan C, Karve J, Blankenburg D, Storen S (2004). Design of an aluminium-based vehicle platform for front impact safety. International J. Impact Eng. 30:1055-1079.

Gouddreau GL, Hallquist JO (1982). Recent developments in large-scale finite element Lagrangian hydrocode technology. Comput. Methods Appl. Mech. Eng. 33:725-757.

Hallquist JO (2006). LS-DYNA 3D: Theoretical manual. Livermore Software Technology Corporation. Livermore.

Herridge JT, Mitchell RK (1978). Development of computer simulation program for collinear car/car and car/barrier collisions. Am. Soc. Mech. Eng. P. 73.

Hofmeister LD (1978). Dynamic analysis of structures containing nonlinear spring. Comput. Struct. 8:609-614.

Jones N (1989). Structural impact. Cambridge University Press. Cambridge.

Jones N, Wierzbicki T (1983). Structural crashworthiness. Butterworth & Co (Publisher) Ltd: London UK.

Jonsén P, Isaksson E, Sundin KG, Oldenburg MM (2009). Identification of lumped parameter automotive crash models for bumper system development. Int. J. Crashworthiness 14(6):533-541.

Kecman D (1997). An engineering approach to crashworthiness of thin-walled beams and joints in vehicle structures. Thin-Walled Struct. 28:309-320.

Kirkpatrick SW, Simons JW, Antoun TH (2000). Development and validation of high fidelity vehicle crash simulation models. Int. Crash. Conf. 16:602-612.

Liaw JC, Walton AC, Brown JC (1988). Structural impact simulation using program KRASH. Cranfield: Cranfield Instit. Technol. pp. 69-80.

Livermore Software Technology Corporation, LS-DYNA Keyword User's Manual (2003). Livermore Software Technology Corporation. California.

Pifko AB, Winter R (1981). Theory and application of finite element analysis to structural crash simulation. J. Comput. Struct. 13:277-285.

Ruan HH, Yu TX (2003). Collision between mass-spring systems. Int. J. Impact Eng. 31:267-288.

Sheh MY, Reid JD. Lesh SM, Cheva M (1992). Vehicle crashworthiness analysis using numerical and experimental method. in International Conference on Vehicle Structural Mechanics and CAE. Warrendale: Society of Automotive Engineering Inc.

Stronge WJ, Yu TXL (1993). Dynamic models for structural plasticity. Springer.

Thomson WT (1988). Theory of Vibration with Application. Prentice Hall. Sydney.

Venkataraman P (2002). Applied optimization with Matlab programming. John Wiley & Sons Inc. New York.

Wu KQ, Yu TX (2001). Simple dynamic models of elastic-plastic structures under impact. Int. J. Impact Eng. 25:735-754.

Wu SR, Cheng J (1997). Advanced development of explicit FEA in automotive applications. Comput. Methods Appl. Mech. Eng. 149:189-199.

6

Influence of fuel in the microwave assisted combustion synthesis of nano α-alumina powder

Ramesh G.[1] , Mangalaraja R. V.[2], Ananthakumar S.[3] and Manohar P.[1]

[1]Department of Ceramic Technology, A.C Tech. Campus, Anna University, Chennai, India.
[2]Department of Materials Engineering, University of Concepcion, Concepcion, Chile.
[3]Materials and Minerals Division, National Institute for Interdisciplinary Science and Technology (NIIST), Trivandrum, India.

Microwave assisted combustion synthesis is used for fast and controlled processing of advanced ceramics. Single phase and sinter active nano crystalline alpha alumina powders were successfully synthesized by different fuel-to-oxidant molar ratios using aluminium nitrate as an oxidiser, glycine as a reducing agents and millipore water as a solvent by microwave assisted combustion synthesis. Thermodynamic modelling of the combustion reaction shows that as the fuel-to-oxidant ratio increases, the amount of gases produced and adiabatic flame temperature also increases. The precursor powders were investigated by thermogravimetry (TG) analyses. The as prepared precursors calcined at 900 to 1200°C in air atmosphere were characterized for their structure and morphology. The thermal analyses (TG/DSC), X-ray diffraction (XRD) and Fourier transform infra red (FT-IR) results demonstrate the effectiveness of the microwave assisted combustion synthesis. The transmission electron microscopy (TEM) observations show the different morphologies of as-prepared powders and shows the particle sizes in the range of <50 nm. The results confirm that the homogeneous, nano scale alumina powders derived by microwave assisted combustion have high crystalline quality and the morphology of the as-prepared precursor powders.

Key words: Synthesis, microwave, glycene-gel, decomposition, alumina, nano crystalline.

INTRODUCTION

Alumina as an excellent ceramic material, finds potential application because of its high melting point (2400°C) with excellent chemical stability, high corrosion resistivity and low volatility in vacuum. Synthesis of fine-grained ceramic products has been the topic of many recent investigations on account of their beneficial properties over coarse-grained ceramics (Yi-quan and Yu-feng, 2001; Edrissi and Norouzbeigi, 2007; Shojaie-Bahaabad and Taheri-Nassaj, 2008; Robert et al., 2009; Mimani and Patil, 2001; Ananthapadmanabhan et al., 2004).

Over the last few decades, a great variety of techniques have been developed to synthesize highly reactive and high-purity materials to fabricate fine-grained ceramic products. The techniques including spray pyrolysis, precipitation (Shaoyan et al., 2008), sol-gel (Mirjalili et al., 2010), hydrothermal and combustion synthesis (Robert et al., 2009; Chyi-Ching et al., 2004; Ganesh and Ferreira, 2007) have been employed to synthesise ultra fine Al_2O_3 powders. Mechanical synthesis of α-Al_2O_3 requires extensive mechanical ball milling and easily introduces impurities. Vapour phase reaction for the preparation fine α-Al_2O_3 powder from a gas phase precursor demands high temperature above 1200°C. Sol-gel method based on molecular precursors usually makes use of metal alkoxides as raw material. However, the high prices of alkoxides and long gelation periods limit the application of this method. The precipitation method suffers from its complexity and time consuming (long washing times and aging time). The direct formation of α-Al_2O_3 via the hydrothermal method needs high temperature and pressure. Most of these techniques involve expensive raw materials and are associated with unmanageable processing steps.

Fortunately, the drawbacks of these methods as mentioned above could be partially eliminated by the so called combustion synthesis (also known as self-propagating high-temperature synthesis). It is emerged as effective powder synthesis route as it is a simple and economic process and yields high-purity powders with excellent homogeneity and fine particle sizes (Singanahally and Alexender, 2008; Edrissi and Mohammad, 2008). Combustion synthesis is not only simple, but safe and rapid process wherein the main advantages are energy and time savings. This quick, straightforward process can be used to synthesize homogeneous, high-purity, crystalline oxide ceramic powders including ultrafine alumina powders with broad range of particle sizes.

The basis for the combustion synthesis comes from the thermo-chemical concepts used in the field of propellants and explosives. This technique involves the exothermic chemical reaction of a fuel (for example, citric acid, urea, glycine, oxalic acid or glycol, etc.) and an oxidizer (for example, nitrates) (Tianyou et al., 2006; Jiang et al., 2007). The exothermicity sometime appears in the form of flame, which temperature can be in excess of 1500°C. The large amount of gases generated during combustion synthesis rapidly cools the product leading to nucleation of crystallites without any substantial growth. The gas generated also can disintegrate the large particles or agglomerates; therefore, the resulted product consists of very fine particles of friable agglomerates or nanoparticles (Xiujing and Yan, 2006).

Actually, the mechanism of the combustion reaction is quite complex. The main parameters influencing the reaction include type of the main fuel, fuel to oxidizer ratio, the amount of oxidizer in excess, ratio of fuels, pH of the solution and rate of calcinations. In general, a good fuel should not react violently nor produce toxic gases, and must act as a complexing agent for metal cations.

In the present research, we have proposed a new method called 'microwave-assisted glycene-gel decomposition technique' to prepare nonocrystalline alumina powders. Compared to conventional heating, microwave heating leads to a more uniform microstructure. In the microwave process, heat will be generated internally within the material, instead of originating from external sources. Because of the rapid heating of the microwave process, lower processing temperatures are needed which in turn provides a suitable condition for the formation of nano-sized powders. Microwave synthesis of alumina powder reduces the sintering temperature and thereby reducing the particle size.

The use of microwave energy for the synthesis of ceramic oxide powders using a few fuels has been earlier reported (Ganesh et al., 2005; Mangalaraja et al., 2009). The powders synthesized by this technique are ultrafine, pure and more homogeneous and less-agglomerated. The present work reports the synthesis of alumina nano powders using varied fuel to oxidant ratio and heating. Microwave assisted glycene-gel decomposition technique under normal atmospheric pressure has not been studied so far thoroughly. Decomposition of the glycene precursor, phase transformations and morphology of the synthesized alumina powder are investigated. Also in this synthesis, Millipore water is used as a solvent to dissolve aluminium nitrate and fuel to synthesis process for producing alumina nano powders.

Generally, natural water contains soft particulates (vegetable debris) and hard particulates (sand, rock) as well as colloidal matters that can interfere in processing factors. Natural water contains dissolved gases such as nitrogen, oxygen and carbon dioxide. The concentrations of oxygen and nitrogen may affect the products. Inorganic ions (such as sodium, calcium, magnesium or iron, and anions, such as bicarbonate, chloride and sulphate) even at trace levels, may affect chemical reactions by acting as catalysts. Milli-Q system has three-step purification in one unit. Secondary purification via high-recovery reverse osmosis cartridge removes 95 to 99% of inorganic ions, 99% of dissolved organics, bacteria, and particulates. The purifications of Millipore process gives the water resistivity 18.2 MΩ-cm at 25°C and the particulate level is <1 particulate/ml.

The extent of conversion to the α-Al_2O_3 phase depends on the temperature and the time of calcinations. Generally, it exists in a number of meta-stable polymorphs before a complete transformation to thermodynamically stable α–alumina. These polymorphs include γ (gamma), Θ (theta), η (eta), δ (delta), χ (chi), k (kappa) and β(beta). The formation of one, two or more of above mentioned transient aluminas before the complete transformation to the final stable α–alumina depends upon the processing conditions, the degree of crystallinity and the presence of impurities in the starting materials. The powders obtained through combustion synthesis were characterized by Fourier transform infra red (FT-IR), X-ray diffraction, thermo gravimetric analysis (TG), and Transmission electron microscopy (TEM).

EXPERIMENTAL

In a typical experiment, a solid mixture containing indispensable quantities of aluminium nitrate (Merck, GR grade), and glycene (Merck, GR grade) was taken in a Pyrex glass dish and was irradiated with microwaves in a modified domestic microwave over (ONIDA, POWER BARBEQUE 28, India, Model No. 28CJS14, microwave 900 W, input range 230 V ac 50 Hz, microwave

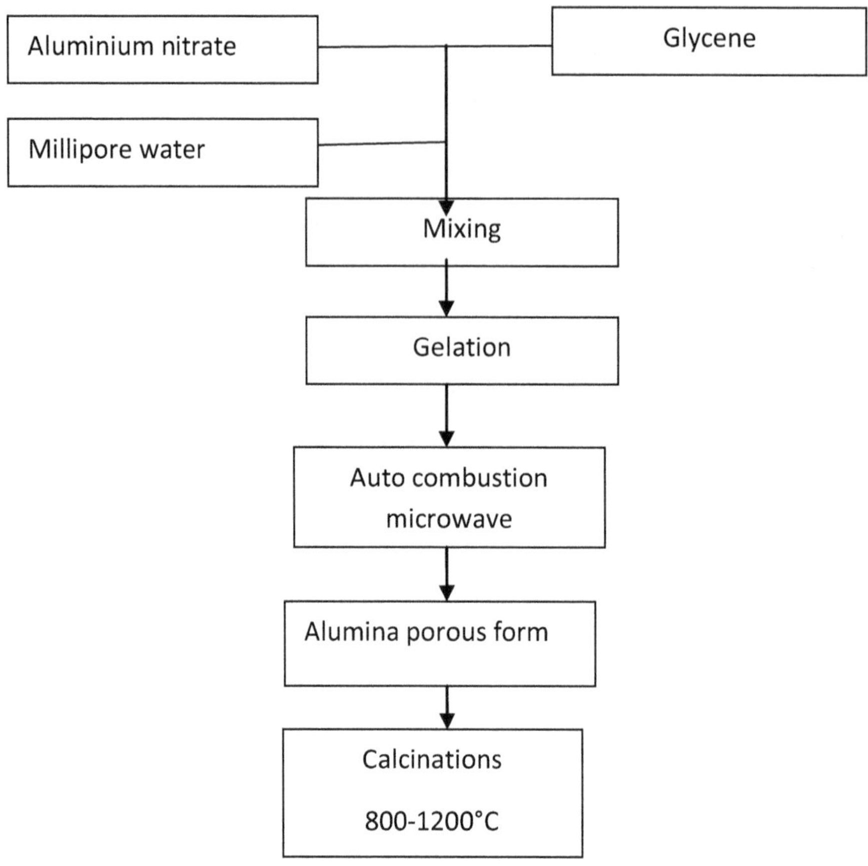

Figure 1. Schematic of the Nitrate-Glycene combustion process.

frequency 2.45 GHz) to produce alumina material. A provision was made for the escape of combustion gases by providing an exhaust of the microwave oven.

In general, the raw materials must be readily available and convenient. In addition, it should react non-violently and produce non-toxic gases. The molar ratio of glycene (combustion aid) to aluminium nitrate (main oxidizer) was selected in the amounts of 0.51 (high level), 0.60 (stiochiometric) and 0.71 (low level). The initial composition of the solution containing aluminium nitrate, $Al(NO_3)_3.9H_2O$ and glycene was derived from the total oxidizing and reducing valences of the oxidizer and fuel using the concepts of propellant chemistry. Carbon, hydrogen and aluminium were considered as reducing elements with the corresponding valences of +4, +1 and +3 respectively. Oxygen was considered as an oxidizing element with the valence of 2, the valence of nitrogen was considered to be 0 because of its conversion to molecular nitrogen (N_2) during combustion. The total calculated valence of metal nitrated by arithmetic summation of oxidizing and reducing valences was -15. The calculated valence of glycene was +9. The stoichiometric composition of the redox mixture demanded that 2(-15) + n (+9) = 0, or n = 3.33 mol. This calculation was done for mixture of aluminium nitrate and glycene.

The products were synthesized by an amount of 15 gm per batch. The meticulous process is given in the form of flowchart as shown in Figure 1. The aluminium nitrate and glycene were dissolved in a required amount of Millipore water and mixed thoroughly to ensure a molecular level mixing to form a homogeneous solution. Metal nitrates possess hygroscopicity; consequently, they easily absorb moisture and become slurry. Therefore, the reactants can be mixed homogeneously during the stirring process. The mixture solution was transferred to a Pyrex glass dish and later introduced into a microwave oven to undergo decomposition in the microwave field of 2.45 GHz. Initially, the solution was boiled to transform to a honey like consistency, that escort to a transparent sticky gel and then the gel was immediately decomposed by a self-combustion accompanied by the evolution of a brown fume and finally yielded a fluffy precursor which did not contain crystalline phases. During decomposition of gel, a large amount of gaseous products N_2, CO_2 and H_2O evolved without the necessity of getting oxygen from outside.

In fact, on all fuel-to-oxidant ratios evaluated, upon auto-ignition, it resulted in a brownish voluminous product identified by XRD as an amorphous structure, which indicates the incomplete combustion probably due to characteristics of fuel employed. Subsequently, these powders were calcined at 800, 900, 1000 and 1200°C at a heating rate of approximately 10°C/min., during 2 h of soaking time. The prepared powders were characterised by X-ray diffraction (XRD), Fourier transform infra red (FT-IR) and thermal analysis (TG and DSC) techniques. The as prepared powders were also investigated by the thermal analysis (TG/DSC) and Fourier transform infra red (FT-IR) studies.

X-ray diffraction was executed on combustion synthesized powders for phase characterization, at a rate of 60/min., using Cu Ka radiation on a General Engineering X-ray diffractometer (model GE-110T). Silicon was also employed as an external standard for correction due to instrumental broadening. The phase structure and crystallite size of the prepared Al_2O_3 powder was estimated from X-ray peak broadening using Scherer formula.

Table 1. Effect of fuel ratio on phase and crystallite size of calcined alumina.

Temperature (°C)	Phase			Crystallite size (nm)			Crystallite size*		
	G/N =0.51	G/N =0.60	G/N =0.71	G/N=0.51(H)	G/N=0.60(S)	G/N=0.71(L)	G/N =0.51	G/N =0.56	G/N =0.69
800	-	-	-	Amorphous	Amorphous	Amorphous			
900	γ , δ	γ	γ , δ , α	49	49	38	90.3	96.8	123.2
1000	γ , δ	γ , δ , α	γ , δ , α	50	46	46			
1200	α	α	α	52	51	45			

*Effect of crystallite size at 1100°C for different glycine-to-nitrate ratios- report of Toniolo et al. (2005).

The theoretical chemical reaction equation of the stoichiometric combustion reaction can be written as follows:

Stoichiometry

$1.0\ Al(NO_3)_3.9H_2O + 1.67\ NH_2CH_2COOH \longrightarrow 0.5\ Al_2O_3 + 2.33\ N_2 + 3.33\ CO_2 + 13.17\ H_2O$

Fuel-rich (+ 17.62%)

$1.0\ Al(NO_3)_3.9H_2O + 1.96\ NH_2CH_2COOH + 1.32O_2 \longrightarrow 0.5\ Al_2O_3 + 4.96\ N_2 + 3.92\ CO_2 + 13.9\ H_2O$

Fuel-lean (-15.54%)

$1.0\ Al(NO_3)_3.9H_2O + 1.41\ NH_2CH_2COOH \longrightarrow 0.5\ Al_2O_3 + 4.41\ N_2 + 2.82\ CO_2 + 12.52\ H_2O + 1.155\ O_2$

RESULTS AND DISCUSSION

Powder X-ray analysis

The average crystallite size values were estimated using the following equation of Scherrer.

$$D = \frac{0.9\lambda}{\sqrt{B2-b2}\cos\theta}$$

Where D is the average crystallite size, λ the wavelength of the radiation, ϴ the Bragg's angle and B and b are the FWHMs observed for the sample and standard, respectively. Silicon powder was used to measure the instrumental peak broadening. Terms B and 2ϴ were obtained from each XRD pattern and PC-APD (X-manager) software that uses information of ICDD (international center of diffraction data) to recognize alumina phases. The average crystallite size for each sample was calculated from the above formula.

Toniolo et al. (2005) studied the crystallite size of the powder for the fuel-to-oxidant ratio varied by the combustion synthesis technique using glycine as fuel and aluminium nitrate as an oxidizer to produce alumina powders through conventional heating process. In this present work, comparison was made between variation in the powder characteristics obtained through microwave assistance for different fuel-to-oxidant ratios with Toniolo et al. (2005). The phase structure and crystallite size of the prepared Al_2O_3 powder estimated from X-ray peak broadening using Scherrer formula is shown in Table 1.

As can be observed from Figure 2 (G/N=0.50), the fuel rich batch powders obtained by the combustion of aluminium nitrate and glycene mixture in a domestic microwave oven for 4 min. exhibited significantly broad XRD peaks indicative of excess quantity of heat generated and radiated from inner atom to outer surface of the material. The sample 1d shows at temperature higher than 1000°C high intensities of α-Al_2O_3 peaks are observed, indicating crystallite growth of the grains. With the increase of calcinations temperature up to 1000°C, the crystallinity of α-Al_2O_3 improved as shown in the samples 1b and 1c. At 1200°C, rather weak peaks of α-Al_2O_3 disappear, indicating γ to α-Al_2O_3 transition. The γ-Al_2O_3 peaks, except for the strongest one disappear, indicating the γ to α-Al_2O_3 phase transition is almost completed. At temperature higher than 1000°C, high intensities of α-Al_2O_3 peaks are observed, indicating crystallite growth of the grains.

As can be observed from Figure 2 (G/N=0.60), the stoichiometric batch powders obtained by the combustion of aluminium nitrate and glycene mixture in a domestic microwave oven for 4 min. exhibited a few broad XRD peaks indicatives of crystallite size growing up and α-Al_2O_3 crystallite rapidly grew up to 51 nm as transformation was completed. XRD patterns show that samples 2a has more amorphous particles than the others. Also the crystallinities of sample 2d are better than those of other samples. Furthermore, samples 2b and 2c are mixtures of δ and γ phases and a sample 2d is a pure α-alumina. The results show that when an XRD profile has broad peaks, the crystallite sizes are small and the crystallinity is poor but when the peaks are narrow, the corresponding crystals are large.

From Figure 2 (G/N=0.71), the fuel rich batch material obtained from the identical reaction mixture in a domestic microwave oven shows in sample 3d, the complete transformation to α-phase took place after calcinations temperature of 1200°C. This result illustrates an advantage of microwave heating since the formation time and temperature are significantly reduced compared to conventional heating. The formation of γ-Al_2O_3 was first detected in sample 3b after increase in calcinations temperature: γ-Al_2O_3 was still the main constituent for this heating amount. Some δ and α-Al_2O_3 the intermediate

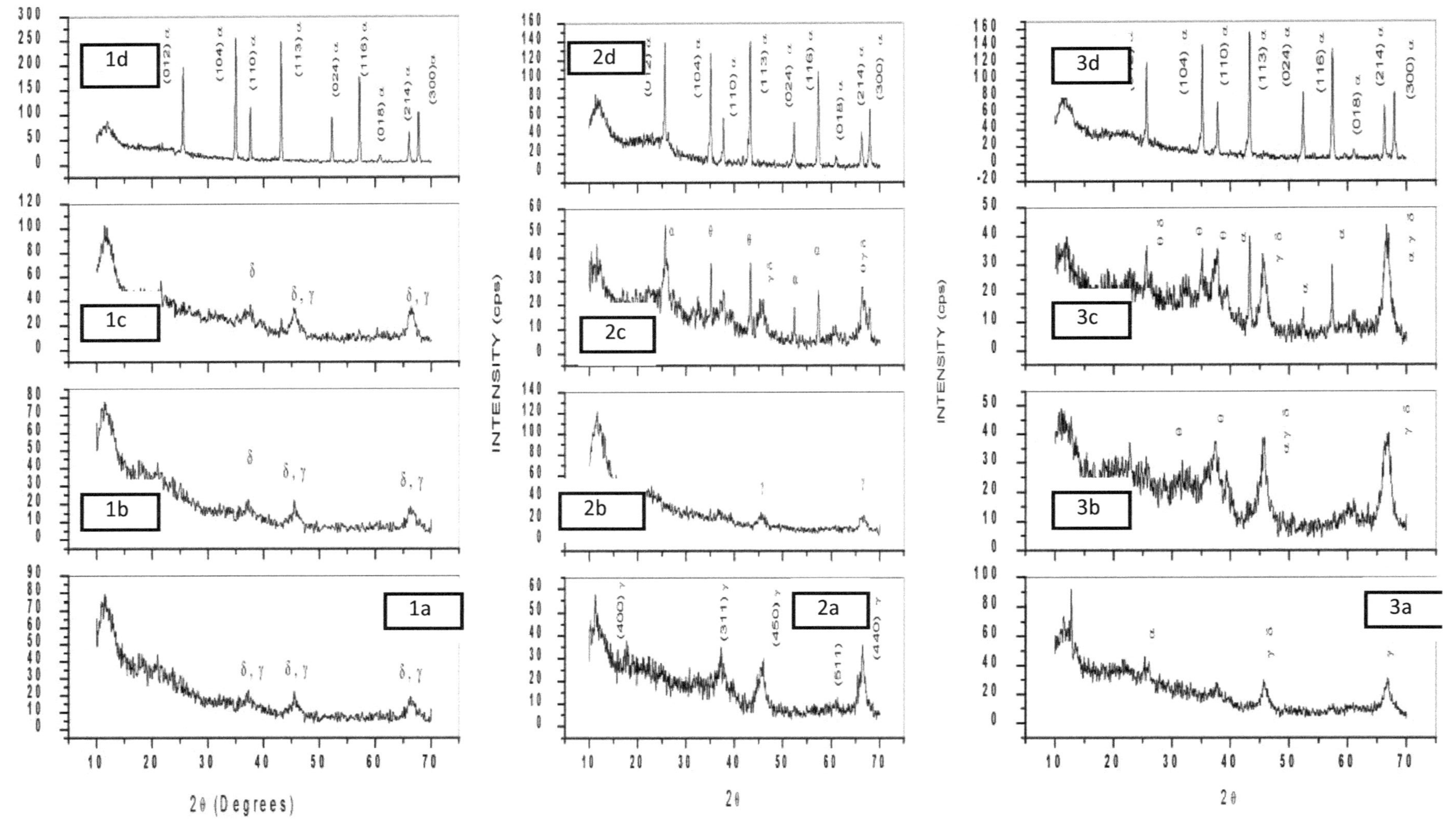

Figure 2. XRD patterns of as-synthesized precursors G/N=0.51,0.60 and 0.71 calcined at a) 800°C (b) 900°C (c) 1000°C (d) 1200°C.

phases were detected before α-Al_2O_3 formation. The amount of gases released during the exothermic reaction is very high, so it cools the reaction environment. These results in poor crystallinity in some prepared powders also control the growth of crystal size. Figure 3 shows the variation of crystallite size with different G/N ratios.

The particle size of α-Al_2O_3 was obtained about

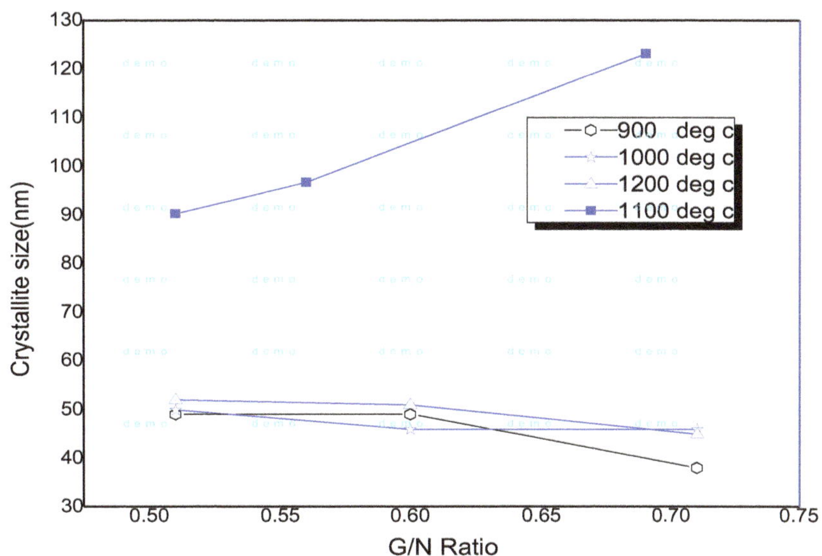

Figure 3. Different molar ratio and crystallite size at different calcined temperature.

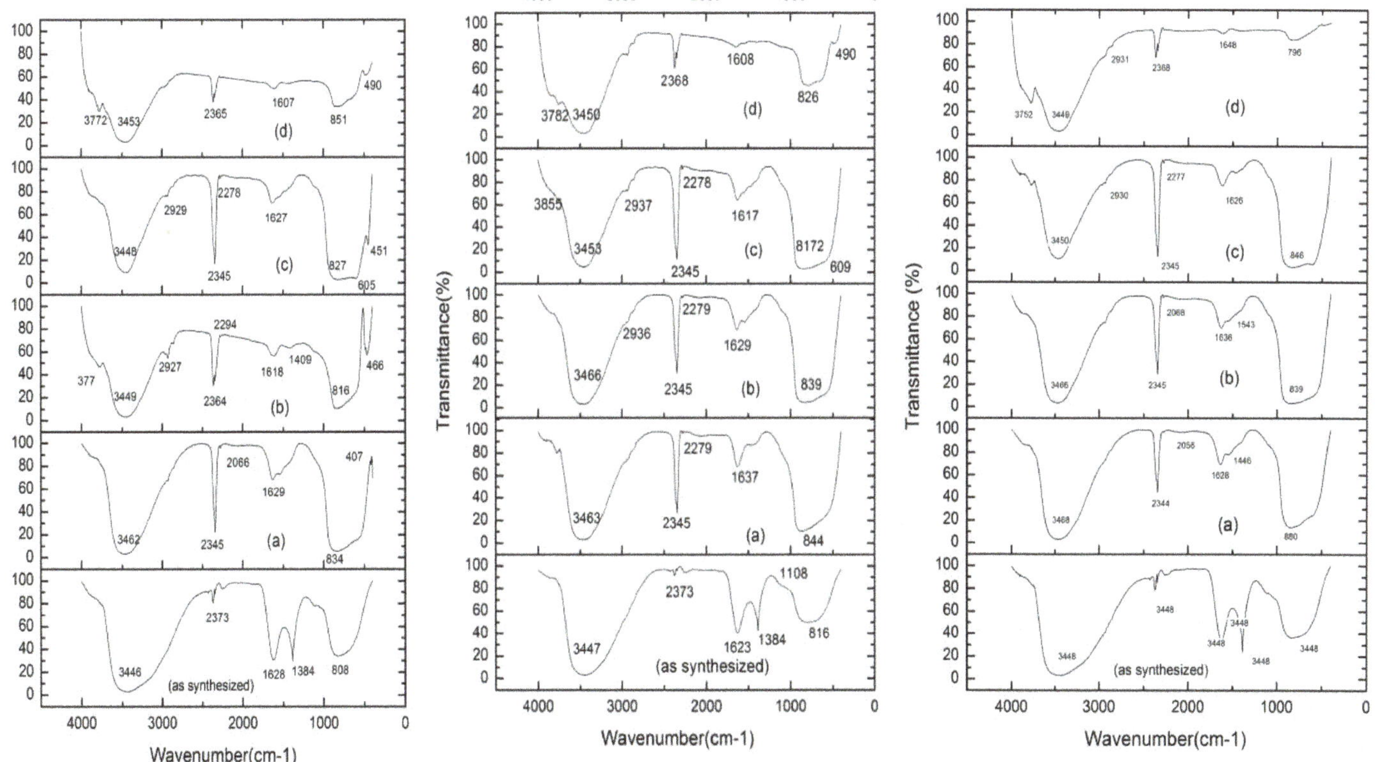

Figure 4. FTIR Spectrum of as-synthesized precursors G/N=0.51,0.60 and 0.71 calcined at 1)as synthesized a) 800°C, (b) 900°C, (c) 1000°C and (d) 1200°C.

38 nm in the case of lean fuel, 46 nm in the case of stoichiometric and 49 nm for fuel rich, and there was prominent change obtained by comparing this microwave assistance and Millipore water usage with respect to conventional process with ordinary distilled water.

FTIR analysis

The FTIR at 4000-400 cm^{-1} for the precursor (G/N=0.51, 0.60 and 0.71) calcined at different temperatures are shown in Figure 4. This clearly shows a broad absorption

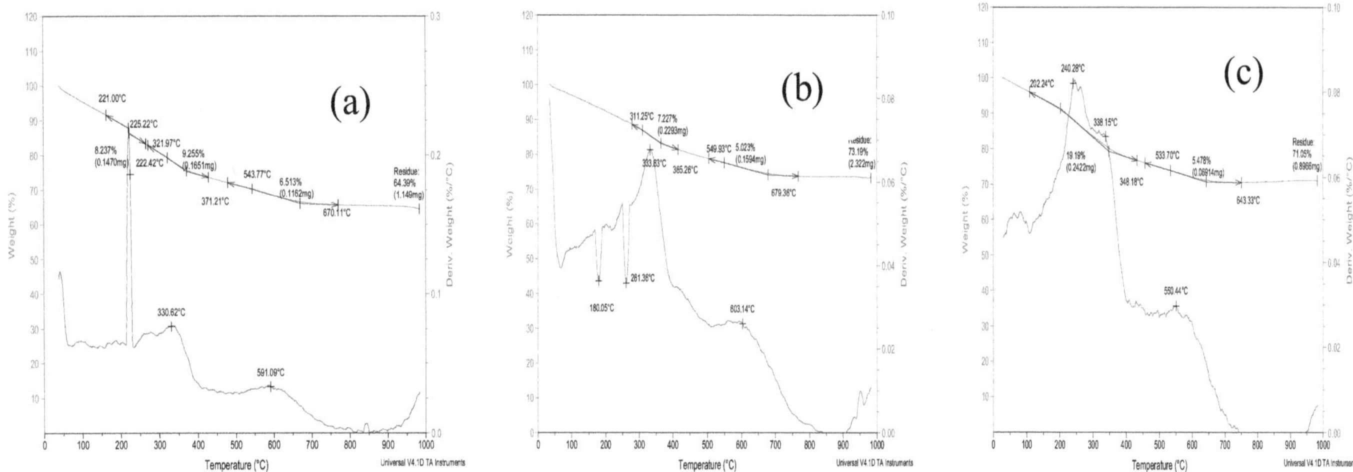

Figure 5. Simultaneous DSC-TGA curves of the as prepared powder of (a)G/N=0.51, (b) G/N=0.60 and (c) G/N=0.71.

at around 3400 cm^{-1} and a small absorption at 654 cm^{-1}, which are characteristic stretching vibration and deformation vibration of hydroxyl group (O-H), respectively. The characteristic bands of nitrate ions in the region of 1385 cm^{-1} indicate its incomplete decomposition during heat-treating process of the gel in the microwave synthesis. Peaks localized at 1623 cm^{-1} are assigned to asymmetrical stretching vibration of carboxyl ions (COO^-). In the FTIR spectra of powder calcined at 800°C and above, the absorption bands of NO_3^- group disappear because of the complete decomposition of nitrate. At different calcined G/N ratios reveals the appearance of ammonium ions in the regions of 2365 cm^{-1}.

DSC-TGA analysis

As shown in Figure 5a (0.51), the most weight loss 9% occurs between 321 to 371°C due to evaporation of volatile components. The weight loss of the as synthesized powder is less than 6.5% between 543 to 670°C. So the volatilization does not occur obviously. From 670 to 983°C, a weight loss occurs due to volatilization of carbon which corresponds to an endothermic reaction. However, there was an exothermic reaction due to crystallization of transition phases of alumina in this range. Above 983°C, an exothermic reaction occurs due to transformation of transition phases of alumina to α-alumina. With TG analysis, it was possible to confirm phase transformation into thermodynamically stable crystallographic alpha alumina which starts at about 983°C. This result agrees with that obtained by XRD analyses.

In Figure 5b (0.6), the TG curve demonstrates two weight loss steps. First, it was verified on a significant fall by about 7.2% at 311°C. The second drop was slight by about 5% at 550 to 750°C. Afterwards the curve became horizontal. In Figure 5c (0.71), the TG curve demonstrates two weight loss steps. First, it was verified on an important fall by about 19.19% at 202°C; it shows exothermic behaviour of auto-ignition. The second drop was slight by about 5.47% between 240 to 534°C. Above 643°C the curve became horizontal.

Between 533 to 643°C is expected and observed weight losses may be attributed to the remained carbon contamination, which can be inferred from the fact that a grayish product was obtained. The total weight loss above 643°C of TGA curve also indicates that the remaining carbon is slowly decomposed. However, compared to three molar ratios processed precursor through microwave synthesis, it inferred that fuel lean is suitable for synthesis of high purity and crystallite particles due to electric induction of atom to produce sufficient temperature for the reactions. The thermal characteristic of the synthesized alumina precursor powder is shown in Table 2.

TEM data

Figure 6 is a TEM photograph taken after the as-synthesized product had been calcined at different temperature of 900, 1000 and 1200°C. The sizes of the spherical particles are dispersed with negligible agglomeration and are homogeneously distributed ranging from 38 to 46 nm, which is about the same as the sizes estimated by using XRD method.

Conclusions

Nanocrystalline Al_2O_3 powders were prepared with crystallite sizes ranging between 38 and 52 nm using

Table 2. Thermal characteristics of the Al_2O_3 precursor materials.

Stages	G/N=0.51		G/N=0.60		G/N=0.71	
	Decomposition Temperature (°C)	Mass loss%	Decomposition Temperature (°C)	Mass loss%	Decomposition Temperature (°C)	Mass loss%
First	222	8.237	311	7.227	202	19.19
Second	371	9.255	550	5.023	534	5.478
Third	670	6.519	---	---	---	--
Stability	983	64.40	983	73.19	983	71.05

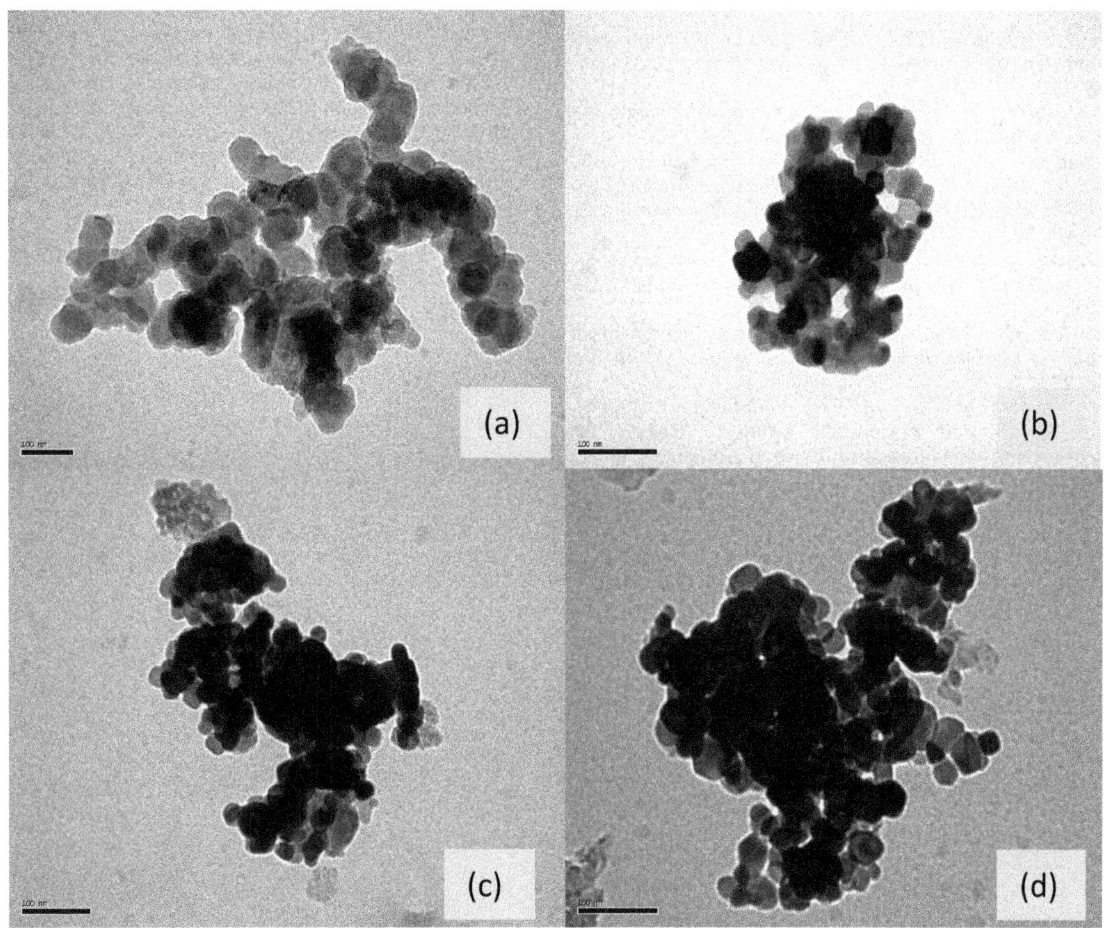

Figure 6. TEM picture of as-synthesized precursors a) G/N= 0.71 calcined at 900°C, (b) G/N= 0.71 calcined at 1000°C, (c) G/N= 0.60 calcined at 1000°C and (d) G/N= 0.71 calcined at 1200°C.

microwave assisted self-ignition process. Microwave assisted glycene – nitrate gel combustion synthesis has outstanding potential for producing nanocrystalline alumina powders. The technique produced pure, homogeneous and nanocrystalline Al_2O_3 particles, which are essential and beneficial to sintering and also to make transparent alumina ceramics.

The γ-Al_2O_3 crystallites appeared at 900°C with particle size of about 38 nm, and there was only a slight increase in grain size with the increase of calcination temperature, γ-Al_2O_3 crystallite almost disappeared and α-Al_2O_3 crystallite rapidly grew up to 52 nm at 1200°C when the transformation was just completed. The increasing molar ratio of G/N was found to be in favour of γ- to α-Al_2O_3 phase transition, whereas the precursor with G/N=0.71 at 900°C yielded a relatively well dispersed ultrafine α-Al_2O_3 powder with particle size of 38 nm. The mechanism of the combustion reaction is quite complex and fuel-to-oxidizer ratios and Millipore water is one of the most important parameters that influence the reaction. The combustion of glycene-nitrate mixture appears to undergo a self-propagating and non explosive exothermic reaction.

REFERENCES

Yi-quan W, Yu-feng Z (2001). "Preparation of plate like nanoalpha alumina particle", Ceramic Int. 27:265-268.

Edrissi M, Norouzbeigi R (2007). "Synthesis and characterization of alumina nanopowders by combustion of nitrate-amino acid gels", Mater. Science-Poland. 25(4).

Shojaie-Bahaabad M, Taheri-Nassaj E (2008). " Economical synthesis of nano alumina powder using an aqueous sol–gel method", Mater. Lett. 62:3364–3366.

Robert I, Ioan L, Cornelia P (2009). "The influence of combustion synthesis conditions on the a-Al_2O_3 powder preparation", J. Mater. Sci. 44:1016–1023.

Mimani T, Patil KC (2001). "Solution combustion synthesis of nanoscale oxides and their composites", Mater. phys. mech. 4:134-137.

Ananthapadmanabhan PV, Thiyagarajan TK, Sreekumar KP, Venkatramani N (2004). "Formation of nano-sized alumina by in-flight oxidation of aluminium powder in a thermal plasma reactor", Scripta Materialia, 50:143–147.

Shaoyan W, Xiaoan L, Shaofeng W, Yang L, Yuchun Z (2008). "Synthesis of γ-alumina via precipitation in ethanol" Mater. Lett. 62:3552–3554.

Mirjalili F, Hasmaliza MB, Chuah AL (2010). "Size-controlled synthesis of nano α- alumina particles through the sol–gel method", Ceramics Int. 36:1253–1257.

Chyi-Ching H, Tsung-Yung W, Jun W, Jih-Sheng T (2004). "Development of a novel combustion synthesis method for synthesizing of ceramic oxide powders", Mater. Sci. Eng. B, 111:49-56.

Ganesh I, Ferreira JMF (2007). "Influence of Phase Composition on Sintered Microstructure of Combustion Synthesized Oxides", Research Letters in Materials Science, Vol.2007, Article ID 91376, P. 5.

Singanahally TA, Alexander SM (2008). "Combustion Synthesis and Nano materials",Current opinion in solid state and Materials Science, 12(3-4):44-50.

Edrissi M, Norouzbeigi R (2008). Taguchi optimization for combustion synthesis of aluminium oxide Nano-particles, Chinese J. Chem. 26:1401-1406.

Tianyou P, Xun L, Ke D, Jiangrong X, Haibo S (2006). " Effect of acidity on the glycine–nitrate combustion synthesis of nanocrystalline alumina powder", Mater. Res. Bulletin. 41:1638–1645.

Jiang L, Yusong W, Yubai P, Jingkun G (2007). "Alumina precursors produced by gel combustion", Ceramics Int. 33:361–363.

Xiujing Z, Yan FGC (2006). Combustion Synthesis of the nano-structured alumina powder, Nano Science. 11(4):286-292.

Ganesh I, Johnson R, Rao GVN, Mahajan YR, Madavendra SS, Reddy BM (2005). Microwave assisted combustion synthesis of nanocrystalline MgAl2O4 spinel powder. Ceramic Int. 31:67-74.

Mangalaraja RV, J. Mouzon, Hedström P, Carlos CP, Ananthakumar S, Odén M (2009). "Microwave assisted combustion synthesis of nanocrystalline yttria and its powder Characteristics", Ceramic Int. 35(3):1173-1179.

Toniolo JC, Lima MD, Takimi AS, Bergmann CP (2005). "Synthesis of alumina powders by the glycine-nitrate combustion process", Mater. Res. Bulletin. 40:561-571.

7

Studies on energy absorption and exposure buildup factors in some solutions of alkali metal chlorides

D. Sardari[1] and M. Kurudirek[2]

[1]Faculty of Engineering, Science and Research Branch, Islamic Azad University, P. O. Box 14515-775, Tehran, Iran.
[2]Department of Physics, Faculty of Science, Ataturk University, 25240, Erzurum, Turkey.

Buildup of photons has been investigated through the solutions LiCl, NaCl and KCl with different salt contents in the energy region 0.015-15 MeV up to a penetration depth of 40 mean free paths. Two types of buildup factors, the gamma ray energy absorption ($EABF$) and exposure buildup factors (EBF) have been calculated simultaneously using the five parameter geometric progression (G-P) fitting formula. The influence of photon energy, penetration depth and chemical composition on the buildup factors has been studied. Also, a comparison has been made between the values of $EABF$ and EBF if any significant variation occurs between them. Moreover, the Monte Carlo simulation study has been made for the purpose of the comparison.

Key words: Gamma ray buildup factor, energy absorption, exposure, solution, Monte Carlo method.

INTRODUCTION

The buildup of photons in various kinds of materials has long been considered as the subject of various investigations in radiation shielding and dosimetry. Brar et al. (1999) have focused on the buildup factor studies of HCO-materials as a function of weight fraction of constituent elements. Brar et al. (1998) have investigated the effect of weight fractions of Fe and Si on buildup factors of some soil samples. Singh et al. (2008) have investigated the variation of energy absorption buildup factors with incident photon energy and penetration depth for some commonly used solvents. Manohara et al. (2010) have studied the energy absorption buildup factors for thermoluminescent dosimetric materials and their tissue equivalence which are of importance in radiation dosimetry, diagnostics and therapy. Recently, chemical composition dependence of exposure buildup factors for some polymers has been studied (Singh et al., 2009). Singh et al. (2010) have studied the buildup of gamma ray photons in fly ash concretes. An experimental investigation based on the effect of finite sample dimensions and total scatter acceptance angle on the gamma ray buildup factor has been made before (Singh and Kumar, 2008). There are different available methods to calculate the buildup factor such as geometric progression (G-P) fitting method (Harima et al., 1986) and invariant embedding method (Sakamoto and Tanaka, 1988; Shimizu, 2002; Shimizu et al., 2004). A reliable document for these methods is American National Standards ANSI/ANS 6.4.3 (1991) which provided buildup factor data for 23 elements, one compound and two mixtures (that is, air and water) and concrete at energies in the range 0.015-15 MeV up to penetration depths of 40 mean free path by using the G-P method. The developed G-P fitting formula is known to be accurate within a few percent errors (Harima et al., 1986; Harima, 1983). Harima (1993) has made an extensive historical review and an assessment for the status of buildup factor calculations and applications.

When it comes to material in dissolved form, it is noteworthy that the use of absorbers in the form of liquids has some definite advantages as compared with solid absorbers: the criterion of homogeneity of the absorber is satisfied instantly, and the strength of the absorber can easily be varied by changing the relative amount of solute and solvent (Kumar et al., 2006). However, there is almost no study based on the calculation of buildup factors for materials in their liquid forms such as solutions with different salt contents. Hence, we embarked on a study including the calculation of energy absorption and exposure buildup factors including their dependence on energy, penetration depth and chemical composition in the energy region of 0.015-15 MeV up to a penetration depth of 40 mean free paths. Also, the calculated buildup factors have been compared with those obtained using Monte Carlo method.

COMPUTATIONAL WORK

To calculate the buildup factors, the G-P fitting parameters were obtained by the method of interpolation from the equivalent atomic number (Z_{eq}). Computations are illustrated step by step as follows:

(i) Calculation of the equivalent atomic number Z_{eq}
(ii) Calculation of geometric progression (G-P) fitting parameters
(iii) Calculation of energy absorption and exposure buildup factors

However, the interaction of gamma rays with materials is based on domination of different partial photon interaction processes in different energy regions, thus Z_{eq} is an energy dependent parameter. Since the buildup effect arises from multiple scattering events, Z_{eq} is derived from the contribution of Compton scattering process. At the first step, the equivalent atomic number for a particular material has been calculated by matching the ratio, $(\mu/\rho)_{Compton}/(\mu/\rho)_{Total}$, of that material at a specific energy with the corresponding ratio of an element at the same energy. Thus, firstly the Compton partial mass attenuation coefficient, $(\mu/\rho)_{Compton}$, and the total mass attenuation coefficients, $(\mu/\rho)_{Total}$, were obtained for the elements of $Z = 4-40$ and for the solutions in the energy region 0.015-15 MeV, using the WinXCom computer program (Gerward et al., 2001a; Gerward et al., 2004b) (initially developed as XCOM (Berger and Hubbell, 1999)). It was reported by Hubbell (1999) that the envelope of the uncertainty of mass attenuation coefficient is in the order of 1-2% in the energy range from 5 keV to a few MeV. In case of the energies of 1 to 4 keV, the discrepancies are known to reach to a value of 25 to 50%. Recently, Chantler (2000) has extended the investigations below 5 keV concluding the presence of huge discrepancies below 4 keV and derived new theoretical results of substantially higher accuracy in near-edge soft X-ray regions in detail. De Jonge et al. (2005) have measured the mass attenuation coefficients and determined the imaginary component of the atomic form factor of molybdenum over the 13.5-41.5 keV energy range. Tran et al. (2005) have measured the X-ray mass attenuation coefficient of silver using the X-ray extended energy range with high accuracy.

At the second step, to calculate the G-P fitting parameters for elements were taken from the ANSI/ANS-6.4.3 (1991) standard reference database which provides the G-P fitting parameters for elements from beryllium to iron in the energy region 0.015-15 MeV up to 40 mfp. The G.P. fitting buildup factor coefficients of the used materials were interpolated according to the given formula (Gupta and Sidhu, 2012).

At the final step, these parameters were used to calculate the energy absorption and exposure buildup factors from the G-P fitting formula (Harima et al., 1986). While the exposure buildup factor, EBF is based on the energy absorption response of air; that is, exposure is assumed to be equivalent to absorbed dose in air as measured by a nonperturbing detector, the energy absorption buildup factor, $EABF$ refers to that absorbed or deposited energy in the attenuating material.

In radiologic sciences, practical problems usually are not legible for analytical solutions. Thus Monte Carlo techniques as strong computational tools are applied in radiation protection and dosimetry. Utilization of Monte Carlo codes such as MCNP (versions 4C, X and 5) has increased over past decade.

The MCNP code being acronym for Monte Carlo N-Particle, was developed at Los Alamos Laboratory (Los Alamos, NM) originally as a neutron and photon transport for reactor analysis in general. This code has been updated and improved repeatedly and their latest version, MCNPX and MCNP5, includes charged particle transport algorithms based on the best available models. This program provides several options for developing spatial and energetic distributions using complex geometric shapes (Briesmeister, 2000).

The code supports a wide variety of scoring options and radiation source modeling. Several variance reduction techniques are also available, allowing performance optimization for a more efficient determination of results. The user must also specify for each problem the *tallies*, or memory regions in which quantities such as energy, flux, etc. are recorded by MCNP. In our problem, we are interested in absorbed energy and/or photon flux either individual after passing through layers of material.

RESULTS AND DISCUSSION

The equivalent atomic numbers of the solutions are given in Tables 1, 2 and 3. The energy absorption and exposure buildup factors for various energies and mixtures are shown in graphical form at specific penetration depths (Figures 1 to 2a, b). In the energy range of interest (0.015-15 MeV), the photon interaction processes namely photoelectric absorption, Compton scattering and pair production partially dominate in different energy regions. Since the buildup of photons arises mainly from multiple scattering, the absorption processes (photoelectric absorption, pair production) reduces the values of $EABF$ and EBF in the low and high energy regions, respectively, and the scattering process (Compton scattering) increases the values of $EABF$ and EBF at the intermediate energy region. The maximum values of $EABF$ and EBF have been observed for LiCl, NaCl and KCl solutions at energy 0.15 MeV except for the KCl solution of highest Z_{eq} (salt content = 0.2) for which the maximum values occur at 0.2 MeV. At this energy the main photon interaction process is Compton scattering. It has been observed that the solutions of high Z_{eq} (that is, KCl (salt content = 0.2) mainly possess the lower values of $EABF$ and EBF whereas the solutions of

Table 1. Equivalent atomic numbers of LiCl solutions.

Energy (MeV)	Equivalent atomic number				
	LiCl solution				
	0.04*	0.08	0.12	0.16	0.2
1.50E-02	8.1	8.7	9.3	9.7	10.1
2.00E-02	8.2	8.8	9.4	9.8	10.3
3.00E-02	8.3	8.9	9.5	10.0	10.4
4.00E-02	8.3	9.0	9.6	10.1	10.5
5.00E-02	8.4	9.1	9.6	10.1	10.6
6.00E-02	8.4	9.1	9.7	10.2	10.6
8.00E-02	8.4	9.1	9.7	10.2	10.7
1.00E-01	8.4	9.2	9.8	10.3	10.8
1.50E-01	8.5	9.2	9.8	10.3	10.8
2.00E-01	8.5	9.2	9.9	10.4	10.9
3.00E-01	8.5	9.3	9.9	10.4	10.9
4.00E-01	8.5	9.3	9.9	10.4	10.9
5.00E-01	8.5	9.3	9.9	10.5	10.9
6.00E-01	8.5	9.3	10.0	10.5	11.0
8.00E-01	8.5	9.3	10.0	10.5	11.0
1.00E+00	8.5	9.3	10.0	10.5	11.0
1.50E+00	7.0	7.4	7.7	8.1	8.5
2.00E+00	6.9	7.2	7.5	7.8	8.2
3.00E+00	6.9	7.2	7.5	7.8	8.1
4.00E+00	6.9	7.2	7.5	7.8	8.1
5.00E+00	6.9	7.2	7.5	7.8	8.1
6.00E+00	6.9	7.2	7.5	7.8	8.0
8.00E+00	6.9	7.2	7.5	7.7	8.0
1.00E+01	6.9	7.2	7.4	7.7	8.0
1.50E+01	6.9	7.1	7.4	7.7	8.0

*refers to the salt content (g/cm^3).

low Z_{eq} (that is, NaCl (weight fraction = 0.04)) mainly dominate the higher values of $EABF$ and EBF. Actually, the reason for higher values of buildup factor in NaCl solutions is its lower Z_{eq} which leads to abundance of Compton scattering events.

Figures 3 and 4a, b, c, d show the influence of penetration depth on buildup factors at the fixed energies. From the above mentioned figures and data in Table 5, it is understood that $EABF$ and EBF values lie between 1 to 1.7, 1 to 1.7 and 1 to1.6 at 0.015 MeV for LiCl, NaCl and KCl solutions, respectively. At this low energy the main interaction process is photoelectric absorption hence the fast removal of photons give rise to the lower values of buildup factors. For photon energies of 0.15 MeV or lower, $EABF$ and EBF increase with the decrease in Z_{eq}. $EABF$ and EBF seem to be independent of variation in chemical composition at 1.5 MeV, thus the buildup factor values remain the same for different solutions. At 15 MeV, an inverse variation occurs as such buildup factors increase with the increase in Z_{eq} after 20 mfp for the given materials. $EABF$ and EBF for the given solutions as a function of weight fractions of Na and H_2O are shown in Figures 5 and 6 a,b. Similar figures have been produced for LiCl and KCl in the course of our studies. It has been shown that for the energies below 1.5 MeV, $EABF$ and EBF increase with the increase in the weight fraction of H_2O whereas they decrease with the increase in weight fraction of Li, Na and K present in solutions. Also, there are significant variations between $EABF$ and EBF where the larger buildup factors occur. In general, $EABF$ have higher values than EBF due to the fact that the materials under investigation have higher Z_{eq} values than that of air. Thus, when the Z_{eq} increases the energy absorption in the medium will be more than that of absorption in air. Figure 7 a, b gives the relative differences (%) between $EABF$ and EBF. In these figures, the positive values

Table 2. Equivalent atomic numbers of NaCl solutions.

Energy (MeV)	Equivalent atomic number				
	NaCl solution				
	0.04	**0.08**	**0.12**	**0.16**	**0.2**
1.50E-02	8.0	8.6	9.0	8.4	9.8
2.00E-02	8.1	8.6	9.1	8.4	9.9
3.00E-02	8.2	8.7	9.2	8.5	10.1
4.00E-02	8.2	8.8	9.3	8.6	10.1
5.00E-02	8.2	8.8	9.3	8.6	10.2
6.00E-02	8.2	8.8	9.4	8.7	10.2
8.00E-02	8.3	8.9	9.4	8.7	10.3
1.00E-01	8.3	8.9	9.4	8.8	10.3
1.50E-01	8.3	9.0	9.5	8.8	10.4
2.00E-01	8.3	9.0	9.5	8.9	10.4
3.00E-01	8.3	9.0	9.5	8.9	10.4
4.00E-01	8.3	9.0	9.6	8.9	10.5
5.00E-01	8.4	9.0	9.6	9.0	10.5
6.00E-01	8.4	9.0	9.6	9.0	10.5
8.00E-01	8.4	9.1	9.6	9.0	10.5
1.00E+00	8.4	9.1	9.6	9.0	10.5
1.50E+00	7.0	7.3	7.7	7.5	8.3
2.00E+00	6.9	7.2	7.5	7.4	8.1
3.00E+00	6.9	7.2	7.5	7.3	8.1
4.00E+00	6.9	7.2	7.5	7.3	8.0
5.00E+00	6.9	7.2	7.4	7.3	8.0
6.00E+00	6.9	7.2	7.4	7.3	8.0
8.00E+00	6.9	7.2	7.4	7.3	8.0
1.00E+01	6.9	7.1	7.4	7.3	8.0
1.50E+01	6.9	7.1	7.4	7.3	8.0

Table 3. Equivalent atomic numbers of KCl solutions.

Energy (MeV)	Equivalent atomic number				
	KCl solution				
	0.04	**0.08**	**0.12**	**0.16**	**0.2**
1.50E-02	8.4	9.2	9.9	10.5	11.0
2.00E-02	8.5	9.4	10.0	10.6	11.2
3.00E-02	8.6	9.5	10.2	10.8	11.4
4.00E-02	8.7	9.6	10.3	10.9	11.5
5.00E-02	8.7	9.6	10.4	11.0	11.6
6.00E-02	8.8	9.7	10.4	11.1	11.6
8.00E-02	8.8	9.8	10.5	11.1	11.7
1.00E-01	8.9	9.8	10.5	11.2	11.8
1.50E-01	8.9	9.9	10.6	11.3	11.8
2.00E-01	8.9	9.9	10.7	11.3	11.9
3.00E-01	9.0	10.0	10.7	11.4	12.0
4.00E-01	9.0	10.0	10.7	11.4	12.0
5.00E-01	9.0	10.0	10.8	11.4	12.0
6.00E-01	9.0	10.0	10.8	11.4	12.0
8.00E-01	9.0	10.0	10.8	11.4	12.0

Table 3. Contd.

1.00E+00	9.0	10.0	10.8	11.4	12.0
1.50E+00	7.2	7.7	8.2	8.7	9.2
2.00E+00	7.0	7.5	7.9	8.4	8.8
3.00E+00	7.0	7.4	7.8	8.3	8.7
4.00E+00	7.0	7.4	7.8	8.2	8.7
5.00E+00	7.0	7.4	7.8	8.2	8.7
6.00E+00	7.0	7.4	7.8	8.2	8.6
8.00E+00	7.0	7.4	7.8	8.2	8.6
1.00E+01	7.0	7.4	7.8	8.2	8.6
1.50E+01	7.0	7.4	7.8	8.2	8.6

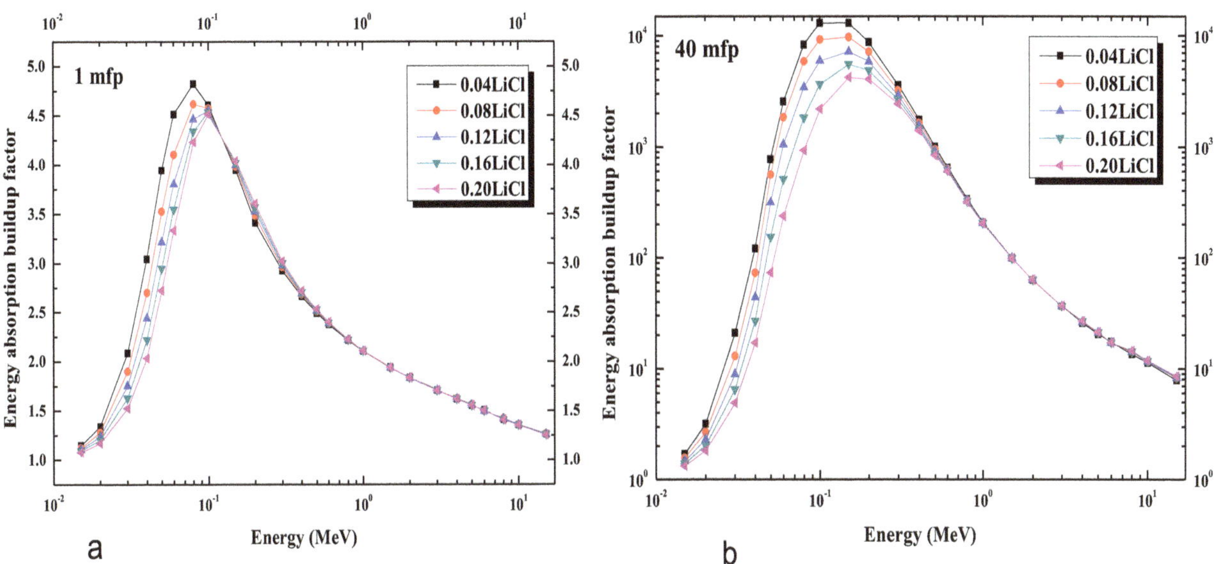

Figure 1(a, b). The energy absorption buildup factor for NaCl solution in the energy region 0.015-15 MeV at 1 and 40 mfp.

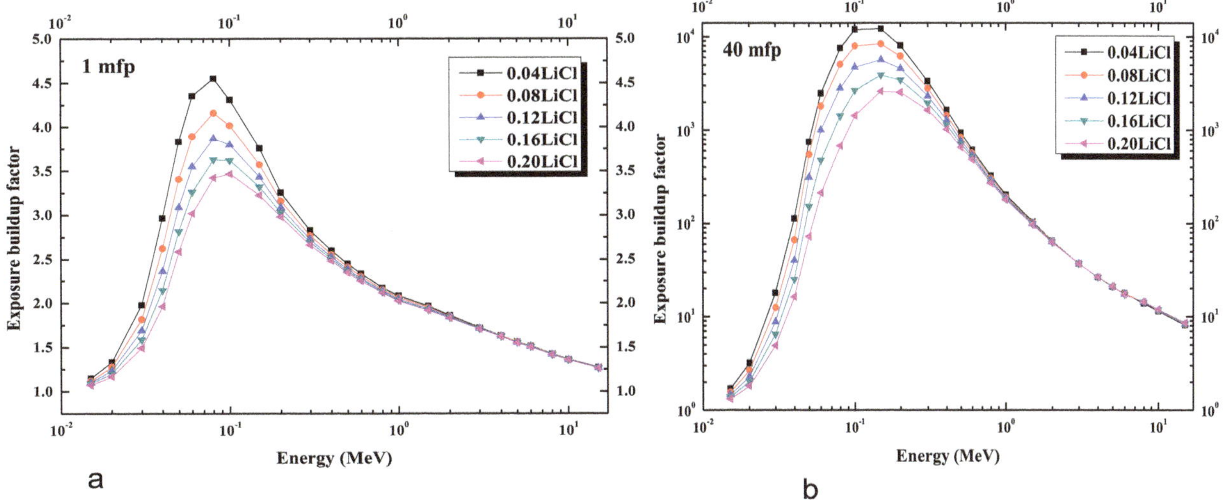

Figure 2(a, b). The exposure buildup factor for NaCl solution in the energy region 0.015-15 MeV at 1 and 40 mfp.

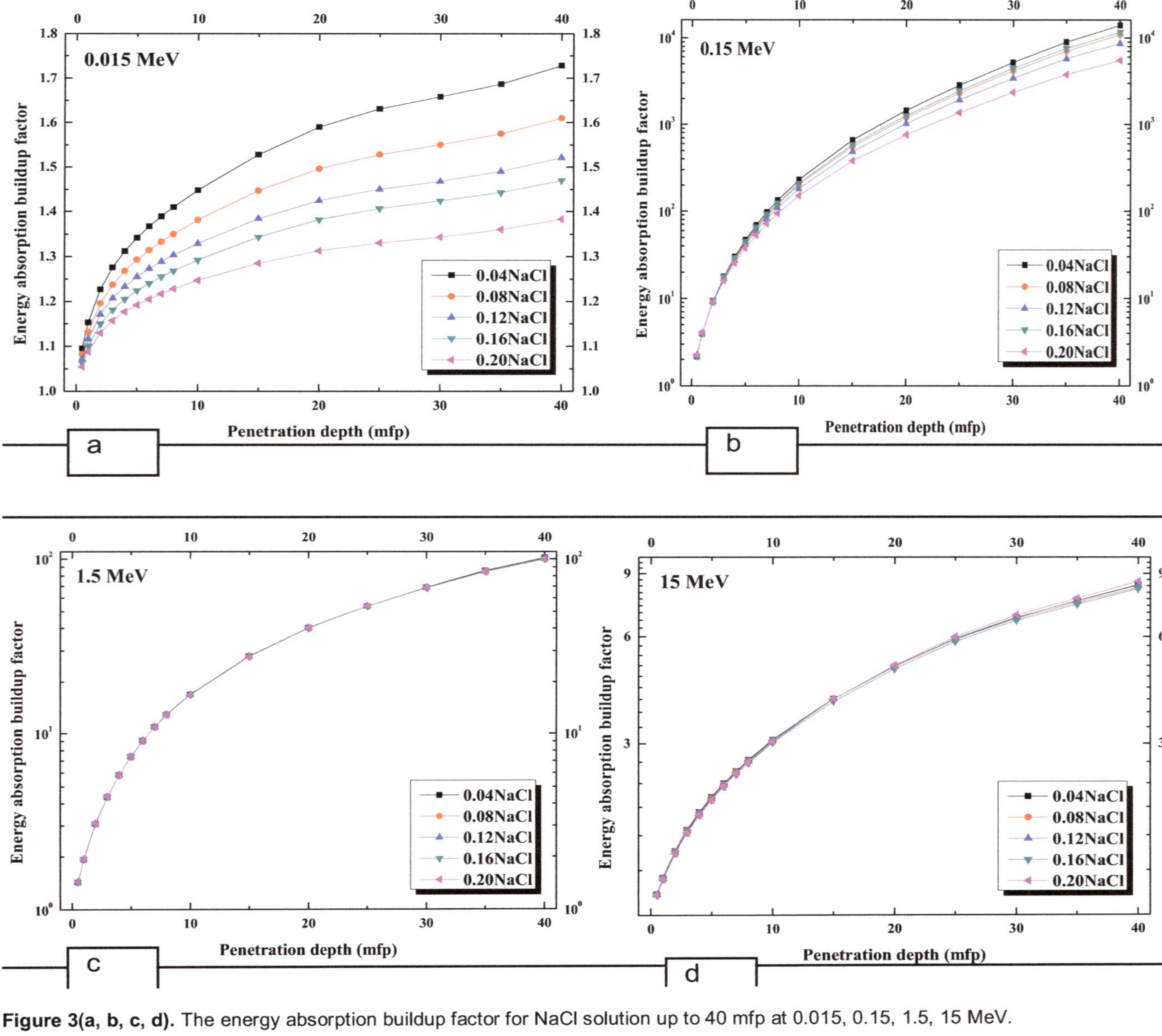

Figure 3(a, b, c, d). The energy absorption buildup factor for NaCl solution up to 40 mfp at 0.015, 0.15, 1.5, 15 MeV.

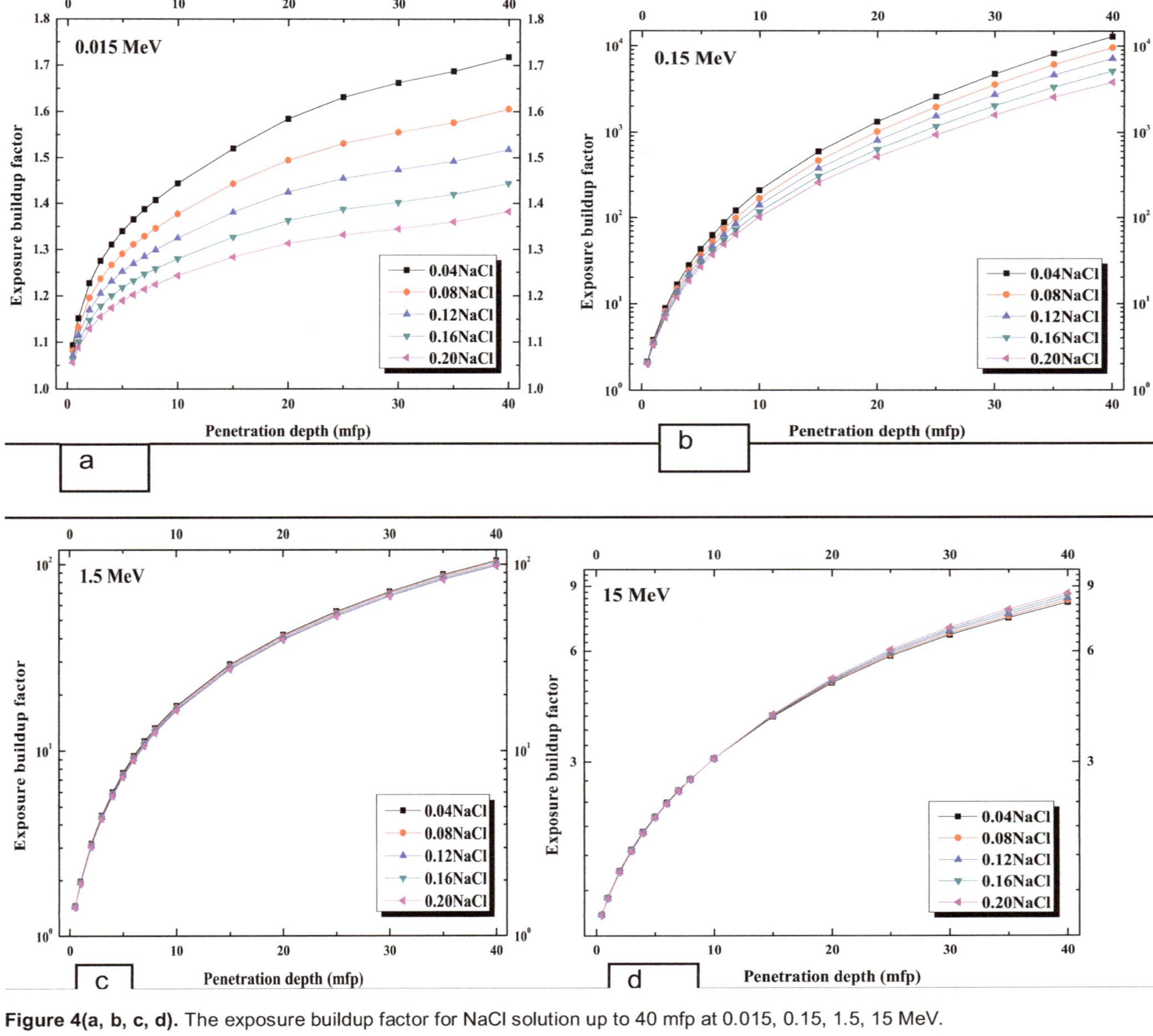

Figure 4(a, b, c, d). The exposure buildup factor for NaCl solution up to 40 mfp at 0.015, 0.15, 1.5, 15 MeV.

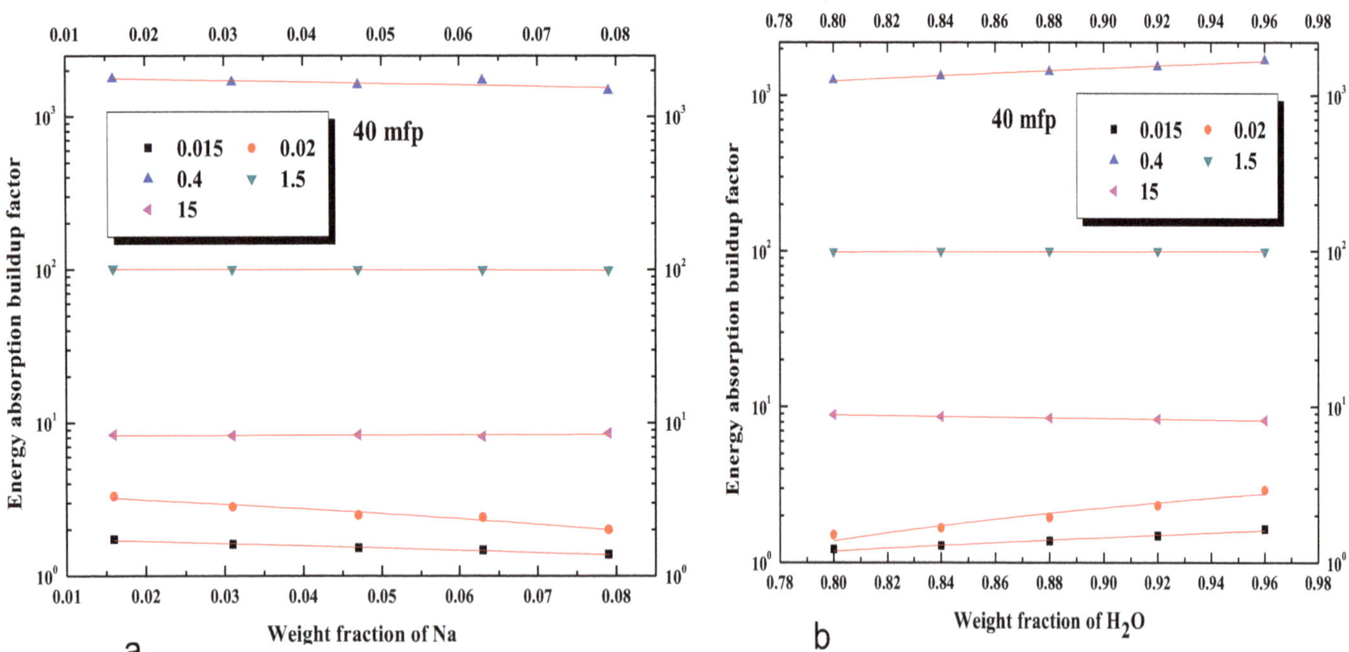

Figure 5(a, b). The energy absorption buildup factors for the given solutions as a function of weight fractions of Na and H_2O.

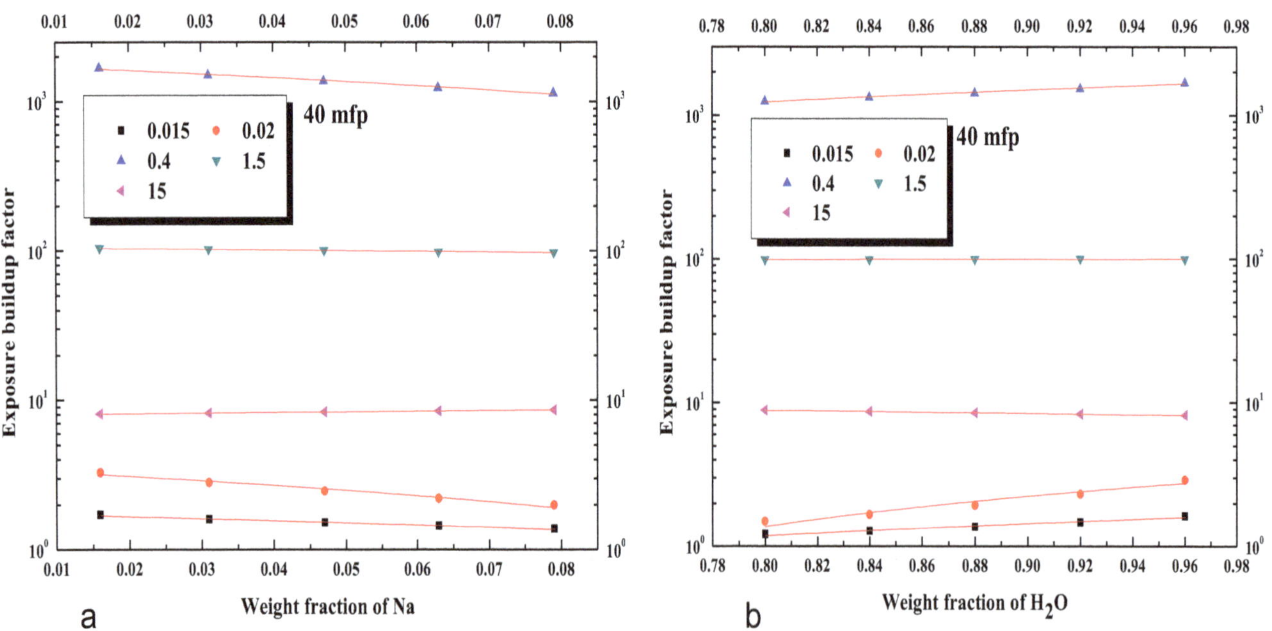

Figure 6(a, b). The exposure buildup factors for the given solutions as a function of weight fractions of Na and H_2O.

of differences (%) refer to the higher values of $EABF$ when compared to EBF.

Over past few years, Asano and Sakamoto (2007) have evaluated the buildup factors of two typical heavy concretes to improve the capability of the various materials for the shielding wall by using the Monte Carlo simulation code, EGS4. They also compared their calculated values by that of concrete in ANSI/ANS-6.4.3 (1991) standard reference database. Both of the calculations are in good agreement except for the slight differences which may be due to (a) the ANSI/ANS data are based on the calculation result data by using the

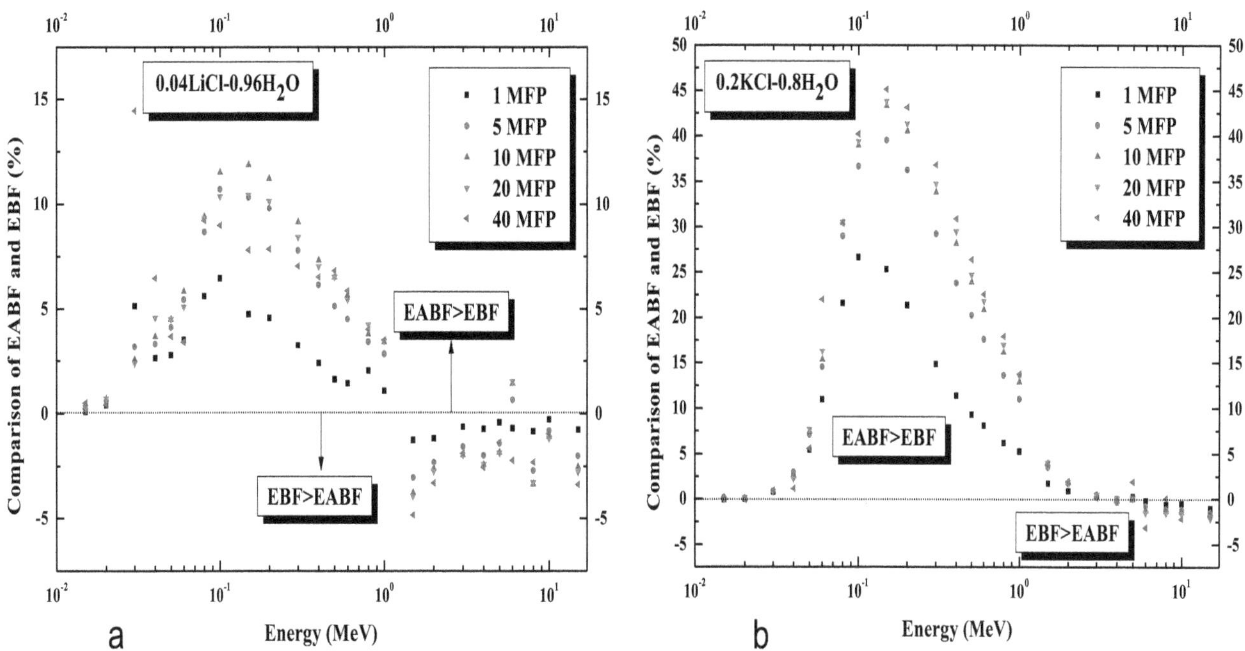

Figure 7(a, b). Difference (%) between $EABF$ and EBF for 0.04LiCl and 0.2KCl solutions in the energy region 0.015-15 MeV up to 40 mfp.

Table 4 (a,b,c,d). Energy absorption (*EABF*) and exposure buildup factors (*EBF*) of some solutions obtained by Monte Carlo simulation.

Energy (MeV) / Solutions	(a) Energy absorption buildup factor 1MFP					
	0.08 LiCl	0.08 NaCl	0.08 KCl	0.2 LiCl	0.2 NaCl	0.2 KCl
0.015	1.10	1.10	1.07	1.07	1.07	1.05
0.15	3.81	3.78	3.39	3.38	3.38	2.96
1.5	2.10	2.09	2.05	2.06	2.04	1.99
15	1.36	1.36	1.36	1.35	1.35	1.35

Energy (MeV) / Solutions	(b) Exposure buildup factor 1MFP					
	0.08 LiCl	0.08 NaCl	0.08 KCl	0.2 LiCl	0.2 NaCl	0.2 KCl
0.015	1.10	1.10	1.06	1.07	1.07	1.04
0.15	3.27	3.26	3.02	3.01	3.00	2.72
1.5	1.86	1.85	1.83	1.83	1.82	1.79
15	1.23	1.23	1.23	1.23	1.22	1.22

Solutions / Energy (MeV)	(c) Energy absorption buildup factor 10MFP					
	0.08 LiCl	0.08 NaCl	0.08 KCl	0.2 LiCl	0.2 NaCl	0.2 KCl
0.015	1.73	1.47	1.27	1.32	1.35	1.21
0.15	138.10	167.71	88.46	86.08	82.23	46.54
1.5	18.12	16.57	22.16	17.28	18.34	22.11
15	2.78	2.50	2.92	2.99	3.50	3.11

Solutions / Energy (MeV)	(d) Exposure buildup factor 10MFP					
	0.08 LiCl	0.08 NaCl	0.08 KCl	0.2 LiCl	0.2 NaCl	0.2 KCl
0.015	1.70	1.47	1.27	1.31	1.35	1.21
0.15	97.45	114.07	68.69	67.97	64.68	40.50
1.5	14.07	12.66	18.08	13.51	13.61	18.03
15	2.32	2.17	2.48	2.45	2.86	2.46

Table 5 (a, b, c, d). Energy absorption (*EABF*) and exposure buildup factors (*EBF*) of some solutions obtained by G-P fitting approximation.

Energy (MeV) \ Solutions	(a) Energy absorption buildup factor 1MFP					
	0.08 LiCl	0.08 NaCl	0.08 KCl	0.2 LiCl	0.2 NaCl	0.2 KCl
0.015	1.13	1.13	1.11	1.08	1.09	1.05
0.15	3.98	3.97	4.00	4.04	4.02	4.09
1.5	1.94	1.94	1.94	1.94	1.94	1.94
15	1.26	1.26	1.26	1.27	1.27	1.25

Energy (MeV) \ Solutions	(b) Exposure buildup factor 1MFP					
	0.08 LiCl	0.08 NaCl	0.08 KCl	0.2 LiCl	0.2 NaCl	0.2 KCl
0.015	1.12	1.13	1.11	1.08	1.09	1.05
0.15	3.57	3.63	3.42	3.22	3.32	3.06
1.5	1.95	1.95	1.94	1.92	1.92	1.91
15	1.27	1.27	1.27	1.26	1.26	1.26

Energy (MeV) \ Solutions	(c) Energy absorption buildup factor 10MFP					
	0.08 LiCl	0.08 NaCl	0.08 KCl	0.2 LiCl	0.2 NaCl	0.2 KCl
0.015	1.36	1.38	1.30	1.21	1.25	1.14
0.15	191.30	200.92	167.24	137.72	151.23	111.54
1.5	16.80	16.81	16.77	16.69	16.70	16.62
15	3.05	3.02	3.03	3.08	3.07	3.03

Energy (MeV) \ Solutions	(d) Exposure buildup factor 10MFP					
	0.08 LiCl	0.08 NaCl	0.08 KCl	0.2 LiCl	0.2 NaCl	0.2 KCl
0.015	1.35	1.38	1.30	1.21	1.24	1.14
0.15	152.01	165.25	120.54	85.56	101.03	63.20
1.5	17.07	17.10	16.82	16.34	16.40	15.96
15	3.06	3.06	3.06	3.06	3.06	3.08

moments method (Eisenhauer and Simmons, 1975) with parallel beam source and the Monte Carlo code, EGS4 with isotropic emission source, (b) the development of the low energy photon computations in EGS4 such as K-X ray, L-X ray and Bremsstrahlung. It was shown by Shimizu et al. (2004) that the methods based on invariant embedding, G.P fitting and Monte Carlo simulation agree for 18 low-Z materials within small discrepancies. In the present study, the solutions have Z_{eq} values ranging from 7 to 12. Hence, the used materials can be considered as low Z_{eq} materials. When compared with other available approximations such as Berger, Taylor and three exponential, the geometric-progression (G.P) fitting seem to reproduce the buildup factors with better accuracy. Harima et al. (1986) have shown that the absolute values of maximum deviations of exposure build factors for water in G. P. fitting is within 0.5-3%, in three-exponential approach is within 0.4 to 9.3%, in Berger approach is within 0.9 to 42.7% and in Taylor approximation is within 0.4 to 53.2%.

Conclusion

It is shown that G-P fitting is a proper method for estimating photon buildup factor in materials with Z < 20. To make a more robust comparison, in the present work the buildup factor of Alkali metals' salt solution in water was computed using MCNP code. This was carried out for penetration depths of 1 and 10 mean free path. The results appear in Tables 4 and 5. For EBF, in short distances from the source, G-P fitting results are in good agreement with MCNP. Namely, the average deviation between two methods is about 5%, the largest one being 10%. In all energies and solution concentration the buildup factor obtained by MCNP at one m.f.p is lower than G-P fitting method. The largest deviation appears at energy 0.15 MeV.

At longer penetration depth, that is, 10 m.f.p the percentage deviation between the results of two methods on average amounts to 20%. Except 0.015 MeV, the outcome of MCNP is lower than G-P fitting. Besides, the agreement between two methods is much better in 0.015 MeV than higher energies.

In case of 1 m.f.p, the answers from MCNP and G-P fitting are consistent. In case of 10 m.f.p, remarkable disagreement is observed in some results but due to increasing number of interactions of photon in the matter, the MCNP results are more reliable. For EABF, again the deviation of results in longer penetration depth is much more pronounced. At 10 m.f.p on average being 20% while for 1 m.f.p. it is 8%. At short distance and energies above 1 MeV the buildup factor data resulting from MCNP is higher than G-P fitting. Since in EABF the absorbed dose is dealt with, this fact might be attributed to better precision of MCNP code in considering detailed interaction after occurring of photoelectric effect.

Regarding the data in Tables 4 and 5, one might say G-P fitting method is consistent with MCNP in following cases:

(i) For EBF at short distances and all energies and all type of solutions,
(ii) For EBF at long distances for energy below 50 keV and all types of solutions,
(iii) For EABF at short distances and energies away from 0.15 MeV, where buildup factor is maximum,
(iv) For EABF at long distances the consistency is poor but in some cases the data are consistent away from 0.15 MeV.

REFERENCES

ANSI/ANS-6.4.3 (1991). Gamma-ray attenuation coefficients and buildup factors for engineering materials. Am. Nuclear Soc. ANSI/ANS-6.4.3-1991. La Grange Park, Illinois.

Asano Y, Sakamoto Y (2007). Gamma ray buildup factors for heavy concretes JAE Data/Code 2007-006.

Berger MJ, Hubbell JH (1999). XCOM: Photon cross sections database. Web Version 1.2, available at http:// physics.nist.gov/xcom, National Institute of Standards and Technology, Gaithersburg, MD 20899, USA.

Brar GS, Sidhu GS, Singh PS, Mudahar GS (1999). Buildup factor studies of HCO-materials as a function of weight fraction of constituent elements. Radiat. Phys. Chem. 54:125-129.

Brar GS, Sidhu GS, Sandhu PS, Mudahar GS (1998). Variation of buildup factors of soils with weight fractions of iron and silicon Appl. Radiat. Isot. 49:977-980.

Briesmeister JF (2000). MCNP–A General Monte Carlo N–Particle Transport Code. LA–13709–M.

Chantler CT (2000). Detailed tabulation of atomic form factors, photoelectric absorption and scattering cross section, and mass attenuation coefficients in the vicinity of absorption edges in the soft X-Ray (Z=30–36, Z=60–89, E=0.1 keV–10 keV). J. Phy. Chem. Ref. Data. 29:597-1048.

Eisenhauer CM, Simmons GL (1975). Point isotropic gamma ray buildup factors in concrete. Nucl. Sci. Eng. P. 56.

Gerward L, Guilbert N, Jensen KB, Levring H (2001a). X-ray absorption in matter. Reengineering XCOM. Radiat. Phys. Chem. 60:23-24.

Gerward L, Guilbert N, Jensen KB, Levring H (2004b). WinXCom- A program for calculating X-ray attenuation coefficients. Radiat. Phys. Chem. 71:653-654.

Gupta S, Sidhu GS (2012). A comprehensive study on energy absorption and exposure buildup factors for some soils and ceramic materials. IOSR J. Appl. Phys. 2(3):24-30.

Harima Y (1983). An approximation of gamma ray buildup factors by modified geometrical progression. Nucl. Sci. Eng. 83:299-309.

Harima Y, Sakamoto Y, Tanaka S, Kawai M (1986). Validity of the geometric-progression formula in approximating gamma ray buildup factors. Nucl. Sci. Eng. 94:24-35.

Harima Y, Sakamoto Y, Tanaka S, Kawai M, Fujita T, Ishikawa T, Kinno M, Hayashi K, Matsumoto Y, Nishimura T (1986). Applicability of geometrical progression approximation (G-P method) of gamma-ray buildup factors. Jpn. Atomic Energy Res. Instit. (JAERI)-M. pp. 86-071.

Harima Y (1993). A historical review and current status of build-up factor calculations and applications. Radiat. Phys. Chem. 41:631-672.

Hubbell JH (1999). Review of photon interaction cross section data in the medical and biological context. Phys. Med. Biol. 44:R1.

De Jonge MD, Tran CQ, Chantler CT, Barnea Z, Dhal BB, Cookson DJ, Lee WK, Mashayekhi A (2005). Measurement of the x-ray mass attenuation coefficient and determination of the imaginary component of the atomic form factor of molybdenum over the 13.5-41.5 keV energy range. Phys. Rev. A71:032702.

Kumar A, Singh S, Mudahar GS, Thind KS (2006). Molar extinction coefficients of some commonly used solvents. Radiat. Phys. Chem. 75:737-740.

Manohara SR, Hanagodimath SM, Gerward L (2010). Energy absorption buildup factors for thermoluminescent dosimetric materials and their tissue equivalence. Radiat. Phys. Chem. 79:575-582.

Sakamoto Y, Tanaka S (1988). Interpolation of gamma ray buildup factors for point isotropic source with respect to atomic number. Nucl. Sci. Eng. 100:33-42.

Shimizu A (2002). Calculations of gamma-ray buildup factors up to depths of 100 mfp by the method of invariant embedding, (I) analysis of accuracy and comparison with other data. J. Nucl. Sci. Technol. 39:477.

Shimizu A, Onda T, Sakamoto Y (2004). Calculations of gamma-ray buildup factors up to depths of 100 mfp by the method of invariant embedding, (III) generation of an improved data set. Nucl. Sci. Technol. 41:413-424.

Singh SP, Singh T, Kaur P (2008). Variation of energy absorption buildup factors with incident photon energy and penetration depth for some commonly used solvents Ann. Nucl. Energy 35:1093-1097.

Singh T, Kumar N, Singh SP (2009). Chemical composition dependence of exposure buildup factors for some polymers Ann. Nucl. Energy 36:114-120.

Singh T, Ghumman SS, Singh C, Thind KS, Mudahar GS (2010). Buildup of gamma ray photons in flyash concretes: A study. Ann. Nucl. Energy 37:681-684.

Singh T, Kumar A (2008). Effect of finite sample dimensions and total scatter acceptance angle on the gamma ray buildup factor Ann. Nucl. Energy 35:2414-2416.

Tran CQ, Chantler CT, Barnea Z, De Jonge MD, Dhal BB, Chung CTY, Paterson D, Wang J (2005). Measurement of the x-ray mass attenuation coefficient of silver using the x-ray extended range technique. J. Phys. B.: At. Mol. Opt. Phys. 38:89-107.

8

Co-digestion of cattle manure with organic kitchen waste to increase biogas production using rumen fluid as inoculums

Tamrat Aragaw[1], Mebeaselassie Andargie[1] and Amare Gessesse[2]

[1]Department of Biology, College of Natural and Computational Sciences, Haramaya University, P. O. Box 138, Haramaya, Ethiopia.
[2]Department of Biology, College of Natural and Computational Sciences, Addis Ababa University, P. O. Box 1176, Addis Ababa, Ethiopia.

Anaerobic co-digestion strategies are needed to enhance biogas production when treating certain residues such as cattle/pig manure. Co-digestion of food waste with animal manure or other feedstocks with low carbon content can improve process stability and methane production. In this study, anaerobic digestion and co-digestion of cattle manure with organic kitchen waste using rumen fluid as inoculums have been experimentally tested to determine the biogas potential. Co-digestion substantially increased the biogas yields by 24 to 47% over the control (organic kitchen waste and dairy manure only). The highest methane yield of 14,653.5 ml/g-VS was obtained with 75% organic kitchen waste (OKW) and 25% cattle manure (CM) additions. In contrast, addition of 75% cattle manure caused inhibition of the anaerobic digestion process, and its cumulative methane yield was 23% lower than that with 25% cattle manure addition.

Key words: Cattle manure, co-digestion, methane, organic kitchen waste, rumen fluid.

INTRODUCTION

Energy is one of the most important factors for human development and to global prosperity. The dependence on fossil fuels as primary energy source has led to global climate change, environmental degradation, and human health problems. 80% of the world's energy consumption still originates from combusting fossil fuels (Goldemberg and Johansson, 2004). Yet the reserves are limited; means do not match with the fast population growth, and their burning substantially increases the greenhouse gas (GHG) concentrations that contributed for global warming and climate change (Schamphelaire and Verstraete, 2009). So, bio-energy (energy production from biomass) can be seen as one of the key options. Among the many bio-energy related processes being developed, those processes involving microorganisms are especially promising, as they have the potential to produce renewable energy on a large scale, without disrupting strongly the environment or human activities (Rittmann, 2008).

Anaerobic digestion (AD) is a technology widely used for treatment of organic waste for biogas production. Anaerobic digestion that utilizes manure for biogas production is one of the most promising uses of biomass wastes because it provides a source of energy while simultaneously resolving ecological and agrochemical issues. The anaerobic fermentation of manure for biogas production does not reduce its value as a fertilizer supplement, as available nitrogen and other substances remain in the treated sludge (Alvarez and Liden, 2007).

Ethiopia has a large population of dairy and beef cattle, generating large amounts of surplus manure that can be used in biogas plants to produce renewable energy. However, the high water content, together with the high content in fibers, are the major reasons for the low methane yields when cattle manure is anaerobically digested, typically ranging between 10 and 20 m^3 CH_4 per tone of manure treated (Angelidaki and Ellegaard, 2003).

Studies demonstrated that using co-substrates in anaerobic digestion system improves the biogas yields due to the positive synergisms established in the digestion medium and the supply of missing nutrients by the co-substrates (Wei, 2000). In a study carried out by Adelekan and Bamgboye (2009) on the different mixing ratios of livestock waste with cassava peels, the average cumulative biogas yield was increased to 21.3, 19.5, 15.8 and 11.2 L/kg TS, respectively for 1:1, 2:1, 3:1 and 4:1 mixing ratios when cassava peel was mixed with cattle waste. In another report, co-digestion of cow dung with pig manure increased biogas yield as compared to pure samples of either pig or cow dung. Comparing to samples of pure cow dung and pig manure, the maximum increase of almost seven and three fold was respectively achieved when mixed in proportions of 1:1 (Muyiiya and Kasisira, 2009). Co-digestion with other wastes, whether industrial (glycerin), agricultural (fruit and vegetable wastes) or domestic (municipal solid waste) is a suitable option for improving biogas production (Amon et al., 2006; Macias-Corral et al., 2008; El-Mashad and Zhang, 2010; Marañón et al., 2012).

Food waste is a desirable material to co-digest with dairy manure because of its high biodegradability (Zhang et al., 2006, 2011; Li et al., 2010; Wan et al., 2011). Study on the biogas production potential of unscreened dairy manure and different mixtures of unscreened dairy manure and food waste using batch digesters at 35°C showed that the methane yield of unscreened manure and two mixtures of unscreened manure and food waste (68/32 and 52/48), after 30 days of digestion, was 241, 282 and 311 L/kg VS, respectively (El-Mashad and Zhang, 2010).

In a study conducted by Zhu et al. (2011), they used different food wastes, including expired creamer; expired beer; slaughterhouse waste (SW); and fat, oil, and grease (FOG), and these food substances were co-digested with dairy manure to determine the methane potential. According to the result, co-digestion substantially increased the methane yields by 2.0 to 4.6 times over the control (dairy manure only).

This study was initiated to investigate the feasibility of biogas production from the different wastes that are generated from Haramaya University and the aims of the present research work were to determine the optimal conditions and mixing ratios for improved production of biogas using co-digestion of cattle manure and solid organic kitchen waste and also identify the key parameters influencing the increase of biogas and methane yield.

MATERIALS AND METHODS

Sample collection and preparation

Fresh cattle manure (CM) from beef and dairy farm, fresh organic kitchen waste (OKW) from staff lounge, and rumen fluid (RF) from the slaughterhouse were collected from Haramaya University compound. 2 kg of fresh cattle manure was collected from eight randomly selected cattle from beef and dairy farms for five consecutive days. In these sites there are special feeds and normal grazing cattle. The special feeds are provided with special type of feeding program that includes silage, concentrate, hay forage, agricultural residues and different grass types, byproducts from Harar Brewery and Hamaressa Food Complex, etc. On the other hand normal grazers are not provided with special type of feeding program rather they graze grasses in the field and get only fodder and agricultural residues. Finally the CM from both types of cattle (special and normal grazers) was sorted and dried separately on a plastic tray using direct sunlight for two days. 3 kg of fresh organic kitchen wastes were also collected from the staff lounge similarly for five consecutive days. The OKW was sorted manually to prevent the inclusion of unwanted and possibly contaminant materials (such as detergents, sand, bones etc.) and then dried with direct sunlight for two days.

Following the methods suggested by Wendland et al. (2006), separately dried cattle manure from special feeds and normal grazers were mixed by weighing equal amount from each source and shredded using shredder (Fritsch- Adam Baumuler model 80a-4S114 type) to an average particle size of 2 mm and kept in a refrigerator at 4°C. The shredded small sized cattle manure and organic kitchen waste were mixed separately with water in 1:5 (solid waste: water) volume ratio, in order to maintain the total solid in the digester between 8 to 15%, which is the desired value for wet anaerobic digestion.

Inoculum preparation

Following the recommendation of Aurora (1983), due to the presence of higher content of anaerobic bacteria in the rumen of the ruminant animals and the abundance of rumen waste disposal from the nearby slaughterhouse, rumen fluid was used as inoculum for anaerobic co-digestion of cattle manure and organic kitchen waste.

Experimental set-up and design

A completely randomized experimental design was used in a 5 × 4 replicated laboratory experiment and it was conducted in a series of five plastic tanks with 2 L capacity which was used as a laboratory scale anaerobic digesters at mesophilic temperature (30 ± 8°C). The working volume of each digester was 1.6 L. In each digester, rumen fluid was used as inoculum. The TS and VS/ TS of the inoculum used were 1.03% (wet basis) and 63.9%, respectively. Each digester was purged for 5 min (300 mL/min) with inert gas (N_2) to create an anaerobic environment. Food waste, cattle manure and their mixtures were separately examined in mono and co-digestion respectively. The characteristics of the different experiments are shown in Table 1. In co-digestion, the amount of organic kitchen waste as well as that of cattle manure in each

Table 1. Properties of organic kitchen waste, cattle manure and rumen fluid (mean ±SD).

Parameter	Organic kitchen waste	Cattle manure	Rumen fluid
pH	5.51±0.129	7.19±0.215	7.45±0.114
MC (%)	82.95±0.169	84.59±0.40	98.98±0.01
TS (Wt %)	17.05±0.169	15.42±0.40	1.03±0.01
VS (Wt %)	15.89±0.52	12.68±0.63	0.66±0.01
VS/TS ratio	93.18±2.54	82.23±2.04	63.9±0.45

Table 2. Properties of cattle manure and organic kitchen waste before digestion (mean value ± SD).

	Parameter (before digestion)				
Mixture	pH	MC (%)	TS (%)	VS (%)	VS/TS (%)
A1	6.95±0.030	86.15±0.128	13.85±0.128	12.85±0.403	92.75±2.398
A2	7.45±0.071	87.46±0.314	12.54±0.314	10.27±0.503	81.9±1.403
A3	7.32±0.065	87.09±0.490	12.83±0.353	10.81±0.470	84.2±1.354
A4	7.19±0.051	86.83±0.358	13.17±0.358	11.34±0.445	86.03±1.159
A5	7.09±0.025	86.42±0.274	13.58±0.274	11.89±0.389	87.5±1.587

digester was varied when it was added. The FW/CM ratios (based on VS) of digestion A3, A4, A5 were designed as 0.3, 1 and 3, respectively, corresponding to the organic kitchen waste and cattle manure amounts of 25:75, 50:50 and 75:25 g-VS/L. In digestion A1, organic kitchen waste was digested alone at the load of 100 g-VS/L, whereas in digestion A2, cattle manure was digested alone at the load of 100 g-VS/L as a control group. Thus, to determine the performance of co-digestion, the co-digestion of A3, A4 and A5 was compared with mono-digestion groups of A1 and A2. In addition, to provide mixing of the digester contents, all digesters were shaken manually for about 1 min once a day prior to measurement of biogas volume.

Measurement of biogas yield

Biogas was collected by water displacement method. In order to prevent the dissolution of biogas in the water, brine solution was prepared. Following the method suggested by Elijah et al. (2009), an acidified brine solution was prepared by adding NaCl to water until a supersaturated solution was formed. Three to five drops of sulphuric acid were added to acidify the brine solution. As biogas production commenced in the fermentation chamber, it was delivered to the second chamber which contained the acidified brine solution. Since the biogas is insoluble in the solution, a pressure build-up and provides the driving force for displacement of the solution. Thus the displaced brine solution was measured to represent the amount of biogas produced. The biogas volume was calculated daily and was transformed into the volume at Standard Temperature and Pressure (STP) condition.

Chemical analysis

The pH, TS and VS of organic kitchen waste and cattle manure samples were measured according to the standard methods (APHA, 1998). The pH values of each digester were monitored in five days interval using digital pH meter (HANNA Model pH-211). Following the method of Radtke et al. (1998) and Yu and Fang (2002), the pH values of the contents of digesters were buffered between 6.8 and 7.4 by the addition of hydrated calcium carbonate. The VS content of the liquor was subsequently measured. The values of VS destructions were calculated based on total mass balances of VS in each digester before and after the digestion test with subtracting the VS contents of the control digesters from that of the testing digester.

RESULTS AND DISCUSSION

Pre-digestion characteristics of substrates

Table 2 summarizes the values obtained in the pre-digestion characteristics of the five feed stocks. As it is shown, there is a considerable amount of variation in the composition of feed mixtures, which is due to the variability in the composition of the samples of the different substrates taken over the experimental period. The content in volatile solids of cattle manure and organic kitchen waste ranged between 9.8-10.8% and 12.4-13.3%, respectively (average values of 10.3 and 12.9%, respectively). On a dry matter (TS) basis, organic kitchen waste contained higher VS than cattle manure. The higher VS content of organic kitchen waste (13 g/kg), compared with that of manure (10 g/kg), means relatively higher energy content, which is desirable from an economic standpoint with regards to biogas energy production. The VS/TS ratios were 82 and 93% for cattle manure and organic kitchen waste, respectively.

Before inoculation the mean pH values of CM and OKW were 7.19 and 5.51, respectively; however, after they are inoculated with rumen fluid, the inoculum mean pH values of the two control groups (A1 and A2) were

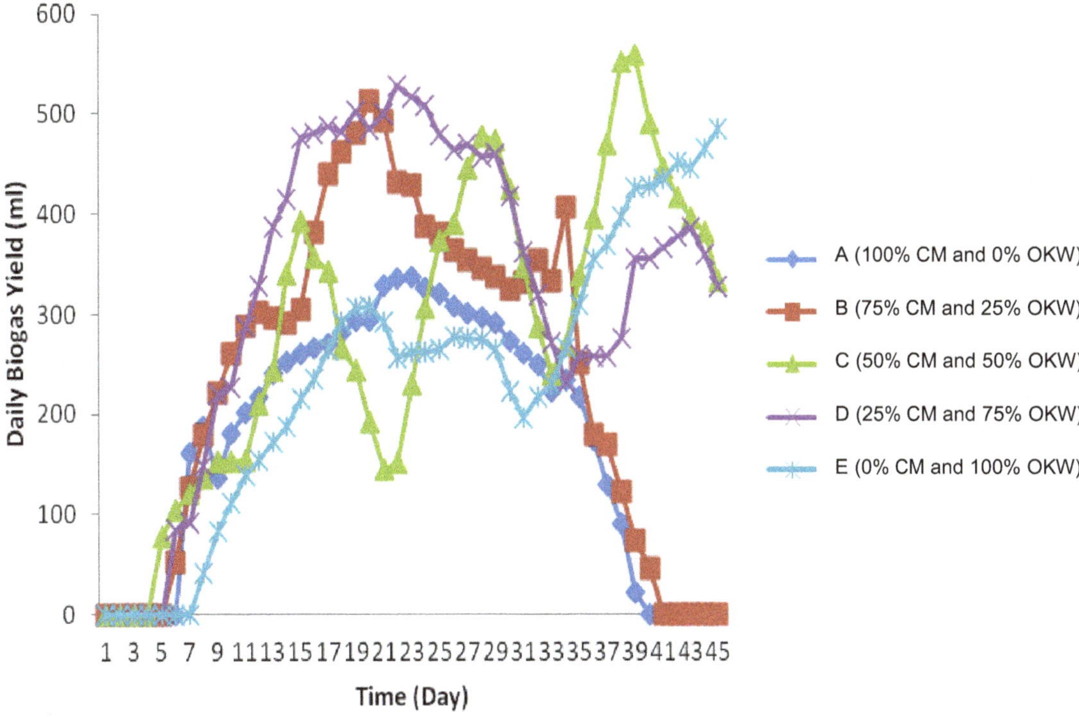

Figure 1. Daily mean biogas yield of digester D in 45 days.

increased. This indicates that the rumen fluid used for this study have had a good buffering capacity as it was also reported earlier (Girma et al., 2004; Forster-Carneiro et al., 2008; Montusiewicz et al., 2008; Uzodinma and Ofoefule, 2008).

Biogas production rate

On average, biogas productions from digesters A2, A3, A4, A5, and A1 were detected on the 7th, 6th, 5th, 6th, and 8th days respectively. The results showed that the co-digestion of samples with the three mix ratios (A4, A5, and A3) produced biogas earlier than the two pure substrates (A1 and A2) that were used as control groups. From the three mix groups, digester A4 produced biogas much faster, followed by digester A5 and A3. This might be due to the attribution of the positive synergetic effect of the co-digestion of CM and OKW in providing more balanced nutrients, increased buffering capacity, and decreased effect of toxic compounds. Digestion of more than one kind of substrate could establish positive synergism in the digester (Mata-Alvarez et al., 2000; Li et al., 2009; Jianzheng et al., 2011). The rapid initial biogas production in digester A4 might be also due to shorter lag phase growth, the availability of readily biodegradable organic matter in the substrate, and the presence of high content of the methanogens.

Biogas production

Biogas production was used mainly as an indication of optimum production and the development of favorable conditions for microbial activity during the digestion process. The daily methane production from the control and digesters are shown in Figure 1. The average daily biogas yield observed from the five digesters (A1, A5, A4, A3, and A2) were 176.77, 237.85, 284.76, 325.63, and 236.18 mL/g-VS, respectively. As compared to digesters A1 and A2, digesters A5, A4, and A3 produced the 1st, 2nd, and 3rd highest volume of biogas on each day during the 45 days of experiment, respectively (Figure 1). The higher biogas production from these mixtures could be due to the balanced (nutrient to microorganism) composition, and stable pH which was attained from the inoculation with rumen fluid and mixing ratios used. On the other hand low average daily biogas production observed from digesters A1 and A2 containing pure 100% OKW and 100% CM, attributed to the unbalanced nutrient to microorganism ratio, and unstable pH value. After the gas production was started and stabilized, digesters A4, A5, and A1 produced the least amount of daily biogas on the 5th, 6th, and 8th, days of the run, respectively. The observed least gas yield from these digesters might be due to the production of volatile fatty acids by the microorganism which hinders the releasing of the biogas. This is in agreement with the report of Budiyono et al. (2010) who also observed low level of

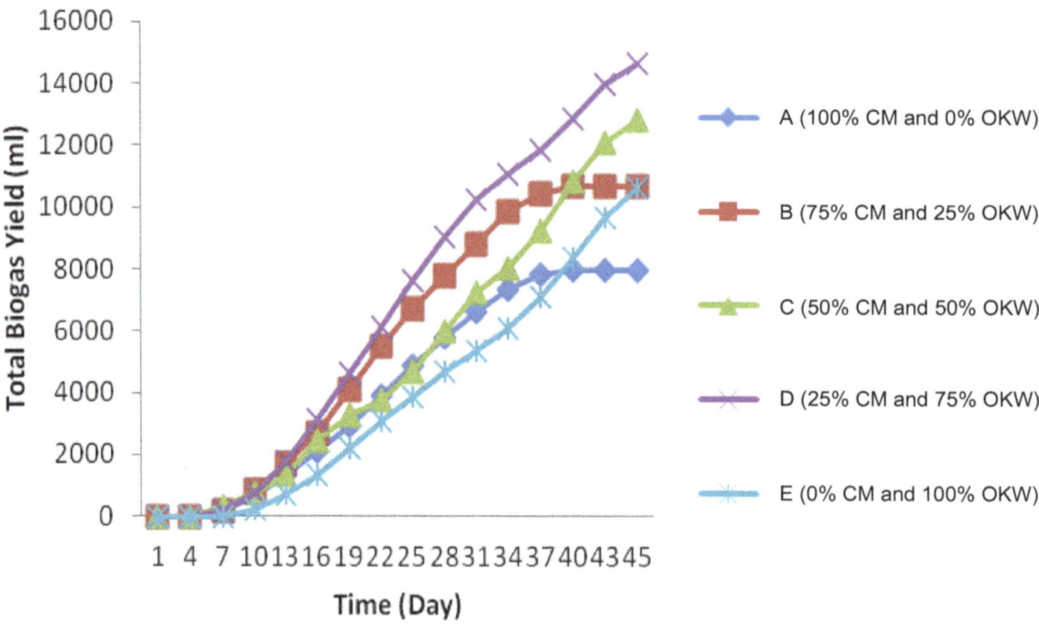

Figure 2. Mean cumulative biogas yield of all samples within 45 days.

biogas production due to the lag phase of microbial growth during these periods of the run.

The cumulative biogas productions of the five samples in all experiments were averaged and the mean cumulative biogas production and total gas production were summarized in Figure 2. As compared to the single anaerobic digestion of the two pure samples, the co-digestion of the three mix ratios produced higher volume of biogas. The total gas produced from the co-digestion of the three mixed samples (A3, A4, and A5) was indicated in Figure 2. From the co-digestion of A3, A4 and A5; 24.12, 37.91 and 47.13% more biogas was produced respectively than the two pure samples used as control. This might be due to mixing of cattle manure with organic kitchen waste provided balanced nutrients, buffering capacity, appropriate C/N ratio and sufficient anaerobic microorganisms. Moreover, the cumulative biogas yield of sample A5 is greater than sample A4 which is greater than sample A3. This might be attributed to the increased content of organic kitchen waste from 25 to 50% and to 75% (Amirhossein et al., 2004; Jianzheng et al., 2011). This result was in accordance with those obtained with co-digestion of 75% brewery waste and 25% sewage sludge (Babel et al., 2009).

Biodegradation during anaerobic digestion

In order to determine which matter in what amount was utilized from the initial feed during the 45 days of retention time and to correlate with the rate and amount of biogas produced, the digestate from each digester were characterized (Table 2). It is important to maintain the pH of an anaerobic digester between 6 and 8; otherwise, methanogen growth would be seriously inhibited (Gerardi, 2003). In this study, the initial pH of all the digesters was in the range of 6.95 to 7.45 even with the addition of acidic food wastes (like injera) indicating the buffering capacity of the cattle manure. But finally the pH showed a significant increase and it was in the range of 7.66 to 8.47. This was predicted because the VFAs produced by acidogens during the start up phase were consumed by methanogens and transferred to the methane. Generally, pH increase accompanies increasing biogas production because methanogens consume VFAs and generate alkalinity. In addition there occurs a decrease in VS and VS/TS ratio and this might be due to the biodegradation and conversion of VS into biogas through the microbial acidogenesis and methanogenesis. At the beginning of the digestion process the average total solids (TS) and volatile solids (VS) content of substrates in all digesters were high (Table 3). But, at the end of the 45 days anaerobic digestion period the contents of both TS and VS were highly reduced and this is attributed to their consumption by fermenting and methanogenic bacteria.

The efficiency of anaerobic co-digestion of cattle manure and organic kitchen waste was evaluated in terms of TS and VS reduction as the amount of dry matter and organic compounds. Table 4 presents the amount of TS, and VS biodegradation and conversion into biogas per mg, TS and VS removed in the anaerobic co-digestion processes of cattle manure with organic kitchen waste at an ambient temperature of 30 ± 8°C.

Table 3. Properties of cattle manure and organic kitchen waste after digestion (mean value ± SD).

	Parameters (after digestion)				
Mixture	**pH**	**MC (%)**	**TS (%)**	**VS (%)**	**VS/TS (%)**
A1	7.66 ± 0.264	94.05 ± 1.067	5.95 ± 1.067	4.71± 0.721	79.16 ± 5.041
A2	8.47 ± 0.173	96.76 ± 0.462	3.24 ± 0.462	0.89 ± 0.307	27.47 ± 6.322
A3	8.27 ± 0.191	96.58 ± 0.486	3.42 ± 0.486	1.28 ± 0.369	37.43 ± 5.824
A4	8.04 ± 0.174	95.96 ± 0.539	4.05 ± 0.539	2.05 ± 0.394	50.62 ± 3.495
A5	7.86 ± 0.236	95.28 ± 0.788	4.72 ± 0.788	2.83 ± 0.568	59.96 ± 4.403

Table 4. Organic matter degradation and biogas yield from each digester.

Treatments	Organic matter composition and its removal						Biogas yield		
	Total solids (mg/vol)			Volatile solids (mg/vol)			Total (ml)	ml/mg TS removed	ml/mg VS removed
	Initial TS (mg)	Removed		Initial VS (mg)	Removed				
		Mg/Vol.	%/Vol.		Mg/Vol.	%/Vol.			
A1	22,160	12,640	57.04	20,560	13,024	63.35	10,628.3	0.84	0.82
A2	20,064	14,880	74.16	16,432	15,008	91.33	7,954.8	0.54	0.53
A3	20,528	15,056	73.34	17,296	15,248	88.16	10,703.3	0.71	0.70
A4	21,072	14,592	69.25	18,144	14,864	81.92	12,814.3	0.88	0.86
A5	21,728	14,176	65.24	19,024	14,496	76.20	14,653.5	1.03	1.01

Biodegradation of TS and VS was high in samples containing high proportion of CM and decreases as the proportion of OKW in the mix ratio increases. With gas production rate of 1.03 ml/mg TS or 1.01 ml/mg VS removed from the biodegradation of 14,176 mg (65.24%) of the initial TS, or 14,496 mg (76.20%) of the initial VS, digester A5 gave the 1st highest cumulative biogas yield of 14,653.50 ml/g-VS. The result showed that in digester A4 and A5 there was a direct relationship between total biogas yield and gas production rate per each milligram of total solids and volatile solids removed. This might be because, the digestion process in these two digesters had more balanced acidogenesis and methanogenesis and the VS removed were utilized for biogas produce more efficiently than the other levels. Similar results were reported by Joung et al. (2008) from the anaerobic co-digestion of swine manure and food waste.

Digester A2 was observed with the highest percentage of TS and VS removal; however, it produced the least cumulative biogas yield of 7,954.75 ml/g-VS. This might be because of the presence of only cattle manure that is inoculated with rumen fluid. Since both cattle manure and rumen fluid are partially digested in the guts of the ruminants less biogas production from cattle manure within short retention period can be attributed to its relatively lower organic content than organic kitchen waste. Generally, it was observed that the TS and VS removal rates were affected by the different mixing ratios of cattle manure with organic kitchen waste and the hydraulic retention time. This suggests that high concentration of anaerobic bacteria content in rumen fluid and cattle manure works effectively to degrade organic matter composed in organic kitchen waste. So the results of this study imply that the biodegradability of organic matter and cumulative biogas yield was improved by co-digesting cattle manure with organic kitchen waste using rumen fluid as inoculum.

Co-digestion performance and synergistic effect

The co-digestion of three mix ratios (75:25, 50:50 and 25:75) of rumen fluid inoculated CM with OKW was performed and biogas productions from the biodegradation of organic matter were compared with pure cattle manure and organic kitchen waste as the controls. As the result indicated, the co-digestions of the three mixes showed improved biogas production rates and achieved higher cumulative biogas production than the two pure samples. This higher biogas production from digesters A3, A4, and A5 with mixed substrates of rumen fluid inoculated cattle manure and organic kitchen waste was due to the increased carbon content of OKW and high concentration of anaerobic bacteria content of cattle

Table 5. Synergistic effect of co-digestion of cattle manure and organic kitchen waste.

Treatments	Percentage of CM /OKW	Cumulative biogas yield				
		Cattle manure (ml)	Co-digestion (ml)	Organic kitchen waste (ml)	Increase (ml)	Increase (%)
A1	0:100	0.00		10,628.25		
A2	100:0	7,954.75		0.00		
A3	75:25	5,966.06	10,703.25	2,657.06	2,080.13	24.12
A4	50:50	3977.38	12,814.25	5314.13	3522.74	37.91
A5	25:75	1988.69	14,653.5	7971.19	4693.62	47.13

manure and rumen fluid. In other words this might be due to synergistic effect of CM to OKW (Table 5). The synergistic effect is mainly attributed to more balanced nutrients, increased buffering capacity, and decreased effect of toxic compounds (Li et al., 2009; Danqi, 2010; Jianzheng et al., 2011). More balanced nutrients in co-digestion would support microbial growth for efficient digestion, while increased buffering capacity would help maintain the stability of the anaerobic digestion system.

As it is shown on Table 5, from the co-digestion of cattle manure and organic kitchen waste with 75:25, 50:50, and 25:75 mix ratios 24.12, 37.91 and 47.13% additional biogas production was obtained, respectively when it is compared with that of the mono-digestions. It is evident from this result that digestion of more than one kind of substrate could establish positive synergism in the digester and provides more balanced nutrients as well as buffering capacity thus enhance the anaerobic digestion process and bio-energy production.

Identification of mix ratio for highest biogas production

As the proportion of OKW in the mix ratio increases from 0 to 25% to 50% and to 75% biogas yield was increased by 24.12, 37.91 and 47.13%, respectively. Thus, digester A5 with mix ratio of 25% CM and 75% OKW produced the highest volume of biogas (Figure 2). This might be due to the high organic content of OKW coupled with the supply of suitable microorganisms and missing nutrients by the rumen fluid and CM make the carbon to nitrogen ratio within the desired range.

Conclusions

Organic kitchen wastes co-digested with cattle manure improved the biogas potential compared to cattle manure alone. The co-digestion of rumen fluid inoculated CM and OKW with mix ratio of 50:50, gives biogas yield earlier and highest average daily and cumulative biogas yield were obtained from the co-digestion of rumen fluid inoculated CM and OKW with 25:75 ratio. The 25:75, 50:50 and 75:25 mix ratios of CM and OKW gave from 24.12 to 47.13% additional biogas yield and cumulative gas production was enhanced by 1.01-1.84 times. Thus, as compared to the mono-digestions of pure CM and pure OKW anaerobic co-digestion of rumen fluid inoculated CM and OKW in 25:75, 50:50, and 75:25 mix ratios enhances both the rate and amount of biogas yield.

ACKNOWLEDGEMENTS

This research was supported by the Ethiopian Ministry of Education. The authors would like to thank Haramaya University Staff Lounge and Animal Farming for their cooperation during sample collection and preparation. The authors also appreciate the support of the Department of Biology Lab Technicians specifically Mr. Samuel Tesfaye and Ms. Yodit.

REFERENCES

Adelekan F, Bamgboye A (2009). Comparison of biogas productivity of cassava peels mixed in selected ratios with major livestock waste types. Afr. J. Agric. Res. 4(7):571-577.

Alvarez R, Liden G (2007). Semi-continuous co-digestion of solid slaughterhouse waste, manure, and fruit and vegetable waste. Renew. Energy. 33:726-734.

Amirhossein M, Noor E, Sharom M (2004). Anaerobic Co-digestion of Kitchen Waste and Sewage Sludge for producing Biogas. Second Int. Conf. Environ. Manag. Bangladesh pp. 129-139.

Amon T, Amon B, Kryvoruchko V, Bodiroza V, Pötsch E, Zollitsch W (2006). Optimising methane yield from anaerobic digestion of manure: effects of dairy systems and of glycerin supplementation. Int. Congr. Ser. 1293:217-220.

Angelidaki I, Ellegaard L (2003). Codigestion of manure and organic wastes in centralized biogas plants. Appl. Biochem. Biotechnol. 109:95-105.

APHA, AWWA, WEF (1998). Standard Methods for the Examination of Water and Wastewater, 20th ed., Washington, DC.

Aurora SP (1983). Microbial Digestion in Ruminants. Indian J. Agr. Res. 2:85-94.

Babel S, Sae-Tang J, Pecharaply A (2009). Anaerobic co-digestion of sewage and brewery sludge for biogas production and land application. Int. J. Environ. Sci. Tech. 6(1):131-140.

Budiyono I, Widiasa S, Johari G, Sunarso T (2010). Increasing Biogas Production Rate from Cattle Manure Using Rumen Fluid as Inoculums. Int. J. Chem. Biol. Eng. 3:1.

Danqi Z (2010). Co-digestion of Different Wastes for Enhanced Methane Production. Msc thesis, The Ohio State University, Ohio, USA.

El-Mashad H, Zhang R (2010). Biogas production from co-digestion of dairy manure and food waste. Bioresource Technol. 101(11):4021-4028.
Elijah T, Ibifuro A, Yahaya S (2009). The Study of Cow Dung as Co-Substrate with Rice Husk in Biogas Production. Sci. Res. Essays. 4(9):861-866.
Forster-Carneiro T, Pérez M, Romero L (2008). Influence of total solid and inoculum contents on performance of anaerobic reactors treating food waste. Bioresource Technol. 99(15):6994-7002.
Gerardi MH (2003). The Microbiology of Anaerobic Digesters. Hoboken, N.J. John Wiley and Sons.
Girma G, Edward J, Peter R (2004). *In vitro* gas production provides effective method for assessing ruminant feeds. Calif. Agric. 58(1):54-58.
Goldemberg J, Johansson TB (2004). World energy assessment overview 2004 update. New York: UNDP. P. 88.
Jianzheng L, Ajay K, Junguo H, Qiaoying B, Sheng C, Peng W (2011). Assessment of the effects of dry anaerobic co-digestion of cow dung with waste water sludge on biogas yield and biodegradability. Int. J. Phys. Sci. 6(15):3679-3688.
Joung D, Sung S, Ki-Cheol E, Shihwu S, SangWon P, Hyunook K (2008). Predicting Methane Production Potential of Anaerobic Co-digestion of Swine Manure and Food Waste. Environ. Eng. Res. 13(2):93-97.
Li R, Ge Y, Wang K, Li X, Pang Y (2010). Characteristics and anaerobic digestion performances of kitchen wastes. Renew. Energ. Res. 28:76-80.
Li X, Li L, Zheng M, Fu G, Lar J (2009). Anaerobic co-digestion of cattle manure with corn Stover pretreated by sodium hydroxide for efficient biogas production. Energ. Fuel 23:4635-4639.
Macias-Corral M, Samani Z, Hanson A, Smith G, Funk P, Yu H, Longworth J (2008). Anaerobic digestion of municipal solid waste and agricultural waste and the effect of co-digestion with dairy cow manure. Bioresour. Technol. 99(17):8288-8293.
Marañón E, Castrillón L, Quiroga G, Fernández-Nava Y, Gómez L, García MM (2012). Co-digestion of cattle manure with food waste and sludge to increase biogas production. Waste Manage. 32:1821-1825.
Mata-Alvarez J, Mace S, Labres P (2000). Anaerobic digestion of organic solid wastes. An overview of research achievements and perspectives. Rev. Paper Bio-resource Technol. 74:3-16.
Montusiewicz A, Lebiocka M, Pawlowska M (2008). Characterization of the biomethanization process in selected waste mixtures. Arch. Environ. Prot. 34(3):49-61.
Muyiiya N, Kasisira L (2009). Assessment of the Effect of Mixing Pig and Cow Dung on Biogas Yield. Agricultural Engineering International: the CIGR E. J. Manuscript PM 1329, Vol. XI.
Radtke D, Wilde F, Davis J, Popowski T (1998). Alkalinity and Acid Neutralizing Capacity. U.S. Geological Survey TWRI Book 94/98.
Rittmann BE (2008). Opportunities for Renewable Bioenergy Using Microorganisms. Biotechnol. Bioeng. 100:203-212.
Schamphelaire W, Verstraete G (2009). Biological Energy Conversion, Biotechnol. Bioeng. 103(2):1.
Uzodinma E, Ofoefule A (2008). Effect of abattoir cow liquor waste on biogas yield of some agro-industrial wastes. National Center for Energy Research and Development. University of Nigeria, Nsukka, Nigeria.
Wan C, Zhou Q, Fu G, Li Y (2011). Semi-continuous anaerobic co-digestion of thickened waste activated sludge and fat, oil and grease. Waste Manage. 31:1752-1758.
Wei W (2000). Anaerobic Co-digestion of Biomass for Methane Production: Recent Research Achievements.
Wendland C, Behrendt J, Elmitwalli T, Al Baz I, Akcin G, Otterpohl R (2006). UASB reactor followed by constructed wetland and UV radiation as an appropriate technology for municipal wastewater treatment in Mediterranean countries.
Yu H, Fang H (2002). Acidogenesis of dairy wastewater at various pH levels. Water Sci. Technol. 45:201-206.
Zhang L, Lee Y, Jahng D (2011). Anaerobic co-digestion of food waste and piggery wastewater: focusing on the role of trace elements. Biosource Technol. 102:5048-5059.
Zhang R, El-Mashad H, Hartman K, Wang F, Liu G, Choate C, Gamble P (2006). Characterization of food waste as feedstock for anaerobic digestion. Bioresource Technol. 98(4):929-935.
Zhu D, Wan C, Li Y (2011). Anaerobic Co-digestion of Food Wastes and Dairy Manure for Enhanced Methane Production. Biol. Eng. 4(4):195-206.

Application of vertical electrical soundings to characterize aquifer potential in Ota, Southwestern Nigeria

A. P. Aizebeokhai[1] and O. A. Oyebanjo[1,2]

[1]Department of Physics, College of Science and Technology, Covenant University, P. M. B. 1023, Ota, Ogun State, Nigeria.
[2]Department of Physics/Telecommunication, Tai Solarin University of Education, Ijagun, Ogun State, Nigeria.

A knowledge of hydrogeophysical parameters of aquifers is essential for groundwater resource assessment, development and management. Traditionally, these parameters are estimated using pumping test carried out in boreholes or wells; but this is often costly and time consuming. Surface geophysical measurements can provide a cost effective and efficient estimates of these parameters. In the present work, geoelectrical resistivity data has been used to characterize and evaluate the aquifer potential at Covenant University, Ota, southwestern Nigeria. Some thirty-five vertical electrical soundings (VESs) were conducted using Schlumberger array with a maximun half-current electrode spacing (AB/2) of 240 m. The geoelectrical parameters obtained were used to estimate longitudinal conductance and transverse resistance of the delineated aquifer. Both the longitudinal conductance and transverse resistance, which qualitatively reflects the hydraulic properties of the aquifer, indicate that the aquifer unit is characterized with high values of hydraulic parameters; consequently a good groundwater potential. Thus, groundwater resource development and management in the area can be effectively planned based on these parameters.

Key words: Hydrogeophysics, geoelectrical parameters, resistivity survey, aquifer potential.

INTRODUCTION

Geophysical methods are increasingly becoming relevant in hydrological applications (Hubbard et al., 1997; Rubin and Hubbard, 2005; Vereecken et al., 2006). Conventional hydrogeologic investigation requires estimates of hydraulic parameters using traditional approaches such as pumping test, slug test and laboratory analyses of core samples. Pumping tests can produce reliable estimates of hydraulic parameters, but the estimates are largely volumetric averages. Laboratory analyses can provide information at a very fine scale, but there are many questions about the reliability of the hydraulic parameters estimates obtained with those analyses. Slug test has the most potential of the traditional approaches for detailed characterization of the variability of hydraulic parameters, but most sites do not have the extensive well network required for effective application of this approach (Butler, 2005). These traditional methods are time-consuming and invasive.

Non-invasive (or minimally invasive) geophysical methods can be used to characterize an image flow and transport processes within the subsurface. Spatial and temporal patterns of hydrological states can be retrieved from the geophysical parameters; thus, estimates of the hydrological and petro-physical parameters that determine flow and transport processes can be made. Geoelectrical resistivity technique is one of the most

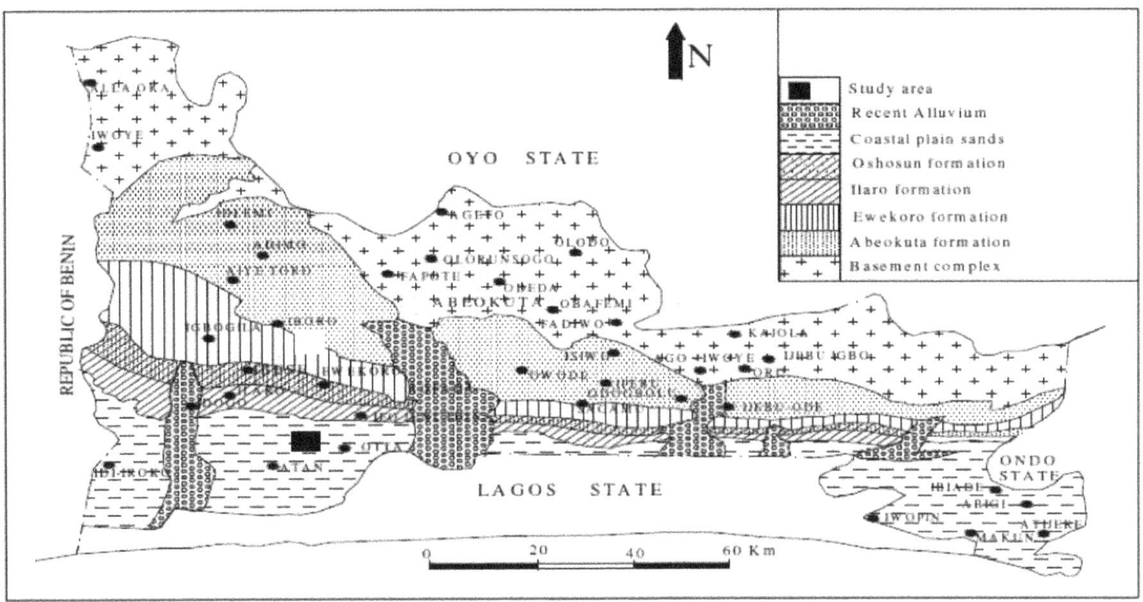

Figure 1. Geological map of Ogun State showing the study area (after Badmus and Olatinsu, 2010).

common geophysical tools used for hydrological investigations. The technique has been widely used in groundwater exploration to determine depth to water-table, aquifer geometry and groundwater quality by analyzing measured apparent resistivity field data. Numerical inversion techniques are often used to obtain the inverse model of the electrical resistivity distribution of the subsurface from the measured apparent resistivity data. This is achieved by solving the nonlinear and mixed-determined inverse problem whose solution is inherently non-unique and sometimes unstable. Typically, the resolution of the inversion result differs spatially, so that some regions may be well resolved while others are prone to exhibit artefacts and interpretation errors (Day-Lewis et al., 2005; Aizebeokhai, 2009).

In general, the inverse geophysical models can be used to estimate the hydraulic properties of aquifer by using analytical relationships between hydraulic parameters and geoelectrical parameters (Niwas and de Lima, 2003). In the present work, some thirty-five vertical electrical soundings (VESs) were conducted in Covenant University campus, Ota, southwestern Nigeria. The survey was carried out between the months of April and May, 2013 as part of the preliminary investigations to evaluate groundwater resource potential in the area. Schlumberger array was used in conducting the measurements with a maximum half-current electrode spacing (AB/2) of 240 m. The geoelectrical parameters obtained from the survey, which characterized the aquifer unit, were used to estimate the longitudinal conductance and transverse resistance of the delineated aquifer. The longitudinal conductance and transverse resistance of a porous medium characterise the hydraulic properties (conductivity and transmissivity) of the medium. The electrical resistivity (or its inverse conductivity) of a porous medium does not directly gives information about the hydraulic conductivity of the medium since the bulk electrical resistivity primarily depends on porosity, water saturation and dissolved ions.

Study area

The study area (Figure 1) falls within the eastern Dahomey (or Benin) Basin of southwestern Nigeria which stretches along the continental margin of the Gulf of Guinea. The area is generally a gently sloping low-lying area characterized by two major climatic seasons namely, dry season spanning from November to March and raining (or wet) season between April and October. Occasional rainfalls are usually witnessed within the dry season, particularly along the region adjoining the coast. Mean annual rainfall is greater than 2000 mm and forms the major source of groundwater recharge in the area.

In general, the rocks are Late Cretaceous to Early Tertiary in age (Jones and Hockey, 1964; Omatsola and Adegoke, 1981; Billman, 1992; Olabode, 2006). The stratigraphy of the basin has been grouped into Abeokuta Group, Imo Group, Oshoshun, Ilaro and Benin Formations (Figure 2). The Cretaceous Abeokuta Group consists of Ise, Afowo and Araromi Formations, and mainly composed of poorly sorted ferruginized grit, siltstone and mudstone with shale-clay layers. Overlying the Abeokuta Group is the Imo Group which is subdivided into the limestone-dominated Ewekoro Formation and the shale-dominated Akinbo Formation. The Akinbo Formation

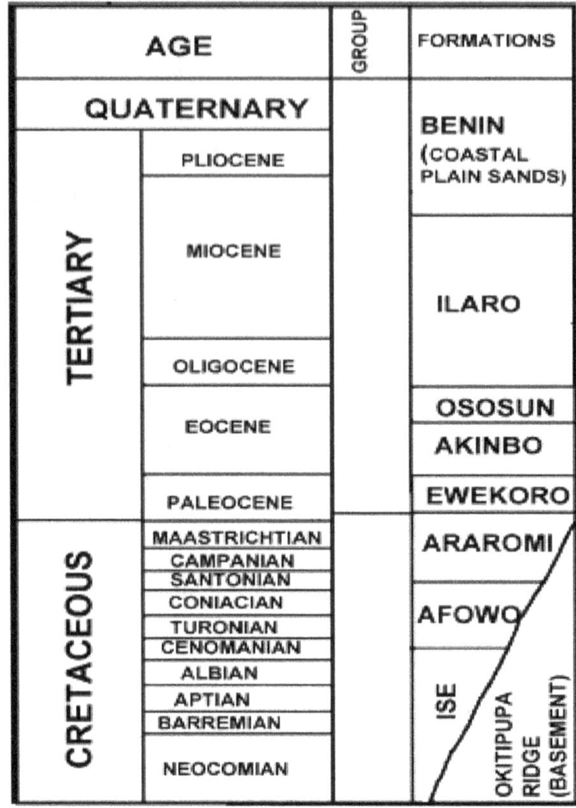

Figure 2. Simplified Cretaceous and Tertiary stratigraphy of the Nigeria part of Dahomey Basin (after Jones and Hockey 1964; Omatsola and Adegoke 1981; Billman 1992).

is overlain by the Oshoshun Fomation and then Ilaro Formation which is predominantly a sequence of coarse sandy estuarine, deltaic and continental beds; the Ilaro Formation displays rapid lateral facies changes. Overlying the Ilaro Formation is the Benin Formation which is predominantly coastal plain sands and Tertiary alluvium deposits. The local geology is predominantly coastal plain sands which are underlain by a sequence of coarse sandy estuarine, deltaic and continental beds largely characterised by rapid changes in facies.

METHODOLOGY

Vertical electrical soundings

A total of thirty-five vertical electrical soundings (VES) were conducted within the study area so as to delineate the subsurface lithological configuration, depth to aquifer(s) and aquifer characteristics. An ABEM Terrameter (SAS 1000 series) was used for the apparent resistivity measurements. Schlumberger electrode configuration was adopted for the resistivity soundings due to its high lateral resolution. The maximum half-current electrode separation (AB/2) used ranges from 130 to 240 m, with an average of 180 m. The spread was sufficient for the effective depth of investigation anticipated. Most of the VESs were conducted along three main profiles (Figure 3). Care was taken to minimize electrode positioning error. A minimum stack of 3 and maximum of 6 were used for measurement. The root-mean squares error associated with the data measurement was minimal, generally less than 0.3%. Measurements with root-mean squares error up to 0.5% or more were repeated after re-checking electrodes contact.

The observed apparent resistivity data were processed by plotting the apparent resistivity values against half-current electrode spacing (AB/2 or half the spread length) at each station on a bi-logarithmic (log – log) graph sheets. Partial curve matching of the field curves with relevant Schlumberger developed master and auxiliary curves was carried out to obtain estimates of the number of layers and their respective resistivities and thicknesses. The geoelectric parameters obtained from this manual interpretation were then used as the initial models for the computer inversion using the Win-Resist code. This computer code uses iterative process by matching the computed data with the observed field data to obtain the inverse models. The iterative process is an attempt to reduce the root-mean squares errors and improve the goodness of fit between the measured data and computed data. The root-mean squares error observed in the inversion range between 1.4 and 2.8%.

Hydraulic parameters estimation

The relationship between the hydraulic conductivity K and geoelectrical resistivity ρ of an aquifer is strongly controlled by the nature of the aquifer substratum (Niwas and Singhal, 1985; Niwas

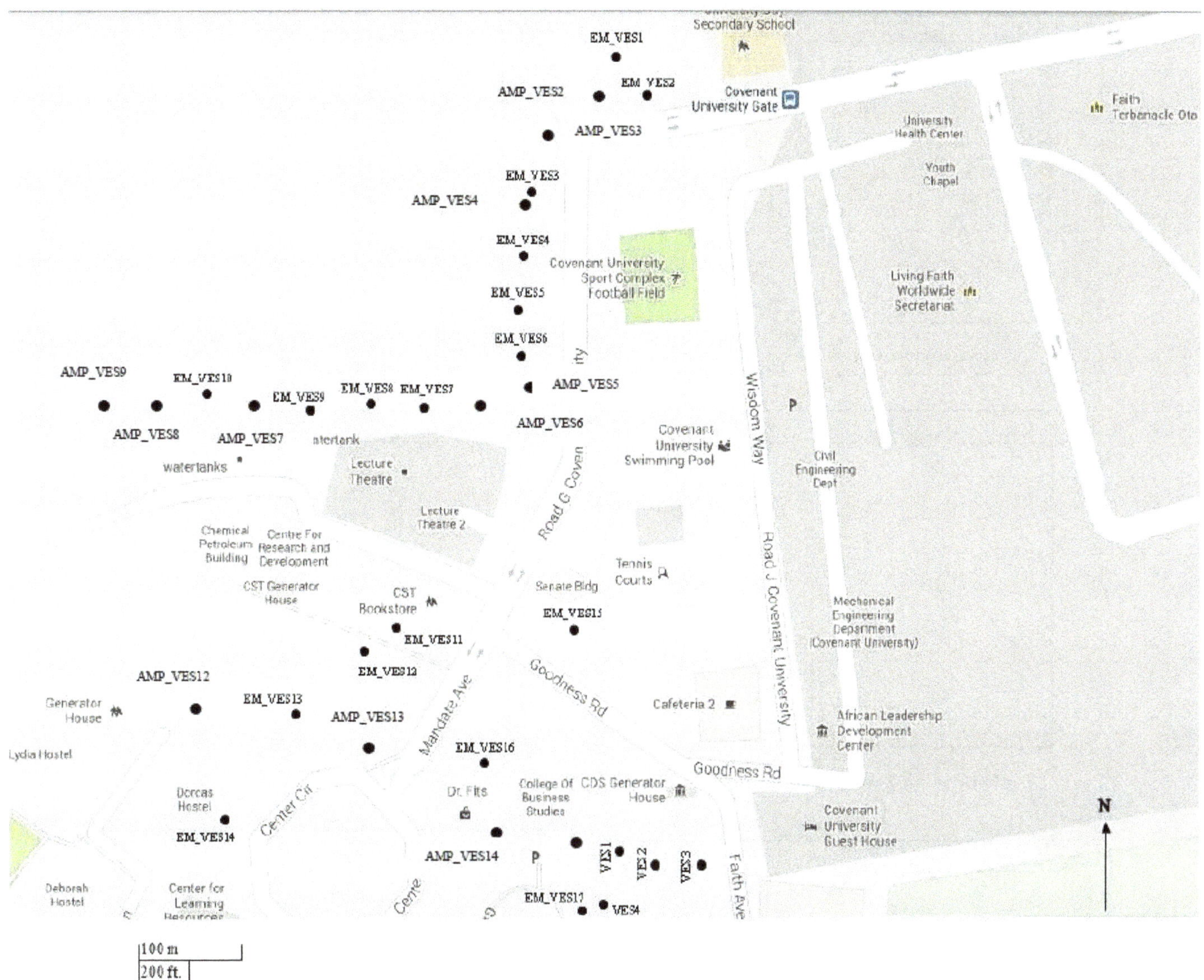

Figure 3. Map showing the study area and locations of VES points.

and de Lima, 2003). For a highly resistive substratum, both the current and the hydraulic flows are dominantly horizontal in a typical unit column of the aquifer, and the relationship between K and ρ, is inverse. If the substratum is highly conductive, the hydraulic flow will still be horizontal while the current flow in a characteristic unit column is dominantly vertical; thus, a direct relation exist between K and ρ. If the aquifer material is cut in the form of a vertical prism of the unit cross-section from top to bottom, fluid flow and current flow in the aquifer material obeys Darcy's law and Ohm's law respectively. Thus, for current and fluid flows in a lateral direction, the transmissivity of the aquifer is given as:

$$T = (K\rho)S \tag{1}$$

where ρ is the bulk resistivity and S is the longitudinal unit conductance of the aquifer material with thickness b given by b/ρ. For a lateral hydraulic flow and current flowing transversely, the transmissivity of the aquifer becomes:

$$T = (K/\rho)R \tag{2}$$

where R is the transverse unit resistance of the aquifer material given by $b\rho$. If the aquifer is saturated with water with uniform resistivity, then the product $K\rho$ or K/ρ would remain constant. Thus, the transmissivity of an aquifer is proportional to the longitudinal conductance for a highly resistive basement where electrical current tends to flow horizontally, and proportional to the transverse resistance for a highly conductive basement where electrical current tends to flow vertically (Niwas et al., 2011). The above equations may therefore be written as:

$$T = \alpha S; \quad \alpha = K\rho \tag{3}$$

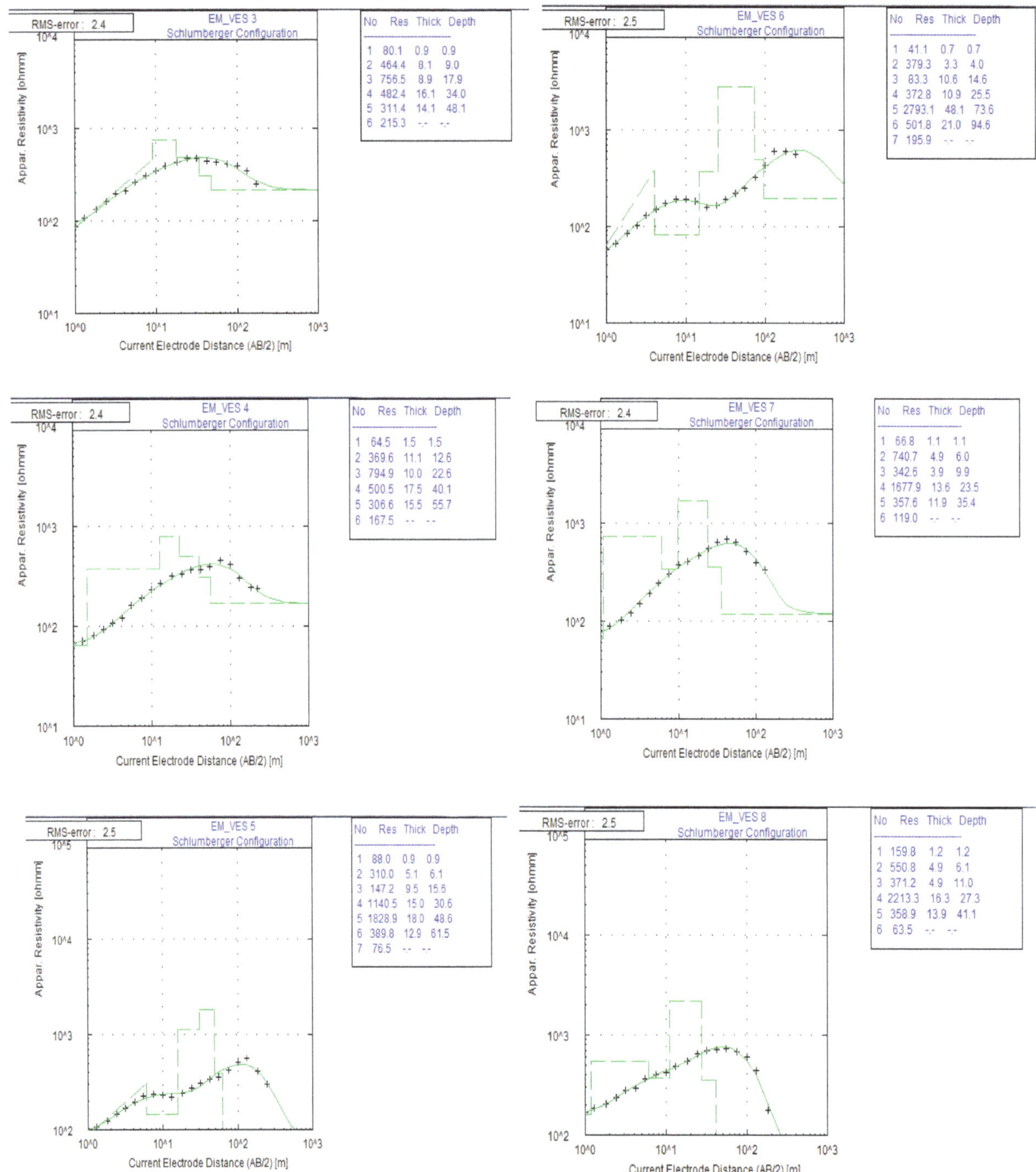

Figure 4. Representative of the iterated VES curves showing the inverse models of the geoelectrical parameters.

and

$$T = \beta R; \quad \beta = K / \rho \tag{4}$$

where α and β are constants of proportionality. From these relations, the model resistivity values obtained from the inversion process were used to estimate the longitudinal unit conductance and transverse unit resistance of the aquifer unit.

RESULTS AND DISCUSSION

Some representative of the output from the computer interpretation of the observed apparent resistivity data are presented in Figure 4. Five to seven layers were generally delineated from the iterated sounding curves. The geoelectrical parameters of the layers correlated for each Traverse are presented in Tables 1 to 3; the

Table 1. Geoelectrical parameters of the VES in Traverse 1.

Layer	1			2			3			4			5			6			7		
Lithology	Top Soil (Sandy Clay)			Lateritic Clay			Lateritic Clay (Compacted)			Clayey/Silty Sand			Laterite (Confining Bed)			Sand (Main Aquifer)			Shale/Clay		
Location	Resistivity (Ωm)	Thickness (m)	Bottom Depth (m)	Resistivity (Ωm)	Thickness (m)	Bottom Depth (m)	Resistivity (Ωm)	Thickness (m)	Bottom Depth (m)	Resistivity (Ωm)	Thickness (m)	Bottom Depth (m)	Resistivity (Ωm)	Thickness (m)	Bottom Depth (m)	Resistivity (Ωm)	Thickness (m)	Bottom Depth (m)	Resistivity (Ωm)	Thickness (m)	Bottom Depth (m)
EM_VES 3	80.1	0.9	0.9	464.4	8.1	9.0	756.5	8.9	17.9	482.4	16.1	34.0			34.0	311.4	14.1	48.1	215.3		
AMP-VES 3	70.8	0.9	0.9	548.8	8.7	9.6	709.5	9.4	19.1	459.0	16.2	35.2			35.2	306.8	15.2	50.4	220.8		
EM_VES 4	64.5	1.5	1.5	369.6	11.1	12.6	794.9	10.0	22.6	500.5	17.5	40.1			40.1	306.6	15.5	55.7	167.5		
AMP-VES 4	41.4	1.1	1.1	405.4	12.4	13.4	443.7	8.7	22.1	304.5	15.4	37.5			37.5	258.4	15.3	52.8	171.5		
EM_VES 5	88.0	0.9	0.9	310.0	5.2	6.1	147.2	9.5	15.6	1140.5	15.0	30.6	1828.9	18.0	48.6	389.8	12.9	61.5	76.5		
AMP-VES 5	37.0	0.7	0.7	446.4	3.6	4.3	76.0	10.5	14.8	422.5	11.6	26.4	4641.6	32.5	58.8	437.1	21.9	80.8	239.4		
EM_VES 6	41.1	0.7	0.7	379.3	3.3	4.0	83.3	10.6	14.6	372.8	10.9	25.5	2793.1	48.1	73.6	501.8	21.0	94.6	195.9		

Table 2. Geoelectrical model parameters of the VES in Traverse 2.

Layer	1			2			3			4			5			6			7		
Lithology	Top Soil (Sandy Clay)			Lateritic Clay			Lateritic Clay (Compacted)			Clayey/Silty Sand			Laterite (Confining Bed)			Sand (Main Aquifer)			Shale/Clay		
Location	Resistivity (Ωm)	Thickness (m)	Bottom Depth (m)	Resistivity (Ωm)	Thickness (m)	Bottom Depth (m)	Resistivity (Ωm)	Thickness (m)	Bottom Depth (m)	Resistivity (Ωm)	Thickness (m)	Bottom Depth (m)	Resistivity (Ωm)	Thickness (m)	Bottom Depth (m)	Resistivity (Ωm)	Thickness (m)	Bottom Depth (m)	Resistivity(Ωm)	Thickness (m)	Bottom Depth (m)
EM_VES 7	66.8	1.1	1.1	740.7	4.9	6.0	342.6	3.9	9.9			9.9	1677.9	13.6	23.5	357.6	11.9	35.4	119.0		
AMP-VES 6	86.9	1.1	1.1	750.7	5.0	6.1	353.2	4.0	10.1			10.1	1726.8	13.4	23.5	347.0	11.9	35.4	118.1		
EM_VES 8	159.8	1.2	1.2	550.8	4.9	6.1	371.2	4.9	11.0			11.0	2213.3	16.3	27.3	358.9	13.9	41.1	63.5		
EM_VES 9	70.8	1.2	1.2	684.4	5.9	7.1	440.1	5.4	12.4			12.4	2659.8	19.5	32.0	368.3	14.9	46.9	91.6		
AMP-VES 7	100.0	0.5	0.5	489.0	3.0	3.5	284.0	3.3	6.9	656.6	5.5	12.4	2770.8	14.2	26.6	341.0	13.1	39.7	25.7		
AMP-VES 8	47.3	0.8	0.8	621.3	4.9	5.7	1033.7	5.4	11.1	979.2	7.6	18.6	2864.2	15.3	34.0	363.8	11.9	45.9	54.1		
EM_VES 10	80.3	1.5	1.5	979.1	4.8	6.3	1509.8	6.9	13.2	738.7	7.7	20.9	2219.9	14.9	35.8	368.0	12.3	48.1	134.7		
AMP-VES 9	72.3	0.9	0.9	509.7	5.2	6.2	1350.8	6.3	12.5	976.4	6.1	18.6	2422.0	14.5	33.1	346.4	11.7	44.7	128.5		

geoelectric parameters are largely consistent among the interpreted sounding curves. The lithologies of the interpreted layers were inferred based on the local geology and available information.

The resistivity of the top soil (sandy clay) varies

Table 3. Geoelectrical model parameters of the VES in Traverse 3.

Layer	1			2			3			4			5			6			7		
Lithology	Top Soil (Sandy Clay)			Lateritic Clay			Lateritic Clay (Compacted)			Clayey/Silty Sand			Laterite (Confining Bed)			Sand (Main Aquifer)			Shale/Clay		
Location	Resistivity (Ωm)	Thickness (m)	Bottom Depth (m)	Resistivity (Ωm)	Thickness (m)	Bottom Depth (m)	Resistivity (Ωm)	Thickness (m)	Bottom Depth (m)	Resistivity (Ωm)	Thickness (m)	Bottom Depth (m)	Resistivity (Ωm)	Thickness (m)	Bottom Depth (m)	Resistivity(Ω m)	Thickness (m)	Bottom Depth (m)	Resistivity (Ωm)	Thickness (m)	Bottom Depth (m)
AMP-VES 12	397.7	0.8	0.8	452.7	3.7	4.5	339.8	4.7	9.2			9.2	1361.7	12.3	21.5	354.8	12.6	34.1	123.5		
EM_VES 13	314.8	0.9	0.9	543.7	4.5	5.4	354.0	5.3	10.7			10.7	1000.9	11.9	22.6	345.9	11.6	34.2	88.0		
AMP-VES 13	413.1	1.2	1.2	566.7	5.3	6.5	483.6	7.8	14.3			14.3	787.7	8.2	22.6	245.1	15.0	37.6	473.1		
EM_VES 15	223.9	0.7	0.7	615.5	3.1	3.9	495.5	4.4	8.3			8.3	1223.2	14.4	22.7	297.3	11.5	34.1	87.9		
AMP-VES 14	306.8	0.6	0.6	606.3	2.2	2.8	344.5	3.4	6.4			6.4	2138.3	11.4	17.7	276.1	10.4	28.1	64.0		
EM_VES 16	181.5	0.8	0.8	547.4	3.3	4.1	456.1	4.1	8.2			8.2	4170.7	15.5	23.7	345.6	11.5	35.2	22.9		
VES 3	502.0	0.8	0.8	835.0	3.2	4.0	571.3	6.2	10.2			10.2	2165.0	21.0	31.2	350.0	12.0	43.2	120.0		
VES 1	418.7	0.6	0.6	814.9	5.4	6.0	156.8	6.1	12.1	288.9	4.3	16.4	4221.5	11.0	27.4	346.0	11.3	38.7	48.2		
VES 4	234.1	0.6	0.6	909.2	3.3	3.9	291.2	4.9	8.8	711.1	7.2	16.0	2150.2	18.2	34.2	345.3	11.6	45.8	103.2		

from $41.1\Omega m$ to $502.0\Omega m$ with mean resistivity of $156.81\Omega m$; the thickness of this layer ranges from 0.5 – 1.5 m. The resistivity of the top soil largely depends on clay volume, moisture content and degree of compaction. The resistivity of the underlying geoelectric layer range from $310.0\Omega m$ to $909.2\Omega m$ with thickness ranging from 2.2 – 13.0 m, while those of the third geoelectric layer are $76.0 - 1509.8\Omega m$ and $3.3 - 10.6m$. The second and third layers are laterally continous and are basically the same lithologic unit, lateritic clay, with different degree of compaction and water saturation. The variability in the resistivity and thickness of these units are shown in Tables 1 to 3. These layers are largely impermeable, especially in areas where they are compacted, and percolation through these layers relatively poor and slow. Consequently, the top soil and possibly the second layer occasionally form parched aquifer; and most parts of the areas are usually flooded due to poor percolation of the underlying layers (Aizebeokhai et al., 2010).

The fourth geoelectric layer, an intercallation of silt, sand and clay, was delineated in all the soundings in Traverse 1 and some of the soundings in Traverses 2 and 3. The range of model resistivity of this layer is $288.9 - 1140.5\Omega m$ with thickness ranging from 4.3 – 17.5 m. This layer is thought to be laterally discontions continous based on the geoeletric layers delineated. However, it may be masked in some cases due to the resistivity contrast between the third and fifth geoelectric layers. Underlying this geoelectric layer is a very high resistive substratum with resitivity ranging from $787.7 - 4641.6\Omega m$ and thickness ranging between 8.2 and 48.1 m.

The sixth geoelectric layer delineated is the main aquifer unit which consists of unconsolidated coarse grain sands. The aquifer unit is confined by the overlying high resistive unit, the depth to the aquifer delineated from the geoelectric parameters ranges from 17.7-73.6 m (Table 4). Its resistivity, ranging between $245.1\Omega m$ and $583.1\Omega m$, and thickness ranging between $10.4m$ and $21.9m$, are more uniform among the geoelectric layers delineated (Table 4). Underlying the aquifer unit is a high conductive clay/shale layer with model resistivity ranging between $22.9\Omega m$ and $239.4\Omega m$. The resistivity of this unit is also largely uniform.

The geoelectric parameters of the aquifer were used to compute the longitudinal conductance and transverse resistance of the aquifer unit (Table 4). These parameters are indicative of the spatial

Table 4. Hydraulic parameters estimated from inverse model resistivity parameters.

S/N	Location	Depth to Aquifer (m)	Aquifer Thickness (m)	Aquifer Resistivity (Ωm)	Longitudinal Conductance (Ω^{-1})	Transverse Resistance (Ωm^2)
1	AMP-VES 2	28.6	12.3	583.1	0.0211	7172.13
2	EM_VES 3	34.0	14.1	311.4	0.0453	4390.74
3	AMP-VES 3	35.2	15.2	306.8	0.0495	4663.36
4	EM_VES 4	40.1	15.5	306.6	0.0506	4752.30
5	AMP-VES 4	37.5	15.3	258.4	0.0592	3953.52
6	EM_VES 5	48.6	12.9	389.8	0.0331	5028.42
7	AMP-VES 5	58.8	21.9	437.1	0.0501	9572.49
8	EM_VES 6	73.6	21.0	501.8	0.0418	10537.80
9	EM_VES 7	23.5	11.9	357.6	0.0333	4255.44
10	AMP-VES 6	23.5	11.9	347.0	0.0343	4129.30
11	EM_VES 8	27.3	13.9	358.9	0.0387	4988.71
12	EM_VES 9	32.0	14.9	368.3	0.0405	5487.67
13	AMP-VES 7	26.6	13.1	341.0	0.0384	4467.10
14	AMP-VES 8	34.0	11.9	363.8	0.0327	4329.22
15	EM_VES 10	35.8	12.3	368.0	0.0334	4526.40
16	AMP-VES 9	33.1	11.7	346.4	0.0338	4052.88
17	AMP-VES 12	21.5	12.6	354.8	0.0355	4470.48
18	EM_VES 13	22.6	11.6	345.9	0.0335	4012.44
19	AMP-VES 13	22.6	15.0	245.1	0.0612	3676.50
20	EM_VES 15	22.7	11.5	297.3	0.0387	3418.95
21	AMP-VES 14	17.7	10.4	276.1	0.0377	2871.44
22	EM_VES 16	23.7	11.5	345.6	0.0333	3974.40
23	VES 1	27.4	11.3	346.0	0.0332	3909.80
24	VES 3	31.2	12.0	350.0	0.0343	4200.00
25	VES 4	34.2	11.6	345.3	0.0336	4005.48

variability of the hydraulic properties (hydraulic conductivity and transmissivity) of the aquifer units. Zones with high longitudinal conductance are generally characterized as areas with low permeability with high clay volume, consequently low hydraulic conducvity. Similarly, areas with low value of longitudinal conductance corresponds to high permeability and hydraulic conductivity. The computed longitudinal conductance for the delineated aquifer unit is generally low, ranging between $0.0211\ \Omega^{-1}$ and $0.0612\ \Omega^{-1}$. This shows that the confined aquifer is characterized with high hydraulic parameters with high permeability and low clay volume. Thus, the aquifer unit is characterized with high hydraulic conductivity and high transmisivity as indicated by the computed longitudinal conductance.

Moreover, many hydrological studies have shown that the transverse resistance parameter can be used to effectively characterize aquifer properties. The transverse resistance of an aquifer increases with increasing transmissivity and yield. The distribution of the transverse resistance range between $2871.44\ \Omega m^2$ and $10537.80\ \Omega m^2$ in the area is presented in Table 4. High values of transverse resistance are generally observed, indicating high transmissivity and high yield of the aquifer units.

Conclusion

Vertical electrical soundings have been used to delineate and characterize the aquifer unit as part of the preliminary investigations to assess groundwater resource potential and development at Covenant University, Ota, southwestern Nigeria. The geoelectrical parameters obtained were used to estimate the longitudinal conductance, and transverse resistance which are reflective of the hydraulic properties of the aquifer were estimated using geoelectric parameters obtained by inverting observed apparent resistivity data. The computed longitudinal conductance indicates high permeabilty and low clay volume in the aquifer unit and thus high hydraulic conductivity for the delineated aquifer unit. Similarly, the computed transverse resistance shows that the aquifer unit is characterized with high transmissivity and yield. Thus, groundwater resource development and management can be effectively planned for.

REFERENCES

Aizebeokhai AP (2009). Geoelectrical resistivity imaging in environmental studies. In: Yanful E. K. (Ed.), Appropriate Technologies in Environmental Protection in Developing World, Springer, pp. 197-305.

Aizebeokhai AP, Alile OM, Kayode JS, Okonkwo FC (2010). Geophysical investigation of some flood prone areas in Ota, southwestern Nigeria. Am. Eur. J. Sci. Res. 5(4):216-229.

Badmus BS, Olatinsu OB (2010). Aquifer characteristics and groundwater recharge pattern in a typical basement complex, southwestern Nigeria. Afr. J. Environ. Sci. Technol. 4(6):328-342.

Billman HG (1992). Offshore stratigraphy and Paleontology of Dahomey (Benin) Embayment. NAPE Bulletin, 70(02):121-130.

Butler JJ (2005). Hydrogeological methods for the estimation of spatial variations in hydraulic conductivity. In: Rubin, Y. And Hubbard, S. (eds) *Hydrogeophysics*, *Water Science and Technology Library*, Chapter 2(50):523. Springer, 23-58.

Day-Lewis FD, Singha K, Binley AM (2005). Applying petrophysical models to radar travel time and electrical resistivity tomograms: resolution-dependent limitations. J. Geophys. Res. Solid Earth. 110:B08206.

Hubbard SS, Peterson JE, Majer Jr. EL, Zawislanski PT, Williams KH, Roberts J, Wobber F (1997). Estimation of permeable pathways and water content using tomographic radar data. Leading Edge. 16:1623-1628.

Jones HA, Hockey RD (1964). The geology of part of southwestern Nigeria. Geological Survey of Nigeria Bulletin, 31:101.

Niwas S, de Lima OAL (2003). Aquifer parameter estimation from surface resistivity data. Ground Water. 41:94-99.

Niwas S, Singhal DC (1985). Aquifer transmissivity of porous media from resistivity data. J. Hydrol. 82:143-153.

Niwas S, Tezkan B, Israil M (2011). Aquifer hydraulic conductivity estimation from surface geoelectrical measurements for Krauthausen test site, Germany. Hydrogeol. J. 19:307-315.

Olabode SO (2006). Siliciclastic slope deposits from the Cretaceous Abeokuta Group, Dahomey (Benin) Basin, southwestern Nigeria. J. Afr. Earth Sci. 46:187-200.

Omatsola ME, Adegoke OS (1981). Tectonic evolution and Cretaceous stratigraphy of the Dahomey Basin. Nig. J. Mining Geol. 18(01):130-137.

Rubin Y, Hubbard S (2005). Hydrogeophysics, Water Science and Technology Library, 50, Springer, Berlin, P. 523.

Vereecken H, Binley A, Cassiani G, Kharkhordin I, Revil A, Titov K (eds) (2006). Applied Hydrogeophysics, Springer-Verlag, Berlin, P. 372.

The efficiency of recyclable energies for building design: Cooling system supply for buildings using traditional wind catchers used in the dry and hot climate of Iran and its combination with solar chimney technology

Hourieh Abna[1], Mohammad Iranmanesh[2], Iman Khajehrezaei[1] and Mohammad Aghaei[3]

[1]Department of Architecture, University of Guilan, Rasht, Iran.
[2]Department of Architecture, University of ShahidBahonar, Kerman, Iran.
[3]Department of Mechanical Engineering, Azad University, Sirjan, Iran.

In recent years, the concept of the usage of new energies for cooling and heating in buildings has attracted the attention of engineers. In this study, the combination of solar chimneys with wind catchers has been suggested to provide natural and pleasant ventilation. It explains the increased efficiency of wind catchers for cooling through the help of solar chimneys. In addition, with the use of solar energy for moving air inside the building by a solar chimney and using the hidden heat of water evaporation, for providing cooling, which is created at the top of wind catcher, we will be able to provide a pleasant environment with suitable temperature and humidity that complies with the standards of pleasant ventilation in hot and dry climates. In continuing this research, attention had been paid to this design for Kerman City, which is located in the hot and dry climate of Iran. Wet channels and walls with radiance are placed together with water injection systems in entrances. In addition, the turbulence model had been used to calculate the buoyancy forces and the radiance model for radiance. The project considers the system performance in a constant manner and the effect of certain parameters, including entrance air temperature, rate of injected water and rate of heat flow field, and temperature and relative humidity.

Key words: Renewable energies, solar building, wind catcher, solar chimney, natural ventilation, economies in fuel usage.

INTRODUCTION

This research which relies on library studies and field research about traditional wind catchers in Iran and also about solar chimneys, is based on the theory that using a solar chimney on the southern part of a building and a wind catcher on the favourable wind side, can increase the ventilation speed in wind catchers. Accordingly, we undertook a numerical study using the limited mass method. In addition, EES software was used for calculating solar energy and fluent software for simulating anumerical solution for theair stream inside the building. Problem geometry is employed using Gambit software with 36796, while the boundary conditions of the problem are based on the actual conditions of the summer in one of the cities of Iran called Kerman.

Due to the desert climate of most cities in Iran, with long warm and dry summers, employing energy optimization in buildings seems necessary. The most important environmental pollutant factor in the world, and, especially in Iran is the consumption of fossil energies in residential, official, and commercial areas for cooling and heating, which has increased the significance of this subject significantly. In this sense, traditional Iranian architecture, due to the shortage of technological facilities had been adopting principles and remedies for the construction of buildings, in which their results implies that the primary cost of the implementation of optimization principles in buildings can be compensated through a reduction of cost in fuel consumption in later years. Accordingly, with a logical insight, building construction based on energy consumption optimization principles can be reasonably well evaluated.

The overbearing heat of the summer in Iran and lack of easy availability to fossil fuels and technological cooling systems in the past caused its people to think about natural and spontaneous ventilation in buildings through the use of a high narrow, four sided, six sided or eight sided, tower-like elements known as wind catchers (Bahadori, 1978; Kalantar, 2005).

The increase in knowledge concerning environmental pollution has led to the increasing attention of scientists for renewable energies and the usage thereof in buildings. Therefore, architects are interested in green features in modern buildings (Liu and Mak, 2007). Green design includes the features of building architecture that are used to decrease air pollution and also control the heat, air transfer and radiation of the sun from outside to inside.

Till date, much research concerning natural ventilation had been undertaken. Afonso and Oliveira (2000) performed some research about solar chimneys, their research showed a wide increase in ventilation speed through the use of the solar chimney (Clito and Armando, 2000). In addition, Chungloo and Limmeechokchai (2007), by spraying water on the roof and the use of a solar chimney for accelerating the internal air stream, concluded that they could make the inside space temperature cooler than the surrounding temperature by about 2 to 6.2°C. In addition, research shows the use of this solar system as an instrument of ventilation in some Italian buildings in the sixteenth-century, which employed underground passages and water for cooling (Chena et al., 2003). Macias et al. (2009) studied cheap cooling systems in buildings in hot and dry climates, in which ventilation was facilitated by a solar chimney and fresh air which was cooled by circulation. This system has been applied and evaluated for living rooms of residential buildings, the results for which show that the use of solar chimneys is favourable as long as the rate of ventilation is not dependent on the outside air speed. In addition, this system is very suitable for places that have significant solar radiance and low wind speed.

Maerefat and Haghighi (2010) studied the natural cooling of a house using asolar chimney and evaporative cooling system. Their results showed that, this system can easily provide good thermal conditions for environmental air with a relative humidity of less than 50% even in environments with high temperatures. Hirunlabh et al. (1999) presented a design, which consist of natural ventilation with a metal solar wall. This wall consisted of a glassy cover, air gap, dark metal board and a layer of fibre and multi-layer boards. The rays of the sun after passing through the glass cover are absorbed by the dark metal board, which causes the internal air in the air gap to become very hot causing acceleration of the air speed. Thus, the speed of ventilation in the room was increased by the increase in the movement of the hot air and its exiting from the top of the glassy gap (Hirunlabh et al., 1999) (Figure 1).

Traditional wind catcher (Baudgeer) introduction

The wind catcher has been a part of traditional buildings in some parts of Iran and has provided good efficiency in spontaneous cooling. This structure is one of the architectural elements that is based on a dry and hot climate and in the humid and hot climate of Iran with attention to the continental condition of the native architecture. Traditional architects, in consideration of the conditions and poor facilities at that time, incorporated this structure into the architecture, thereby creating spontaneous cooling in buildings in hot areas, particularly in the hot and dry desert areas in Iran. Such a feature provided the buildings a special beauty.

In a manner, it can be said that this soil-oriented element blends with the dusty nature of the region in such a way that these wind catchers have come out of the earth naturally. However, nowadays, this critical component of traditional architecture has been forgotten, and has been substituted with mechanical devices that work with fossil fuels. One of the most important reasons for the current situation is due to the construction of buildings with more than one floor, since wind catchers cannot supply cooling to all floors. Another reason is that, most of the time; the air flow speed in the wind catcher is low and results in its very low efficiency on some days of the year. Thus, some arrangements and strategies need tobe applied in order to make use of them again.

Studying the different sections of wind catcher (baudgeer)

A wind catcher is a vertical channel that can be observed in plain form as a square, rectangle, octagon or circle. Wind catchers consist of two main sections: one is the inner section of the channel, which starts from the ceiling and continues towards the underground, and the other is the outer part located on the roof, including the openings through which the wind enters (Figures 2, 3, 4, and 5). They

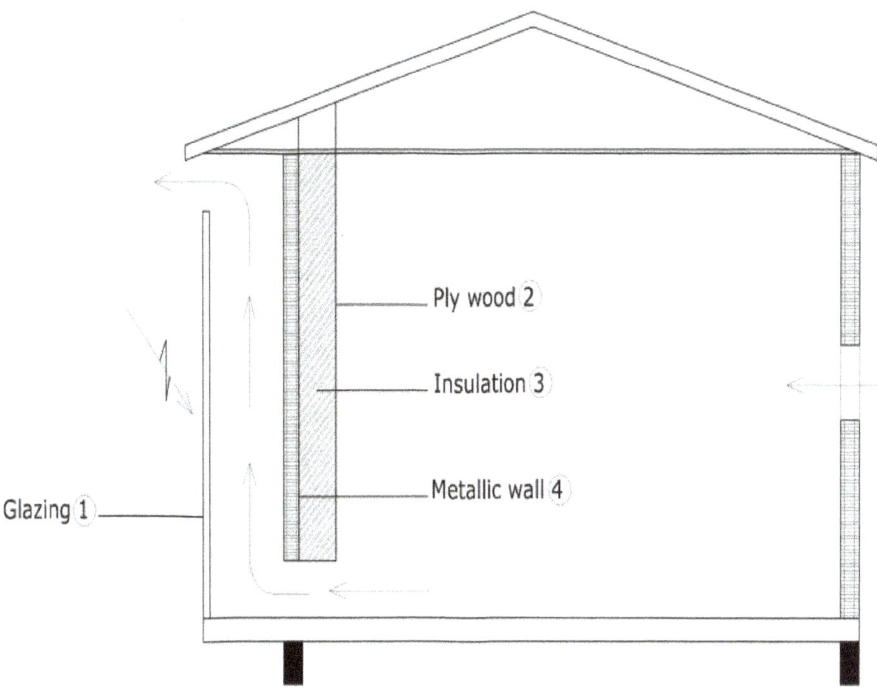

Figure 1. Natural ventilation using metal solar wall. 1, glassy cover; 2, multi-layer boards; 3, insulator; 4, metal wall.

Figure 2. An instance of wind catchers in the warm and dry regions of Iran.

Figure 3. An instance of wind catchers in warm and dry regions of Iran.

Figure 4. Section-perspective of a traditional house in Iran (Yazdcity).

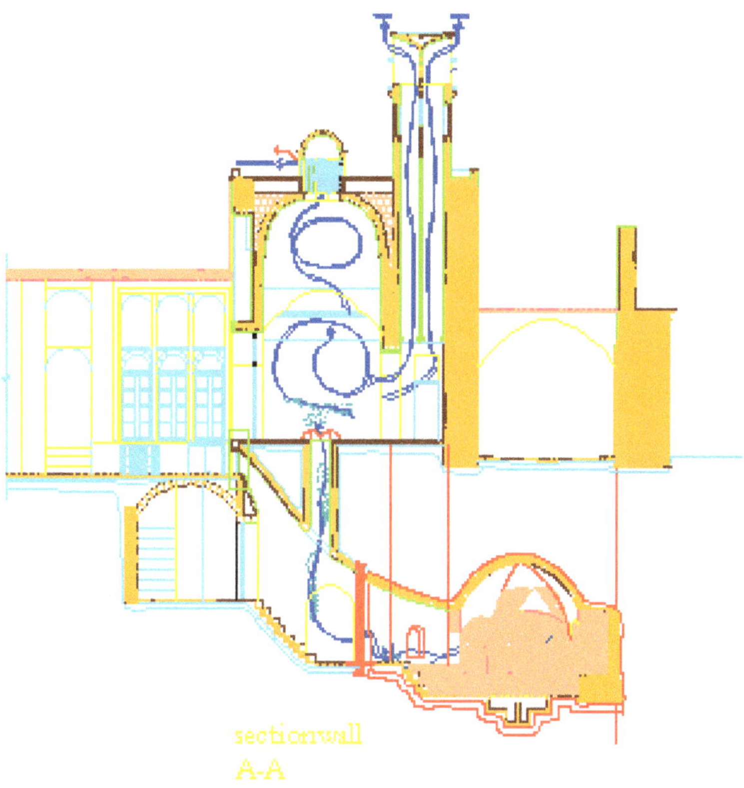

Figure 5. Section- air circulation by air entrance and suction through the wind catcher and roof hatch in the previous figure.

facilitate natural cooling operation in two ways; first, by driving the air, and, second, by cooling through water evaporation. This traditional structure consists of different components that have been considered by the architects not only due to their operational role, but also from the aesthetic point of view.

The main components of the wind catcher are the shelf, body, partitions (including main and minor partitions) and the openings between the partitions of the wind shield. The shelf refers to the upper section of the wind catcher in which the air passage channel is located. The body refers to the section beneath the shelf, which is the separating distance between the shelf and the roof. The partitions are the walls, which divide the air entrance channel to some smaller parts (main partitions extend up to the centre of the tower while minor partitions just go up to the width of the outer walls).

Solar chimney

Some researchers suggest the use of a solar chimney on the other side of the building in order to increase the efficiency of automatic air-conditioning and cooling. Applying this method indicates the completion of the idea for making maximum use of natural energies. A solar chimney is a tool designed for making use of the natural energy of the sun, which operates by increasing the pressure power of the air mass as the result of driving air flow along the chimney channel due to transformation of the heat energy to kinetic energy of air movement. Its structure usually consists of an absorbing wall, air divider, and a glass cover with high sun radiation-conducting power. Solar chimneys are mainly of two types: first are the chimneys installed on the roof, which are inclined, and, second are the chimneys connected to the southern wall of the building.

Applying solar chimneys in order to speed the air movement in the cooling operation of wind catchers

Thus, as was mentioned, by combining the solar chimney system and wind catcher, that is, using a solar chimney on the southern side of the building and applying a wind catcher in an appropriate wind-face direction, the air-conditioning in the wind catcher can be quickened. In this research, it has been done by a numerical study using the limited volume method. A 4-storey building has been considered in which the height of each of its floors is 2.5 m (Figure 6). As can be observed, the inside of the solar chimney grows warm as the result of the sun's radiation and the passage of sunlight through the glass. This warm air moves upwards and forms an air flow by the suction it

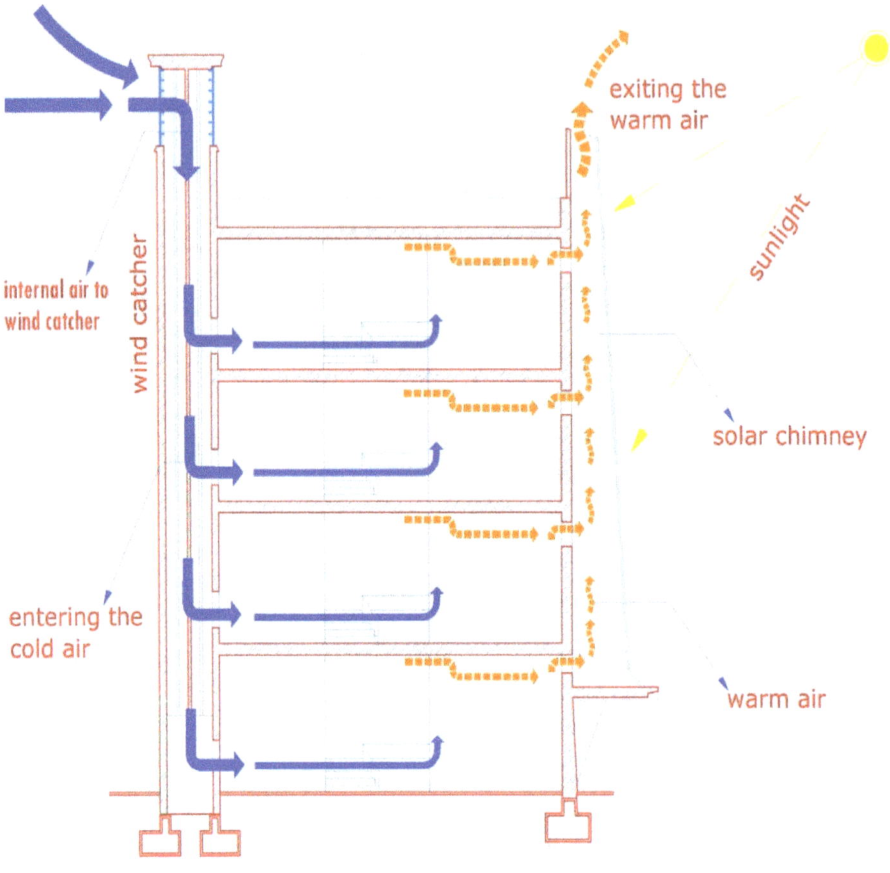

Figure 6. Geometry of input air to wind catcher and output air from solar chimney.

creates in the building. In fact it undertakes the role of a fan.

Equations governing the flow

The equations governing the flow inside the building as control volume or accounting region are as follows:

$$\frac{\partial \rho u_i}{\partial x} = s_m \qquad (1)$$

This is the source phrase, which can be added to or omitted from the main flow; it is related to the water evaporation or condensation phenomena. In the above equation, ρ is the mass density of the air, which is at its lowest at the time of entrance and increased by sprinkling water into it making it heavy. This desired phenomenon creates a sort of negative floatation, which results in the downward movement of heavy air. The total mass density or mass density of the water steam or dry air components can be calculated by calculating the pure moisture and also the total and minor pressure of the water steam and dry air for any node (Kalantar, 2009; Ashare, 1981).

Momentum equations

The averaged equation of Navir Stocks, which governs the flow, is as follows:

$$\frac{\partial}{\partial x_i}(\rho u_i u_j) = \frac{\sigma p}{\partial x_i} + \frac{\partial}{\partial x_j}\left[\mu\left(\frac{\partial u_i}{\partial x_j} + \frac{\partial u_j}{\partial x_i}\right)\right] + \frac{\partial}{\partial x_j}(-\rho\overline{\acute{u}_i\acute{u}_j}) + f_i \quad (2)$$

The phrase introduces the Reynolds stresses, which have been created in relation to the speed fluctuations around the average amount by using the Boussinesq's hypothesis; and f_i introduces the fluid weight power in the given direction (User's Guide, 2000).

Energy equation

In order to calculate the temperature distribution inside the building, an energy equation is required, which is as follows:

$$\frac{\partial}{\partial x_j}[u_i(\rho E + p)] = \frac{\partial}{\partial x_j}[(\lambda + \frac{c_p \mu_t}{pr_t})\frac{\partial T}{\partial x_j} - \sum_i h_i j_i] + s_h$$

In this equation E is:

$$E = \sum h_i y_i + \frac{U^2}{2} \quad (3)$$

λ is fluid heat conduction coefficient

is the penetrating flux of component j; and is related to any energy source.

Transfer equation

The transfer equation for water steam and sprinkling it into the air is as follows:

$$\frac{\partial}{\partial x_i}\left(\rho Y_{H_20} U_i\right) = \frac{\partial}{\partial x_i}\left[\left(\rho D_{H_20} + \frac{\mu_t}{sc_t}\right)\frac{\rho Y_{H_20}}{\partial x_i} + s_{H_20} \quad (4)$$

Radiation equations

Total daily radiation is the sum of scattered daily radiation and direct daily radiation; thesolar radiation data are usually available as total daily radiation. In order to obtain the rate of energy received by the collector, the instantaneous amount of direct and scattered radiation should be calculated. To do this, first the daily scattered radiation is separated from the total radiation, and then the instantaneous amounts of the scattered and direct radiation are determined. In this research, the Miguel method is used for obtaining the daily scattered radiation which is one the most recent methods presented in this are (Miguel et al., 2001; Tsilingiris, 1993).

$$\begin{cases} f_d = 0.952 & k_t < 1.3 \\ f_d = 0.868 + 1.335K_t - 5.782k_t^2 + 3.721k_t^3 & \\ & 1.3 < k_t < 0.8 \\ f_d = 0.141 & k_t > 0.8 \end{cases} \quad (5)$$

K_t is air clearness coefficient, which is equal to the ratio of the total radiation on the earth's surface to the ultra-atmosphere radiation.

f_d is the daily scattered radiation fraction, which is equal to the ratio of the scattered radiation to the total daily radiation on the horizon.

After determination of f_d based on the air clearness coefficient and by having the total daily radiation, the daily scattered radiation can be calculated. After obtaining the daily scattered radiation, the direct radiation can be calculated as the difference in the total and scattered radiations. In order to obtain the instantaneous amounts of the direct and scattered radiation, the Jain et al. (1988) method is used, which has the best conformity with reality.

$$g(t, \alpha, \beta) = \frac{\Gamma(\alpha + \beta)}{\Gamma(\alpha)\Gamma(\beta)N^{\alpha+\beta-1}} \quad (6)$$

$$\left[\frac{N}{2} + (t - 12)\right]^{\alpha-1} \left[\frac{N}{2} + (t - 12)^{\alpha-1}\right]^{\beta-1} \quad (7)$$

For total radiation: $\alpha = \beta = 2.061 + 0.0385N$
For diffused radiation: $\alpha = \beta = 1.969 + 0.0153N$

$$\Gamma(a) = (a-1)\,(a-2)K(b!)\; 0 \le b \le 1 \quad (8)$$

The Jin relation g (t,α, β) determines the amounts of r_G, that is, the ratio of total instantaneous radiation to the total daily radiation, and r_D, that is, the ratio of instantaneous scattered radiation to daily scattered radiation. In addition, the instantaneous amounts of direct radiation can be calculated as the difference of these two amounts. The r_G and r_D amounts are calculated for hourly amounts. In this research, in order to study the received energy amount more precisely, the instantaneous radiation amounts were calculated in time intervals of 6 min. In order to calculate these amounts in time intervals of 6 min, these ratios are multiplied by 0/1.

Determining the amount of absorbed radiation

By having the instantaneous data of direct and scattered radiation, the amount of the radiation absorbed by the collector can be determined using Equation 6. This equation calculates the absorbed radiation, which consists of three types of radiation: direct radiation, scattered radiation, and reradiated scattered radiation, hour by hour (Sukhatme, 1984; Duffie and Beckman, 1980).

$$S = I_b R_b(\tau\alpha)_b + I_d(\tau\alpha)_d\left(\frac{1+\cos\beta}{2}\right) + \rho_g(I_b + I_d)(\tau\alpha)_g\left(\frac{1-\cos\beta}{2}\right) \quad (9)$$

Rb is the ratio of the intensity of direct sun radiation on an inclined slab to the intensity of direct sun radiation to the same slab when it is horizontal.

is the vision coefficient of the collector to the sky and is the vision coefficient of the collector to the earth. In this research, heat fluxes and radiation equations were calculated by EES program at different hours, and, finally, the temperature of 305°C and heat flux of 1000(w/m^2) were obtained.

Introducing the selected numerical method and determining the material for the absorbents and walls

As was mentioned before, the Fluent Software was used for numerical calculation of the air flow inside the building, and marginal conditions of the related issue were considered based on the real summer conditions in one of

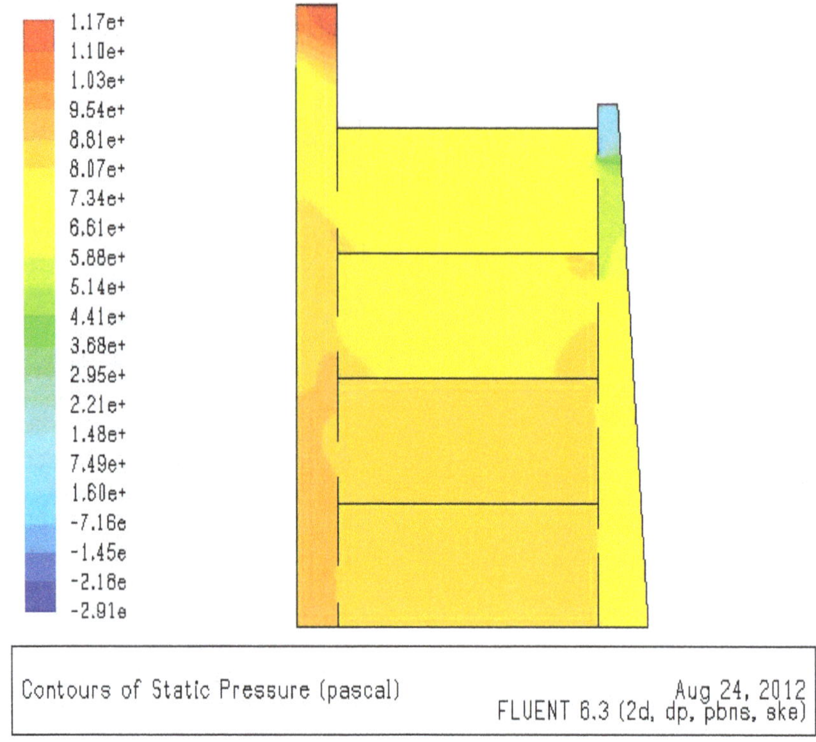

Figure 7. Calculating network.

the cities of Iran, named Kerman. Copper was considered for the material for the absorbents so that itcan operate the best for natural air-conditioning, and the walls were considered to be of brick type. The flow was obtained in the control volume and based on the k-ε model, and the pressure base was solved using simple algorithm. The temperature difference between the radiation absorbance and the air inside the chimney results in a difference in air density, which creates the natural flow movement in the chimney, and, in turn, contributes to the gas exiting.

Model simulation in the fluent software

Figure 7 shows the accounting network of the building and wind catcher in Gambit software. Studying the fine mesh and mesh-free cases results in the following geometry; by applying this geometry in fluent, the effects of the solar chimney on the air-conditioning of the building can be observed. In Figure 8, one can see the pressure counters, and, as can be observed, the pressure is high at the entrance and reduces gradually towards the outlet. These factors drive the air towards the outlet. In addition, an approximate high pressure difference can be observed in the upper and lower parts of the chimney, which is due to the dropping density as a result of the solar heating in the chimney. Figure 9 indicates the pattern of flow lines. The vortexes created in each floor contribute to the better air-conditioning in the building. The direction of these vortexes is counter-clockwise, which contributes to air circulation. The height difference between the air entrance and outlet in each floor is also a factor, which creates these vortexes.

Conclusion

Clean energies or so-called natural energies, are completely compatible with nature and do not harm the environment, which in as much as fossil energies are limited, expensive, and unrenewable, has encouraged people to move towards sustainable development. One of the important branches of sustainable development is sustainable architecture. This sustainability can be observed clearly in Iran's traditional architecture and also that of some other countries. In addition, such architecture can be revived by implementing certain changes in its use through logical insight and new technologies. Wind catchers are among the sustainable factors in Iran's traditional architecture; the integration of which with solar chimneys provides the possibility of their employment in multi-storey buildings. Accordingly, the integration of solar chimneys and wind catchers in buildings was investigated in this research.The results showed that employing traditional air conditioning systems in modern systems, demonstrates a decrease in the consumption of fossil fuels of about 50 to 60%, which

Figure 8. Pressure counter.

Figure 9. Flow route line.

is an important and effective contribution to a clean environment. Employing a solar chimney on the south side, in order to increase the upward movement of the air through it, causes an increase in the wind catcher's efficiency for cooling and conditioning the air.

The results of this simulation demonstrate that inside the solar chimney, due to the radiation of sun and passing the light through the glass, the air is warmed and moves upward. Accordingly, by means of the suction produced in the building, a flow is formed, which simulates the action of a fan.Therefore, by increasing the internal channel temperature, the density of this part is reduced, and, since the internal room temperature is lower than the internal channel temperature, the density of this part rises above the internal channel density, which results in a difference of densities between the two regions (channel and room space) causing the flow. As this temperature difference increases, the flow speed and the magnitude of air conditioning also increases.

Furthermore, according to the results obtained from running the program for different cases, it was observed that by employing water sprayers near the entrance of the air to the wind catcher (inside the blades) and other points along the route, had a crucial effect on various parameters, such as temperature, relative humidity, flow speed, mass density etc. The temperature is decreased significantly, while the relative humidity and mass density increased. It was observed that as the amount of injected water increases, the water temperature decreased until the saturation state was achieved. In this case, the water is not vaporized anymore and the addition of water had no effect and is just gathered as liquid at the bottom of the wind catcher.

REFERENCES

Ashare Handbook (1981). “Fundamentals”, American Society of heating, refrigerating and air_ conditioning Engineers, Inc., Atlanta, Georgia.

Bahadori MN (1978). Passive cooling systems in Iranian architecture. Scientific American. pp. 144-145.

Chena ZD, Bandopadhayaya P, Halldorssonb J, Byrjalsenb C, Heiselbergb P, Lic Y (2003). An experimental investigation of a solar chimney model with uniform wall heat flux. Build. Environ. 38:893-906.

Chungloo S, Limmeechokchai B (2007). Application of passive cooling systems in the hot and humid climate: The case study of solar chimney and wetted roof in Thailand. Build. Environ. 42:3341-3351.

Clito A, Armando O (2000). Solar chimneys: simulation and experiment. Energy Build. 32(1):71.

Duffie JA, Beckman WA (1980). Solar engineering of thermal processes Wiley, New York.

Hirunlabh J, Kongduang W, Namprakai P, Khedari JS (1999). Study of natural ventila-tion of houses by a metallic solar wall under tropical climate. Renew. Energy 18(1):109-119.

Jain PC, Jain S, Ratto CF (1988)."A new model for obtaining horizontal instantaneous global and diffuse radiation from the daily values." Solar Energy 41(5):397-404.

Kalantar V (2005). Natural ventilation the building with wind tower and renewable energy without using fuel oil, the third conference on fuel conservation in building, '13-14 Mar. Tehran-Irans. pp. 1566-1577.

Kalantar V (2009). "Numerical simulation of cooling performance of wind tower (Baud-Geer) in hot and arid region.” Renew. Energy 34(1):246-254.

Liu L, Mak CM (2007). “The assessment of the performance of a wind catcher system using computational fluid dynamics.” Build. Environ. 42:1135-1141.

Macias M, Gaona JA, Luxan JM, Gomez G (2009). “Low cost passive cooling system for social housing in dry hot climate.” Energy Build. 41:915-921.

Maerefat M, Haghighi AP (2010). “Natural cooling of stand-alone houses using solar chimney and evaporative cooling cavity”, Renew. Energy 35:2040-2052.

Miguel A, Bilbao J, Aguiar R, Kambezidis H, Negro E (2001). "Diffuse solar radiation model evaluation in the nprthMediteranean belt area." Solar Energy 70(2):143-153.

Sukhatme SP (1984). "Solar Energy", Principles of thermal collection and storage, New Dehli, India, Tata MacGraw-hill.

Tsilingiris PT (1993). "Theoreticalmodeling of a solar air conditioning system for domestic applications." Energy Conversion Manag. 34(7):523-53l.

User's Guide (2000). FLUENT 6.1.

Comparative study of four methods for estimating Weibull parameters for Halabja, Iraq

Salahaddin A. Ahmed

Department of Physics, Faculty of Science and Science Education, School of Science, University of Sulaimani, Iraq.
E-mail: salahaddinahmed@gmail.com.

The Weibull distribution is the standard function used by the wind energy community to model the wind speed frequency distribution. In this study, four methods are presented for estimating Weibull parameters (Shape and Scale), namely, Maximum likelihood method (MLM), Rank regression method (RRM), Mean-standard deviation method (MSD), and Power density method (PDM). To compare the methods, a period of 4 years (2001 - 2004) of monthly time series data of Halabja city was considered. Two distinct analytical methods are studied to determine the parameter estimation accuracy of these methods; coefficient of determination and root mean square error (RMSE) are used as measurement tools. The Rank regression and MSDs are recommended to estimate the shape parameter; also the Rank regression is recommended for use with our time series wind data to estimate the scale parameter.

Key words: Weibull distribution, parameter estimation, energy pattern factor, accuracy.

INTRODUCTION

Weibull has been recognized as an appropriate model in reliability studies and life testing problems such as time to failure or life length of a component or product. Over the years, estimation of the shape and scale parameters for a Weibull distribution function has been approached through Maximum likelihood method (MLM), linear method, and several versions of regression analysis. In recent years, Weibull distribution has been one of the most commonly used, accepted, recommended distribution to determine wind energy potential and it is also used as a reference distribution for commercial wind energy softwares such as Wind Atlas Analysis and Application Program (WAsP). The two-parameter Weibull distribution function is commonly used to fit the wind speed frequency distribution.

The preferred method of estimating the Weibull parameters was a graphical way using the cumulative wind speed distribution, plotting it on special Weibull graph paper. Estimation of the two-parameter Weibull distribution occurs in many real-life problems. The Weibull distribution is an important model especially for reliability and maintainability analysis.

Weibull distribution can be used to model the wind speed distribution at a particular site and hence, it can help in wind resource assessment of a site. By calculating the two parameters (shape and scale) for Weibull distribution the wind speed frequency curve for a site can be made (Prasad et al., 2009) and the key to perform wind turbine and wind farm energy calculation. Several methods have been proposed to estimate Weibull parameters (Marks, 2005; Rider, 1961; Kao, 1959; Pang et al., 2001; Pandey et al., 2011; Seguro and Lambert, 2000; Stevens and Smulders, 1979; Bhattacharya and Bhattacharjee, 2010). In literature about wind energy, these methods are compared several times and in different ways (Akdag and Ali, 2009; Silva et al., 2004; Yilmaz et al., 2005; Gupta, 1986; Rahman et al., 1994; Lei, 2008; Kantar and Senoglu, 2007), however, results and conclusions of the previous studies are different. Several of fit tests are used in literature. A method for estimating parameters of mixed distributions using sample moments has been outlined by Paul (1961) who considered compound Poisson, binomial, and a special case of the mixed Weibull distribution. A graphical method for estimating the mixed Weibull parameters in life testing of electron tubes is proposed by John (1959). For these reasons, according to the results of the studies, it might be concluded that suitability of the method may vary with the sample data size, sample data distribution, sample data format and goodness of fit test (Akdag and Ali, 2009).

The present work is based on the time series wind data collected over a period of 4 years (2001 - 2004) (hourly). The location concerned in this study named Halabja is situated in east Sulaimani/North Iraq 35° 11' 7" North latitude, 45° 58' 42" East longitude and it is at an elevation of 692 m above sea level. There is no obstacle around wind speed measuring location, the wind data recorded from a mechanical cup type anemometer at height of 2 m above the ground level.

In present study, four methods for estimating the parameters of the Weibull wind speed distribution are presented [MLM, Rank regression method (RRM), Mean-standard deviation method (MSD), and the Power density method (PDM)] by Akdag and Ali (2009). The aim of this work was to select a method that gives more accurate estimation for the Weibull parameters at this location in order to reduce uncertainties related to the wind energy output calculation from any Wind Energy Conversion Systems (WECS).

WEIBULL DISTRIBUTION

The Weibull distribution is characterized by two parameters, one is the scale parameter c (m/s) and the other is the shape parameter k (dimensionless). In Weibull distribution, the variations in wind speed are characterized by two functions which are the probability density function (PDF) and the cumulative distribution function (CDF). The PDF, $f_{v,k,c}$ indicates the fraction of time (or probability) for which the wind is at a given speed V. It is given by Bhattacharya and Bhattacharjee (2010) and Weisser (2003).

$$f_{(v,k,c)} = \frac{k}{c}\left(\frac{V}{c}\right)^{k-1} e^{-\left(V/c\right)^k} \tag{1}$$

Where v > 0, and k, c > 0

The CDF of the speed V gives the fraction of the time (or probability) that the wind speed is equal or lower than V, thus, the cumulative distribution $f_{v,k,c}$ is the integral of the PDF, given by

$$F_{(V,k,c)} = \int_0^\infty f(V)dV = 1 - e^{-\left(V/c\right)^k} \tag{2}$$

The average wind speed can be expressed as:

$$\overline{V} = \int_0^\infty V f(V) dV \tag{3}$$

$$\overline{V} = \int_0^\infty V \frac{k}{c}\left(\frac{V}{c}\right)^{k-1} e^{-\left(V/c\right)^k} dV \tag{4}$$

This can be rearranged as:

$$\overline{V} = k\int_0^\infty \left(\frac{v}{c}\right)^k e^{-\left(v/c\right)^k} dV \tag{5}$$

Taken

$$x = \left(\frac{v}{c}\right)^k, \; dV = \frac{c}{k} x^{\left(\frac{1}{k}-1\right)} dx \tag{6}$$

Equation 5 can be simplified as:

$$\overline{V} = c\int_0^\infty e^{-x} x^{1/k} dx \tag{7}$$

This is the form of the standard gamma function, which is given by

$$\Gamma n = \int_0^\infty e^{-x} x^{n-1} dx \tag{8}$$

From Equations 7 and 8, let $n = 1 + \frac{1}{k}$ the average speed can be expressed as:

$$\overline{V} = c\Gamma\left(1 + \frac{1}{k}\right) \tag{9}$$

The standard deviation of wind speed V is given by

$$\sigma = \sqrt{\int_0^\infty (v - \overline{v})^2 f(v) dv} \tag{10}$$

$$\sigma = \sqrt{\int_0^\infty v^2 f(v) dv - 2\overline{v}\int_0^\infty v f(v) dv + \overline{v}^{2}}$$

or

$$\sigma = \sqrt{\int_0^\infty v^2 f(v) dv - 2\overline{v}\,\overline{v} + \overline{v}^{2}} \tag{11}$$

Using

$$\int_0^\infty v^2 f(v) dv = \int_0^\infty v^2 \frac{k}{c}\left(\frac{v}{c}\right)^{k-1} dv = \int_0^\infty c^2 x^{2/k} \frac{k}{c}\left(\frac{v}{c}\right)^{k-1} dv$$

Equating to

$$\int_0^\infty c^2 x^{2/k} e^{(-x)} dx \tag{12}$$

And putting $n = 1 + \frac{2}{k}$, then the following equation can be obtained. Hence, get the standard deviation

$$\sigma = \left[c^2\Gamma\left(1 + \frac{2}{k}\right) - c^2\Gamma^2\left(1 + \frac{1}{k}\right)\right]^{1/2} \tag{13}$$

or

$$\sigma = c\sqrt{\Gamma\left(1 + \frac{2}{k}\right) - \Gamma^2\left(1 + \frac{1}{k}\right)} \tag{14}$$

METHODS FOR ESTIMATING WEIBULL PARAMETERS

Maximum likelihood method (MLM)

Maximum likelihood technique, with many required features is the most widely used technique among parameter estimation techniques. The MLM method has many large sample properties that make it attractive for use; it is asymptotically consistent, which means that as the sample size gets larger, the estimate converges to the true values.

Let $v_1, v_2, v_3 \ldots\ldots\ldots\ldots v_n$ be a random sample size n drawn from a PDF $f(v,\theta)$ where θ is an unknown parameter. The likelihood function of this random sample is the joint density of the n random variables and is a function of the unknown parameter. Thus, (Yilmaz et al., 2005; Nilsen, 2011),

$$L = \prod_{i=1}^{n} f_{v_i}(v_i, \theta) \tag{15}$$

The maximum likelihood estimator of θ say $\overline{\theta}$ is the value of θ that maximizes L or, equivalent, the logarithm of L. Often but not always, the MLM of θ is a solution of $\frac{d \log L}{d\theta} = 0$.

Now, we apply the MLM to estimate the Weibull parameters, k and c. Consider the Weibull PDF given in Equation 1, then likelihood function will be (Yilmaz et al., 2005; Nilsen, 2011):

$$L(v_1, v_2, \ldots\ldots v_n, k, c) = \prod_{i=1}^{n}\left(\frac{k}{c}\right)\left(\frac{v_i}{c}\right)^{k-1} e^{-\left(\frac{v_i}{c}\right)^k} \tag{16}$$

On taken the logarithms of Equation 16, differentiating with respect to k and c in turn, and equating to zero, one can obtain the estimating equations

$$\frac{\partial \ln L}{\partial k} = \frac{n}{k} + \sum_{i=1}^{n} \ln v_i - \frac{1}{c}\sum_{i=1}^{n} v_i^k \ln v_i = 0 \tag{17}$$

$$\frac{\partial \ln L}{\partial c} = \frac{-n}{k} + \frac{1}{c^2}\sum_{i=1}^{n} v_i^k = 0 \tag{18}$$

In eliminating c between Equations 17 and 18 and simplifying, one can get

$$\frac{\sum_{i=1}^{n} v_i^k \ln v_i}{\sum_{i=1}^{n} v_i^k} - \frac{1}{k} - \frac{1}{n}\sum_{i=1}^{n} \ln v_i = 0 \tag{19}$$

This may be solved to get the estimate of k. This can be accomplished by the use of standard iterative procedures (that is, Newton-Raphson method), which can be written in the form

$$x_{n+1} = x_n - \frac{f(x_n)}{f'(x_n)} \tag{20}$$

Where

$$f(k) = \frac{\sum_{i=1}^{n} v_i^k \ln v_i}{\sum_{i=1}^{n} v_i^k} - \frac{1}{k} - \frac{1}{n}\sum_{i=1}^{n} \ln v_i \tag{21}$$

And

$$f'(k) = \sum_{i=1}^{n} v_i^k (\ln v_i)^2 - \frac{1}{k^2}\sum_{i=1}^{n} v_i^k (k \ln v_i - 1) - \left(\frac{1}{n}\sum_{i=1}^{n} \ln v_i\right)\left(\sum_{i=1}^{n} v_i^k \ln v_i\right) \tag{22}$$

The shape parameter k can be estimated using Equations 21 and 22 with Equation 20 as:

$$k = \left(\frac{\sum_{i=1}^{n} v_i^k \ln v_i}{\sum_{i=1}^{n} v_i^k} - \frac{1}{n}\sum_{i=1}^{n} \ln v_i\right)^{-1} \tag{23}$$

Once k is determined, c can be estimated using Equation 18 as follows:

$$c = \left(\frac{1}{n}\sum_{i=1}^{n} v_i^k\right)^{1/k} \tag{24}$$

Rank regression method (RRM)

The second estimation technique, we shall discuss is known as the least squares method. This is, in essence, a more formalized method of the manual probability plotting technique, in that it provides a mathematical method for fitting a line to plotted failure data points.

It is so commonly applied in engineering and mathematics problems that are often not thought of as an estimation problem. With the help of this method the parameters are estimated with regression line equation by cumulative density function. From Equation 1, the cumulative density function of Weibull distribution function with two parameters can be written as (Justus et al., 1978):

$$F(v_i) = 1 - e^{-\left(\frac{v}{c}\right)^k} \tag{25}$$

This function can be arranged as:

$$\{1 - F(v_i)\}^{-1} = e^{-\left(\frac{v}{c}\right)^k} \tag{26}$$

If we take the natural logarithm of Equation 26

$$-\ln\{1 - F(v_i)\} = \left(\frac{v_i}{c}\right)^k \tag{27}$$

And then retake the natural logarithm of Equation 27, we get the following equation:

$$\ln[-\ln\{1 - F(v_i)\}] = -k \ln c + k \ln v_i \tag{28}$$

Equation 28 represents a direct relationship between ($\ln v_i$) and $-\ln\{1 - F(v_i)\}$ which should be minimized

$$\sum_{i=1}^{n}\{\ln[-\ln(1 - F(v_i)] - \ln[-\ln(1 - E(F(v_i))]\}^2 \tag{29}$$

Parameters of Weibull distribution with two parameters are

estimated by minimizing with Equation 29. The two parameters c and k are intersecting by the following equations:

$$k = \frac{n\sum_{i=1}^{n}\ln v_i \ln[-\ln\{1-F(v_i)\}] - \sum_{i=1}^{n}\ln v_i \sum_{i=1}^{n}\ln[-\ln\{1-F(v_i)\}]}{n\sum_{i=1}^{n}\ln v_i^2 - \{\sum_{i=1}^{n}\ln v_i\}^2} \quad (30)$$

$$c = \exp\{\frac{k\sum_{i=1}^{n}\ln v_i - \sum_{i=1}^{n}\ln[-\ln\{1-F(v_i)\}]}{nk}\} \quad (31)$$

From Equations 30 and 31, k and c can be estimated, respectively.

Mean-standard deviation method (MSD)

The Weibull factors k and c can also be estimated from the mean and standard deviation σ of wind data, consider the expression for average and standard deviation given in Equations 9 and 14, from these, one has (Fung et al., 2007; Weisser and Foxon, 2003):

$$\left(\frac{\sigma_v}{\bar{v}}\right)^2 = \frac{\Gamma\left(1+\frac{2}{k}\right)}{\Gamma^2\left(1+\frac{1}{k}\right)} - 1 \quad (32)$$

Where

$$\bar{v} = \frac{1}{n}\sum_{i=1}^{n} v_i \quad (33)$$

And

$$\sigma^2 = \frac{1}{n}\sum_{i=1}^{n}\left(v_i - \bar{v}\right)^2 \quad (34)$$

n is the number of wind observation. Once σ_v and $\bar{v}$ are calculated for a given data set, then *k* can be determined by solving Equation 32 numerically, once *k* is determined, *c* is given by

$$c = \frac{\bar{v}}{\Gamma\left(1+\frac{1}{k}\right)} \quad (35)$$

In a simpler approach, an acceptable approximation for k is (Akhlaque et al., 2006):

$$k = \left(\frac{\sigma_v}{\bar{v}}\right)^{-1.086} \quad (36)$$

Power density method (PDM)

This is a new method suggested by Akdag and Ali (2009). It is used to estimate the two-Weibull parameters, depends on the energy pattern factor method; it is related to the averaged data of wind speed. This method has simpler formulation, easier implementation and also requires less computation. According to the Weibull probability distribution, the mean wind speed of Equation 35 can be written as (Silva et al., 2004; Paula et al., 2012):

$$\bar{v} = c\Gamma\left(1+\frac{1}{k}\right) \quad (37)$$

And hence the cubic mean wind speed is given as:

$$\bar{v}_{cu} = c^3\Gamma\left(1+\frac{3}{k}\right) \quad (38)$$

To determine the energy pattern factor (E_{pf}) one can write Equations 37 and 38 as:

$$E_{pf} = \frac{\overline{v_{cu}^3}}{\bar{v}^3} = \frac{\Gamma\left(1+\frac{3}{k}\right)}{\Gamma^3\left(1+\frac{1}{k}\right)} \quad (39)$$

Equation 39 is known as energy pattern factor (Epf) method. Weibull parameters can be estimated with solving energy pattern factor Equation 39 numerically or approximately by power density technique using the simple formula as follows:

$$k = 1 + \frac{3.69}{(E_{pf})^2} \quad (40)$$

Once k is determined, c can be estimated using Equation 37.

COMPARISON AND ACCURACY OF THE METHODS

Four methods for estimating the parameters of the Weibull wind speed distribution for wind energy analysis for Halabja city are presented. The application of each method is demonstrated using a sample wind speed data set, and a comparison of the accuracy of each method is also performed with the actual time series data for the our case study (Halabja city). In order to compare the methods, monthly mean wind data used for Halabja region is obtained from meteorological automatic station which covers the period of 4 years (2001 - 2004).

Two tests were employed to determine the accuracy of the four methods given in this article, first is the coefficient of determination R2 of Equation 41 used to how well the regression model describes the data, and second is root mean square error (RMSE) of Equation 42.

$$R^2 = 1 - \frac{\sum_{i=1}^{N}(X_i - x_i)^2}{\sum_{i=1}^{N}(X_i - \underline{X})^2} \quad (41)$$

$$RMSE = \sqrt{\frac{1}{N}\sum_{i=1}^{N}(X_i - x_i)^2} \quad (42)$$

Where N is the total number of intervals, Xi the frequencies of observed wind speed data, xi the frequencies distribution value estimated with Weibull distribution, X the mean of Xi values.

RESULTS AND DISCUSSION

Once coefficient of determination and RMSEs are computed the difference methods can be compared in accuracy as shown in Tables 1 and 2. Weibull parameters have been estimated monthly according to the four methods with the actual time series data for all the years (2001 - 2004). Figure 1 shows the histogram of the actual frequency distribution of diurnal wind speed for all these years with the Weibull function for fitting a wind data probability distribution. Figures 2 and 3 show the

Table 1. Monthly estimated Weibull parameters with actual data.

Month	Mean V (m/s)	Actual Data		ML method		RR method		MSD method		PD method	
		K	$C_{(m/s)}$	K	$C_{(m/s)}$	K	$C_{(m/s)}$	K	$C_{(m/s)}$	K	$C_{(m/s)}$
Jan	1.49	2.3551	1.6115	1.6072	1.6868	2.9191	1.5669	1.3648	1.6291	1.9102	1.6798
Feb	1.86	1.8691	2.0018	1.4568	2.0841	2.2292	1.9420	1.2725	2.0082	1.8971	1.8688
Mar	1.87	2.3688	2.0723	2.0169	2.1260	2.6248	2.0487	2.0558	2.1113	2.8403	2.0985
April	1.97	2.5514	2.1995	2.2075	2.2420	3.0264	2.1463	2.1368	2.2244	2.9989	2.2053
May	2.16	2.6301	2.4000	2.2757	2.4431	3.1939	2.3338	2.3462	2.4370	3.0965	2.4144
Jun	2.45	3.4864	2.7189	3.1190	2.7445	4.4182	2.6468	3.2168	2.7343	3.6739	2.7146
July	2.43	4.0289	2.6602	3.2868	2.7031	4.6771	2.6293	3.7930	2.6868	3.7933	2.6868
Aug	2.35	3.3551	2.6115	3.1590	2.6240	4.4598	2.5417	3.5062	2.6090	3.7152	2.6038
Sep	2.01	3.0376	2.2200	2.5361	2.2657	3.4402	2.1931	2.6195	2.2622	3.3178	2.2398
Oct	1.75	2.4456	1.9556	2.1106	1.9840	3.1626	1.8745	2.0821	1.9758	2.9119	1.9614
Nov	1.79	2.2320	1.9837	1.9703	2.0281	2.5358	1.9473	1.8807	2.0169	2.7973	2.0087
Dec	1.86	2.1564	2.0682	1.8891	2.1184	2.4237	2.0338	1.7736	2.0908	2.7021	2.0915
Mean	1.99	2.709	2.2086	2.3029	2.2542	3.2592	2.1587	2.3373	2.2322	2.9712	2.2145

Table 2. Statistical analysis for all the methods with actual data.

Method	Variance	Standard deviation	Coefficient of variation	Coefficient of determination		Root mean square error	
				K	C (m/s)	K	C (m/s)
Actual data	0.5977	0.7731	0.4094				
ML	0.8459	0.9197	0.4605	0.6290	0.4270	0.3235	0.4464
RR	0.4263	0.6529	0.3374	0.6401	0.4335	0.6328	0.1157
MSD	0.8081	0.8989	0.4544	0.6201	0.4232	0.2534	0.3702
PD	0.5251	0.7246	0.3666	0.5307	0.4088	0.3448	0.3089

estimated parameters c and k, respectively versus the months of years, the similarity can be seen among the methods with the true data almost for all the months for parameter c, while for parameter k, the divergence of the methods with the actual data obtained due to the difference in the estimated values, these also have appear in the RMSE results as shown in Table 2. The rank regression and mean standard deviation methods give satisfactory results for the shape parameter estimation, while rank regression method give satisfactory result for the scale parameter estimation. Graphically, Figure 4 shows the annual mean wind data of the probability density function for Weibull distributions using the estimated Weibull parameters by all methods have been compared with the annual mean wind true time series data, as a result the power density method is the most fitted method to estimate the Weibull parameters in our study case. We can also see that all methods are similar enough to show that each method would be sufficient for determining our parameter estimates.

Conclusion

According to the results, it might be concluded that suitability of these methods may vary with the sample such as data size, sample data distribution (months), sample data format, and of fit tests. When wind data is available in time series format, according to the R2 and RMSE tests both the RRM and MSD, respectively are the recommended methods for estimating the shape parameter, while for the scale parameter and for the both tests, the RRM is recommended method to estimate. Graphically, the curves of the methods show that the best way to estimate the two Weibull parameters is the PDM. This fact is also been supported by means of the RMSE and R2 statistical tests (Table 2). From this comparative study, it is observed that the values of RMSE and R2 have magnitudes that are almost similar for all the methods.

ACKNOWLEDGEMENT

The author would like to thank Dr. Samira Mhamad

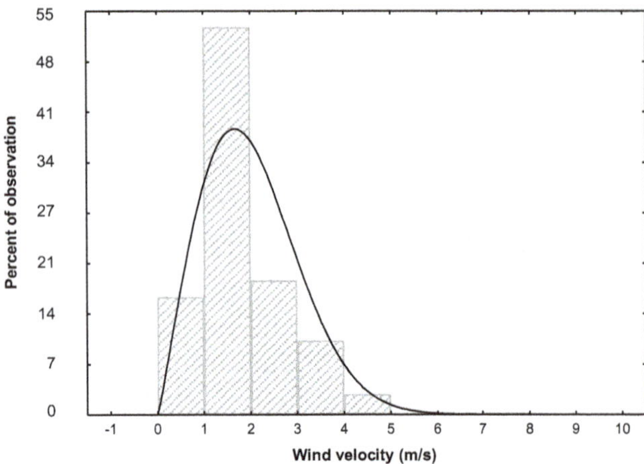

Figure 1. Show the histogram of the time series distribution of the actual wind data.

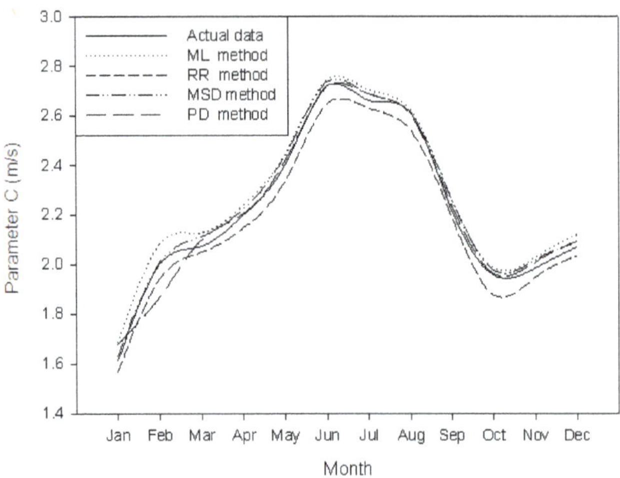

Figure 2. Estimated Weibull parameter C (m/s) versus the months.

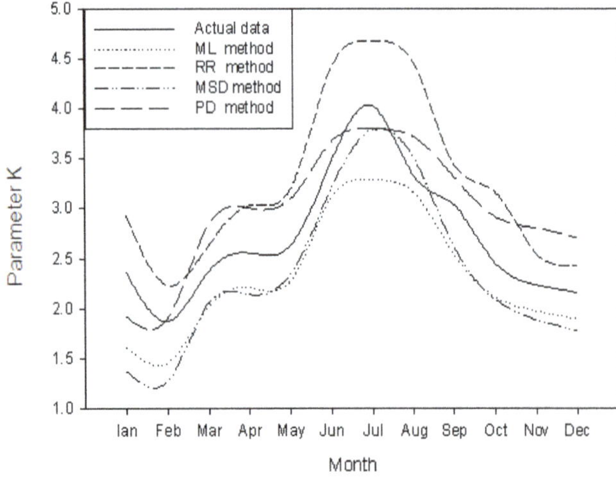

Figure 3. Estimated Weibull parameter K versus the months.

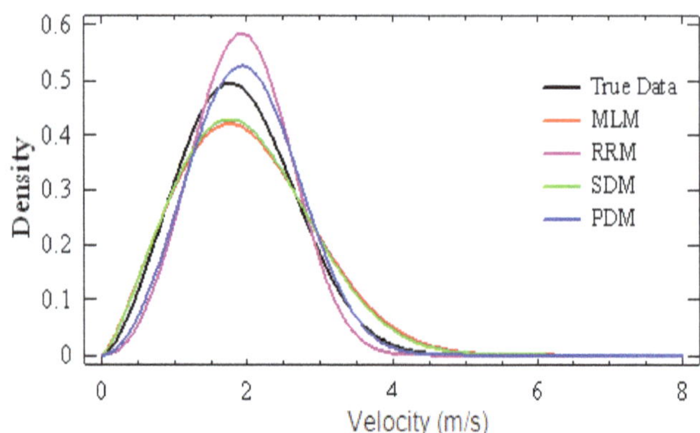

Figure 4. The Weibull probability density function using all methods with actual wind data.

(Department of Statistics) for providing her information to this paper.

REFERENCES

Akdag SA, Ali D (2009). A new method to estimate Weibull parameters for wind energy applications. Energy Convers. Manag. 50:1761-1766.

Akhlaque AM, Firoz A, Wasim AM (2006). Assessment of wind power potential for coastal areas of Pakistan. Turk. J. Phys. 30:127-135.

Bhattacharya P, Bhattacharjee R (2010). A study on Weibull distribution for estimating the parameters. J. Appl. Quant. Methods 5(2):234-241.

Fung CH, Yim HL, Lau KH, Kot SC (2007). Win enegy potential in Guangdong providence. J. Geophys. Res. 112:1-12.

Gupta BK (1986). Weibull parameters for annual and monthly wind speed distributions for five locations in India. Sol. Energy 37(6):469-471.

Justus CG, Hargraves W, Amir M, Denise R (1978). Methods for estimating wind speed frequency distribution. J. Appl. Meteorol. 17:350-353.

Kantar YM, Senoglu B (2007). Estimating the location and scale parameters of the Weibull distribution: an application from engineering. Proceedings of the 16th IASTED International Conference, Applied Simulation and Modelling. Palma de Mallorca, Spain.

Kao JH (1959). A graphical estimation of mixed Weibull parameters in life testing of electron tubes. Technometrics 1:389-407.

Lei Y (2008). Evaluation of three methods for estimating the Weibull distribution parameters of Chinese pine (*Pinus tabulaeformis*). J. For. Sci. 54(12):566-571.

Marks NB (2005). Estimation of Weibull parameters from common percentiles. J. Appl. Stat. 32(1):17-24.

Nilsen MA (2011). Parameter estimation for the two-parameter Weibull distribution. Msc. Thesis. Department of Statistics, Brigham Young University.

Pandey BN, Nidhi D, Pulastya B (2011). Comparison between Bayesian and maximum likelihood estimation of scale parameter in Weibull distribution with known shape under linex loss function. J. Sci. Res. 55:163-172.

Pang WK, Jonathan JF, Marvin DT (2001). Estimation of wind speed distribution using markov chain Monte Carlo techniques. J. Appl. Meteorol. 40:1476-1484.

Paula AC, Ricardo CS, Carla FD, Maria EV (2012). Comparison of seven numerical methods for determining Weibull parameters for wind energy generation I the northeast region of Brazil. Appl. Energy 89:395-400.

Prasad RD, Bansal RC, Sauturaga M (2009). Wind Modeling based on wind input data conditions using Weibull distribution. Inter. J. Glob.

Energy 32(3):227-240.
Rahman S, Halawani TO, Husain T (1994). Weibull parameters for wind speed distribution in Saudi Arabia. Sol. Energy 53(6):473-479.
Rider PR (1961). Estimating the parameters of mixed Poisson, Binomail and Weibull distribution by the method of moments. Bull. Institute Int. Destatistque 2:38.
Seguro JV, Lambert TW (2000). Modern estimation of parameters of the Weibull wind speed distribution for wind energy analysis. J. Wind Eng. Ind. Aerodynamic 85:75-84.
Silva G, Alexandre P, Daniel F, Everaldo F (2004). On the accuracy of the Weibull parameters estimators. European Wind Energy Conference and Exhibition (EWEC). London UK.
Stevens MJ, Smulders PT (1979). The estimation of parameters of the Weibull wind speed distribution for wind energy utilization purpose. Wind Eng. 3(2):132-145.
Weisser D, Foxon TJ (2003). Implications of seasonal and diurnal variations of wind velocity for power output estimation of a turbine: A case study of grenada. Int. J. Energy Res. 2:1165-1179.
Weisser DA (2003). Wind energy analysis of Grenada, an estimation using Weibull density function. Renew. Energy 28:1803-1812.
Yilmaz V, Aras H, Celik HE (2005). Statistical analysis of wind speed data. Eng. & Arch. Fac. Eskisehir Osmangazi University. 98(2):1-10.

12

Markov chain model for the dynamics of cooking fuel usage: Transition matrix estimation and forecasting

Garba S. Adamu[1] and Danbaba A.[2]

[1]Department of Mathematics, Waziri Umaru Federal Polytechnic, Birnin Kebbi, Nigeria.
[2]Department of Statistics, Usmanu Danfodiyo University, Sokoto, Nigeria.

Markov chain models are valuable tools for modeling data that vary over time. They are suitable models to use when modeling the transitions of variables between discrete states over time. In this paper, Markov chain model was applied to the data obtained from 300 households living in the headquarters of three Local Governments (Argungu, Arewa, and Augie) of Kebbi State, Nigeria. The data was information concerning the main source of fuel for cooking used by each of the households. The types of fuels were; fuel-wood, gas, kerosene and electricity. The initial distribution of the households was obtained based on the information in December, 2010 which was used as a baseline. Thereafter, the subsequent data for 2011, 2012 and 2013 were labeled as Periods 1, 2 and 3, respectively. Using this procedure, a transition matrix for the households was obtained and analyzed using Markov chain model. The model was implemented using R-statistical software version 3.0.2. The results obtained indicate a high probability of increase in the use of wood as fuel for cooking by the households. The probability of using other alternative fuels diminishes over time.

Key words: Households, cooking fuel, Markov chains, transition probability.

INTRODUCTION

According to Thierauf and Klekamp (1975), Markov chains originated with the studies of Markov (1906 - 1907) on the sequence of experiments connected in a chain, and with attempts to describe mathematically the physical phenomenon known as Brownian motion. In many real world problems, it is convenient to classify individuals or items into distinct categories or states. We can then analyze the transitions of these individuals or items from one state to another over time (Elwood and James, 1978). Markov chains is a method of studying changes in state of variables with respect to changes in time, in an effort to predict the future state of those variables (Richard and Charles, 1975). According to Sung et al. (2004) and Welton and Ades (2005), Markov chain models are useful tools for modeling data that vary over time. Markov models are appropriate for the analysis of problems in marketing, income tax auditing, car rental services, inventory, machine maintenance and replacement, stock market analysis and hospital administration (Kannan and Lakshmikanthan, 2002). Markov chain model is a suitable model to use when modeling the transitions of patients between discrete health states over time especially the progression over stages of a disease (McDonnel et al., 2002;

Meredith, 1976). Sonnenberg and Beck (1993) applied Markov model in the prognosis of clinical problems. They modeled the events of interest as transition from one state to another.

Fuel is indispensable to human existence for warmth and food preparation. There are different ways a man obtains his required fuel; among these is the fuel-wood. It was approximated that 2.5 to 3.0 billion people rely on wood for fuel. Wood accounts for up to 58% of all energy requirements in African savanna areas (Williams, 2003). Personal interviews and observations indicate that majority of households in Argungu, Arewa and Augie Local Government headquarters are dependent on fuel-wood for cooking. This may not be unconnected to lack of inter-fuel substitution for household choice and the use of a given source of fuel, depend on socio-economic (e.g. family income), demographic (e.g. family size, household composition, life style and culture) and location attributes (e.g. proximity to sources of modern and traditional fuels) (Ayotebi, 2000; Adebaw, 2007).

The use of fuel-wood in Nigeria greatly contributes to desert encroachment and consequently has implications with regard to climate change. Upon this, a little comes to light about the drives and dynamics of fuel-wood consumption in Nigeria (Abebaw, 2007). For the purpose of this research, a household is defined as group of persons living together and maintaining unique eating arrangement. The head of a household is responsible for proving the necessities in the household (NBS, 2012).

MATERIALS AND METHODS

Data collection

In this research, the population was the entire households in the three Local Government headquarters (Argungu, Arewa, and Augie) of Kebbi State, Nigeria. A convenience sampling technique was used to select a sample of 300 households that use one of the common types of fuel as their major source of energy for cooking were interviewed for the purpose of determining and predicting the dynamics of fuel use for the period of 3 years (2010 to 2013). However, the selection of the sample was guided by National Bureau of Statistics Report (NBS, 2012). The detail of the sample was Argungu 185, Augie 50, and Arewa 65 households. The respondents were asked questions pertaining their main source of fuel for cooking among the following categories; A (fuel-wood), B (cooking gas), C (kerosene) and D (electricity).

Analytical technique

For the purpose of this research, Markov chain is used in the prediction of households' choice of means of fuel for cooking in the three Local Government headquarters of Kebbi State, Nigeria. A stochastic process is regarded as a sequence of random variables over time. A random variable taking one of the values 1, 2, 3 … *k* is associated with each point and the sequence is determined by Markov chain with transition matrix *P* (Tijms, 2003; Hsu, 1997; Schuss, 2010; Cox and Isham, 1980; Norris, 1997; Bailey, 1964). The sequence of number of household that use a particular fuel type is considered to be a realization of a stochastic process. If X_t denotes the number of households that maintain the use of a particular fuel for a given period, X_t is a random variable describing the outcome of the fuel usage on the tth period and is termed as "the state" of the process. In Markov process, the probability of moving from one state to another depends only on the present state and not the history.

According to Sung et al. (2006), they defined $\{s_{m0}, s_{m1}, \ldots,\}$ as a sequence of random variables indexed by time, taking finite values in $\varepsilon = \{1, \ldots, J\}$. Assume that the sequence $\{s_{m0}, s_{m1}, \ldots,\}$ forms a first order Markov chains as the conditional probability distribution of s_{mt} given $s_{m,t-1}, \ldots, s_{m,0}$ depend only on the value of $s_{m,t-1}$. Let $X_{ij}(t)$ represents the transition from state *i* at time ($t-1$) to state *j* at time *t*. Let a matrix of state transition probabilities be defined where each row entry represents an initial state and each column entry represents a destination state. That is,

$$X(t) = \begin{bmatrix} x_{11}(t) & \cdots & x_{1M}(t) \\ \vdots & \ddots & \vdots \\ x_{M1}(t) & \cdots & x_{MM}(t) \end{bmatrix} \tag{1}$$

Where

$$\sum x_{ij} = 1. \tag{2}$$

And x_{ij} is defined as

$$x_{ij} = \mathbf{Pr}(X_n = j \mid X_{n-1} = i) \tag{3}$$

More generally, let n_{ij} denote the number of individuals who were in state *i* in period *t*-1 and are in state *j* in period *t*. The probability of an individual being in state *j* in period *t* given that they were in state *i* in period *t*-1, denoted by x_{ij}, can be estimated using the following formula:

$$x_{ij} = \frac{n_{ij}}{\sum_j n_{ij}} \tag{4}$$

Thus, the probability of transition from any given state *i* is equal to the proportion of individuals that started in state *i* and ended in state *j* as a proportion of all individuals in that started in state *i*. According to Hsu (1997), Bailey (1964) and Ross (2007), the Markov chain described above has an initial probability vector

$$X_0 = (i_1, i_2, i_3, \ldots i_n) \tag{5}$$

i's are the states and transition matrix $X_{ij} = P$, the probability vector after *n* repetitions of the experiment is

$$V = X_0 P^n \tag{6}$$

That is, for any regular transition matrix *P*, there is a unique vector *V* such that for any probability vector X_0 and for large value of *n*,

Table 1. Initial distribution of households over the states (t = 0).

State	Number of households	Percentage
A	278	92.7
B	11	3.6
C	9	3.0
D	2	0.7

$V = X_0 P^n$. Vector V is called the equilibrium vector of the Markov chain. From the above fact, and for large value of n,

$$X_0 \cdot P^n \cdot P = V \cdot P \qquad (7)$$
$$X_0 \cdot P^{n+1} = VP$$

as $n \to \infty, X_0 \cdot P^n \to V$ so that $X_0 \cdot P^{n+1} \to VP \to V$, (Danbaba and Isah, 2002). Moreover, at equilibrium, Equation 8 represents the proportion of the households in each state.

$$\lim_{n\to\infty} P^n = W\left(\lim_{n\to\infty} \Lambda^n\right)W^{-1} = L \qquad (8)$$

Where P is the matrix of transition probabilities, W and Λ are matrices of the eigenvectors and eigenvalues of P, respectively and the rows of L are all the same. (Burley and O'sullian, 1986; Ross, 2007).

Basic assumptions

It is assumed in this paper that;

a) Markov process is homogeneous and finite,
b) The number of fuel types (states) remain constant, that is, no new type of fuel used by the selected households,
c) Households used only one of the fuels at a regular interval, that is, yearly in this case.
d) No household leave the system throughout the periods of this research.

RESULTS AND DISCUSSION

At the beginning of data collection, at t = 0, there were a total of 300 households, out of which 278 or 92.7% were in Atate A (fuel-wood), 11 household or 3.6% were in State B (gas), 9 households or 3.0% were in State C (kerosene) and only 2 households or 0.7% use electricity as main source of fuel for cooking, that is State D. (Table 1). Table 2 summarizes the flow of fuel users from one type of cooking fuel to another from December, 2010 to December, 2013.

Transition matrix

$$P = \begin{array}{c|cccc} & A & B & C & D \\ A & 0.9882 & 0.0083 & 0.0035 & 0 \\ B & 0.4643 & 0.3214 & 0.1786 & 0.0357 \\ C & 0.3636 & 0.1818 & 0.4546 & 0 \\ D & 0.2500 & 0.5000 & 0 & 0.2500 \end{array} \qquad (9)$$

The information in Table 2 is more useful when transformed into a transition probability matrix (9). To calculate the entries in the matrix, we sum up (say A to A for example) values and divide it by the row total ($\frac{836}{846} = 0.9882$). Continuing in this manner for other transition routes, we obtained a one-step transition matrix (9). The probabilities of the household moving from state A to States B, C and D are 0.0083, 0.0035 and 0, respectively. In other words, after one-step, the chance of making transition from fuel-wood to gas and kerosene is low. There is no chance of moving from fuel-wood to electricity. The probability of the household remaining in State A (continue using wood for cooking) is as high as 0.9882. The probability of making a forward transition ($\text{i.e., Wood} \to \text{gas} \to \text{kerosene} \to \text{electricity}$) is low while the probability of making backward transition (that is, $\text{Wood} \leftarrow \text{gas} \leftarrow \text{kerosene} \leftarrow \text{electricity}$) is high. The chance of abandoning the use of cooking gas for wood is relatively high (0.5). The chances of abandoning kerosene and electricity for wood are 0.4 and 0.3, respectively. Also, the chance of leaving electricity for gas is good (0.5). The tendency of the household to continue with the use of cooking gas, kerosene and electricity is 0.3, 0.5 and 0.3, respectively. Most importantly, the probability of leaving the use of wood for cooking for its alternative is approximately 0.

Transition diagram

Figure 1 shows a one step transition diagram. It shows the movement of households from one type of cooking fuel to another.

In 2011, 4 households migrate from wood to gas, one from wood to kerosene and none from wood to electricity. Five (5) moved from gas to wood, 1 each from gas to kerosene and electricity. Four households move from using kerosene to fuel-wood, 1 kerosene to gas, and none move from kerosene to electricity. One household moves from electricity to gas, and no other movement from electricity to other fuel types. The same explanation follows in 2012 and 2013. At the end of the data collection, the distribution of the households over the states changed as follows; 290 or 96.6% wood, 5 or 1.7% gas, and 5 or 1.7% kerosene.

To compute fuel shares of the households for a

Table 2. Flow of households from one type of cooking fuel to another (2010 - 2013)

State	December 2010	2011				2012				2013				December 2013
		A	B	C	D	A	B	C	D	A	B	C	D	
A	278	273	4	1	0	278	2	2	0	285	1	0	0	290
B	11	5	4	1	1	5	2	3	0	3	3	1	0	5
C	9	4	1	4	0	2	2	2	0	2	1	4	0	5
D	2	0	1	0	1	1	1	0	0	0	0	0	0	0

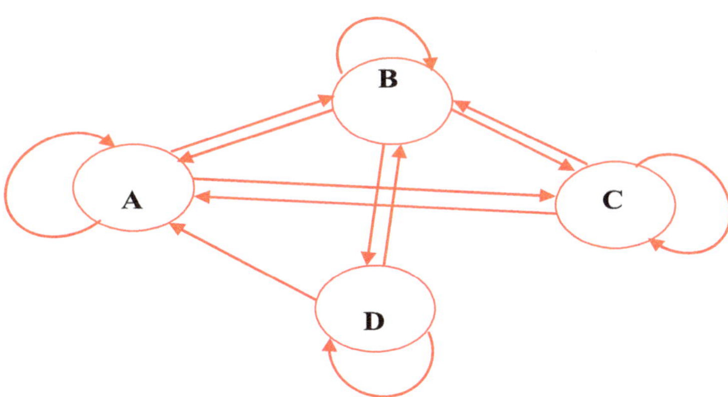

Figure 1. Transition diagram showing the possible transitions.

Table 3. Fuel shares from December, 2010 to December, 2022.

Year	Period	Fuel shares of the households (%)			
		A	B	C	D
2010	1	98.82	0.83	0.35	0.00
2011	2	98.17	1.15	0.65	0.03
2012	3	97.79	1.31	0.85	0.05
2013	4	97.57	1.41	0.96	0.06
2014	5	97.43	1.47	1.03	0.07
2015	6	97.36	1.50	1.07	0.07
2016	7	97.31	1.52	1.10	0.07
2017	8	97.29	1.53	1.11	0.07
2018	9	97.27	1.54	1.12	0.07
2019	10	97.26	1.54	1.12	0.07
2020	11	97.25	1.54	1.13	0.07
2021	12	97.25	1.54	1.13	0.07
2022	13	97.25	1.55	1.13	0.07

Particular year, matrix P and the fuel shares of the preceding year are required. The household-shares of the four competing fuel types for the periods of December, 2010 to December, 2022 have been summarized in Table 3. The table indicates that if the present trends continue for instance, fuel-wood will have 97.25% of the households in the year 2020, while gas, kerosene and electricity will have 1.54, 1.13 and 0.07%, respectively.

Prediction of fuel usage by the households

Applying equation 6 to the initial probability vector

Table 4. Projection of transition probabilities for 13 years (2010 to 2022).

State	2010	2011	2012	2013	2014	2015	2016	2017	2018	2019	2020	2021	2022
AA	0.988	0.982	0.978	0.975	0.974	0.973	0.973	0.973	0.973	0.973	0.973	0.973	0.973
AB	0.008	0.012	0.013	0.014	0.015	0.015	0.015	0.015	0.015	0.015	0.015	0.015	0.015
AC	0.004	0.007	0.008	0.010	0.010	0.011	0.011	0.011	0.011	0.011	0.011	0.011	0.011
AD	0.000	0.000	0.001	0.001	0.001	0.001	0.001	0.001	0.001	0.001	0.001	0.001	0.001
BA	0.464	0.682	0.803	0.873	0.914	0.938	0.952	0.961	0.966	0.968	0.970	0.971	0.971
BB	0.321	0.157	0.092	0.059	0.040	0.031	0.024	0.020	0.018	0.017	0.016	0.016	0.016
BC	0.179	0.140	0.094	0.062	0.042	0.029	0.022	0.018	0.015	0.014	0.013	0.012	0.012
BD	0.036	0.020	0.011	0.006	0.004	0.002	0.002	0.001	0.001	0.001	0.001	0.001	0.001
CA	0.363	0.609	0.758	0.846	0.898	0.929	0.947	0.957	0.964	0.967	0.969	0.971	0.971
CB	0.182	0.144	0.098	0.066	0.046	0.033	0.026	0.022	0.019	0.018	0.017	0.016	0.016
CC	0.455	0.240	0.137	0.083	0.052	0.035	0.025	0.020	0.016	0.014	0.013	0.012	0.012
CD	0.000	0.007	0.007	0.005	0.004	0.003	0.002	0.001	0.001	0.001	0.001	0.001	0.001
DA	0.250	0.542	0.722	0.826	0.887	0.923	0.944	0.955	0.962	0.967	0.969	0.971	0.971
DB	0.500	0.288	0.154	0.088	0.055	0.038	0.028	0.023	0.020	0.018	0.017	0.016	0.016
DC	0.000	0.090	0.094	0.073	0.052	0.036	0.026	0.020	0.017	0.014	0.013	0.012	0.012
DD	0.250	0.080	0.030	0.013	0.006	0.003	0.002	0.002	0.001	0.001	0.001	0.001	0.001

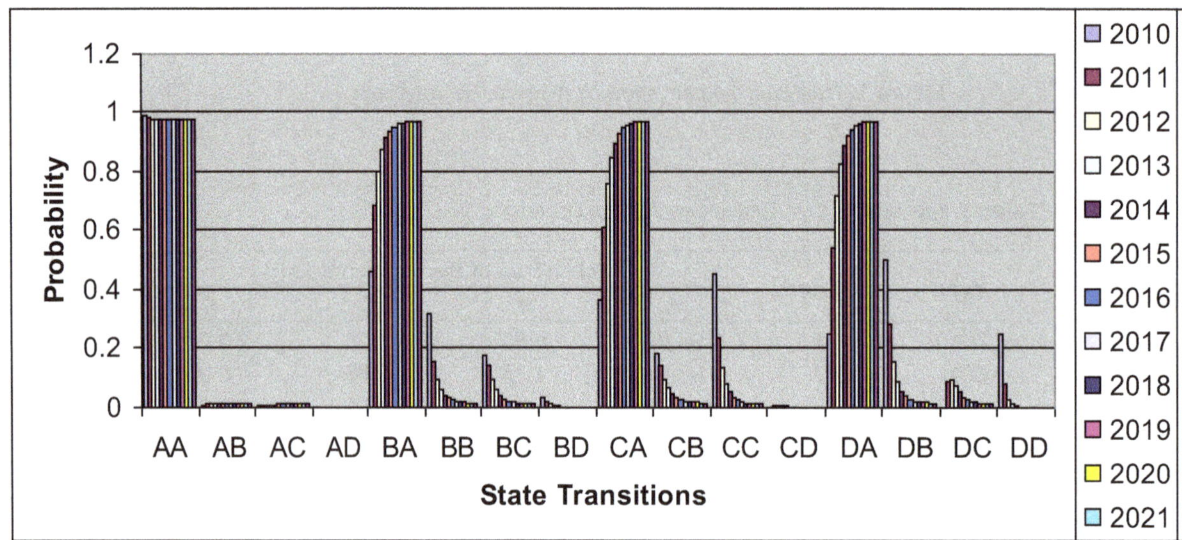

Figure 2. Projection of transition probabilities of moving to each state from 2010 to 2022.

described in Table 1, that is, $X_0 = (0.927 \quad 0.036 \quad 0.030 \quad 0.007)$ and Transition matrix (9), the projected transition probabilities in Table 4 were obtained. Table 4 gives the projection of transition probabilities for 13 years (that is, 2010 to 2022).

Figure 2 is a graphical representation of the four competitive fuel type's transition probabilities for the periods under study. The transition probability of the fuels decreases over the years in favor of the fuel-wood. That is, the probability of the households using gas, kerosene and electricity reduces steadily over the years. The probability of preference of the fuel-wood over other types of fuels for cooking increases across the years. In other words, the probability of the households moving from gas to wood, kerosene to wood and electricity to wood increases steadily over the years. The hope for the households changing cooking fuel from fuel-wood to its alternative is very little. Eventually, all the transition probabilities remain constant in 2021 that is after 12 years.

Conclusion

This paper used Markov chain model to analyze the

behavior of the households in respect of using four types of fuel for cooking in three local government headquarters of Kebbi State, Nigeria. Three years observation periods were used and transition probabilities were calculated up to equilibrium stage. At that state, it is predicted that the probability of households using wood as fuel for cooking is 0.971. This means 97.1% of the household will use wood as their main source of fuel for cooking in the year 2022. This is a sharp increase of fuel-wood users on the initial figure of 92.7%. The probability of the households using gas for cooking is 0.016 or only 1.6% of the households will use gas for cooking,

indicating a sharp decline from the initial figure 3.6%. The same situation was observed for kerosene and electricity declining from 3.0 and 0.7% to 1.2 and 0.1%, respectively. These finding indicates that the high demand of wood for cooking will continue to linger among the households. In view of the above, fuel wood accounted for major part of the fuel sources for cooking in the three local governments. As more and more households depend on the use of fuel wood as a source of fuel, the demand for its exploitation has continued to increase. As a result, fuel-wood exploitation has thus gone beyond mere gathering of dead wood to deliberate and indiscriminate cutting of live trees. The disturbing aspect of fuel-wood extraction is that it can hardly be replaced. Therefore, utilization of fuel-wood in these local government areas will certainly contribute greatly to desert encroachment, and consequently has implications on the climate change and other ecological problems. Hence, the rate of rising exploitation of fuel-wood calls for serious and urgent concern at national and local levels.

Conflict of Interests

The author(s) have not declared any conflict of interests.

REFERENCES

Ayotebi O (2000). Overview of environmental problems in Nigeria. National Centre for Economic Management and Administration (NCEMA) Paper presented at the Conference on Environment and Sustainable Development: Ibadan, 17-18 August.

Abebaw DA (2007). Household determinants of fuelwood choice in urban Ethiopia: a case study of Jimma town. J. Dev. Areas. 41(1):117-126.

Bailey NTJ (1964). The Elements of Stochastic Processes. Wiley, New York, pp. 15–35.

Burley TA, O'sullian G (1986). Operation Research. Macmillan Press Ltd, London. P. 178.

Cox DR, Isham V (1980). Point. Chapman & Hill, London: pp. 54-55.

Danbaba A, Isah GA (2002). Projection of Soft Drinks Consumption in Sokoto State. An Application of Markov Chain Model. Nig. J. Basic. Appl. Sci. 11:73-79.

Elwood SB, James SD (1978). Essentials of Management Science/Operations Research. John Wiley & Sons Inc. USA: pp. 173-193.

Hsu HP (1997). Theory and problems of probability, random variables, and random processes. Schaum's Outline Series, McGraw Hill New York: pp. 161-172.

Kannan D, Lakshmikanthan V (2002). Handbook of Stochastic Analysis and Applications. Marcel Daker Inc. New York: pp. 4-11. http://dx.doi.org/10.1081/SAP-120014691

McDonnel J, Govarde AJ, Rutten FH, Vermeiden JPW (2002). Multivariate Markov chain analysis of the probability of pregnancy in infertile couples undergoing assisted reproduction. Hum. Reprod. 17(1):103-106. http://dx.doi.org/10.1093/humrep/17.1.103

Meredith J (1976). Program evaluation techniques in the Health Services. Am. J. Pub. Health 66(11):1069-1073. http://dx.doi.org/10.2105/AJPH.66.11.1069, PMid:824961 PMCid:PMC1653491

National Bureau of Statistics, (2012).Social Statistics in Nigeria, http://www.nigerianstat.gov.ng accessed 12 February 2013. 11:25 A.M.

Norris JR (1997). Markov Chains. Cambridge University Press U.K: pp. 40-47. http://dx.doi.org/10.1017/CBO9780511810633

Richard IL, Charles AK (1975). Quantitative Approaches to Management. McGraw-Hill, New York, pp. 436-459.

Ross SM (2007). Introduction to Probability Models 9th Edition, Academic Press Elsevier. London: pp. 185-280. PMid:17396482

Schuss Z (2010). Theory and Applications of Stochastic Processes, an Analytical Approach. Springer, New York. P. 207.

Sonnenberg FA, Beck MJ (1993). Markov Models in Medical Decision Making: A Practical Guide. Medical Decision Making, Hanley and Belfus, Inc. Philadelphia, PA, pp. 323-329. http://dx.doi.org/10.1177/0272989X9301300409

Sung M, Erkanli A, Angold A, Castello EJ (2004). Effects of age at first substance use and psychiatric co morbidity on the development of substance use disorders. Drug and Alcohol Dependence 75:287-299. http://dx.doi.org/10.1016/j.drugalcdep.2004.03.013, PMid:15283950

Sung M, Soyer R, Nhan N (2006). Bayesian Analysis of Nonhomogeneous Markov Chains: Application to Mental Health Data. http://www.gwu.edu. Accessed 12 October 2013. 10:32 P.M.

Thierauf RJ, Klekamp RC (1975). Decision Making Through Operations Research, 2nd edition John Wiley& Sons Inc. New York, P. 283.

Tijms HC (2003). A First Course in Stochastic Models. John Wiley & Sons Ltd. The Atrium West Sussex, England: 81-119. http://dx.doi.org/10.1002/047001363X

Welton NJ, Ades AE (2005). Estimation of Markov chain transition probabilities and rates from fully and partially observed data: uncertainty propagation, evidence synthesis and model calibration. Medical Decision Making. 25(633):633-645. http://dx.doi.org/10.1177/0272989X05282637PMid:16282214

Williams M (2003). Deforesting the earth from prehistory to Global Crisis. American Forests. University of Chicago Press.

Numerical modeling of transients in gas pipeline

Mohand KESSAL[1], Rachid BOUCETTA[2], Mohammed ZAMOUM[2] and Mourad TIKOBAINI[1]

[1]Laboratoire Génie Physique des Hydrocarbures, Faculté des hydrocarbures et de la chimie, Université M'Hamed Bougara de Boumerdés-35000- Algérie.
[2]Laboratoire Génie Physique des Hydrocarbures, Faculté des Sciences, Université M'Hamed Bougara de Boumerdés-35000-Algérie.

A set of equations governing an isothermal compressible fluid flow is analytically and numerically analyzed. The obtained equations are written in characteristic form and resolved by a predictor-corrector lambda scheme for the interior mesh points. The method of characteristics (MOC) is used for the boundaries. Advantages of explicit form of these schemes and the flexibility of the MOC are used for an isothermal fast transient gas flow in short pipeline. The results, obtained for a simple practical application agree with those of other methods.

Key words: Gas transients, gas pipeline, lambda scheme, method of characteristics.

INTRODUCTION

Gas transients equations in pipelines can be linear or generally non linear. They also may be parabolic or hyperbolic of the first or second order. As a rule, simple models are an alternative which presents a reasonable compromise between the description accuracy and the cost of solution. These models are obtained by neglecting some terms in the basic set of equations, as a result of a quantitative estimation of the particular elements of the equations for given operating conditions of the pipeline.

Several relatively new numerical schemes were tested to integrate equations of conservation. These include Godunov and TVD schemes (Leveque and Yee, 1990). These second order schemes have the advantage that shock waves problems and other discontinuities can be treated with relatively good accuracy.

In their studies, many authors have considered fast gas transients employing numerical techniques as method of characteristics (MOC) (Kameswara and Eswaran, 1993; Greyvenstein, 2001) and finite differences (Greyvenstein, 2001; Gato and Henriques, 2005), with a relatively good agreement each other. In the last decades Behbahani-Nejad and Bagheri (2010a) have used transfer functions of a single pipeline in order to develop a mathematical Simulik library. Obtained results are satisfactory with the classical methods.

With a reduced order modeling approach, Behbahani-Nejad and Shekari (2010b) have compared their results with the conventional numerical techniques for a simple gas transient example. A good agreement was observed. By the use of time space least square spectral method with a technique based on hierarchical interpolations in space and time, on numerical examples, including fast gas transients, particularly in the case of severe conditions flowing, Dorao and Fernandino (2011) have successfully handled the problem of strong shock wave.

Simulating gas transients in pipes, Ebrahimzadeha et al. (2012) have used an orthogonal collocated method technique to solve the corresponding governing equations.

Its performances are verified and tested for two practical examples corresponding to isothermal and non isothermal cases.

In this work an, old idea, relative to a physically meaningful simples schema, have been developed by Moretti (1979), Zannetti and Colasurdo (1981) and Gabutti (1983). They are known as the lambda schemes. An important advantage of these schemes is related to the concept of "non reflecting boundary condition" :a characteristics form of the boundary conditions equations is applied in order to avoid the use of the improperly reflecting technique. An explicit form of them is adapted and applied to an isothermal fast transient gas flow in short pipeline.

THEORETICAL MODELING

If we consider an isothermal flow in pipeline with variable cross-sectional area in which one dimensional continuity equation is:

$$\frac{\partial \rho}{\partial t} + \frac{\partial}{\partial x}(\rho V) = 0 \tag{1}$$

and momentum equation is given by :

$$\frac{\partial}{\partial t}(\rho V) + \frac{\partial}{\partial x}\left[(\rho V^2 + P)\right] = -\frac{f_g \rho V|V|}{2D} - \rho g S\frac{\partial z}{\partial x} \tag{2}$$

Writing equation of state for natural gas as:

$$P = \rho\frac{ZRT}{\mu} \tag{3}$$

and taking into account the isothermal conditions, the acoustic wave speed becomes:

$$a = \left(\frac{ZRT}{\mu}\right)^{0.5} \tag{4}$$

In the absence of field data, steady state variable distributions constitute the initial conditions. These steady state initial conditions are obtained by the use of an appropriate analytical equation (Zhou and Adewumi (1996) for Z=1:

$$\bar{\rho} = \frac{f_g m_i^2}{Da^2 \rho_i^2}\left(\frac{D}{f_g}\ln\bar{\rho} - \Delta l\right) + 1 \tag{5}$$

Where $\bar{\rho} = \left(\frac{\rho}{\rho_i}\right)^2$ and $m = \rho V$. This equation, which is implicit in $\bar{\rho}$, is well suited for iterative method to determine density or pressure distribution.

In earlier work, considering above equation set, two applications of fast and slow fluid flows have been developed by using two explicit finite-difference schemes (Kessal, 2000). Then, in the same way, a one-dimensional lambda scheme is proposed to study the first case.

NUMERICAL SCHEME

In order to analyze the gas transients phenomena in short pipelines the characteristic method is used to convert the initial partial differential equation set (1) and (2) into ordinary differential equations (Lister, 1960). The physical interpretation is that the waves travel with the speed "a", given by the relation (4), propagating in this way the effect of the initial boundary conditions. Then the transformation of Equations (1) and (2) yields:

$$\frac{1}{\rho a}\frac{dP}{dt} + \frac{dV}{dt} + \frac{f_g}{2D}V|V| + g\frac{dz}{dx} = 0 \tag{6}$$

$$\frac{1}{\rho a}\frac{dP}{dt} - \frac{dV}{dt} - \frac{f_g}{2D}V|V| - g\frac{dz}{dx} = 0 \tag{7}$$

These equations are associated respectively with the following characteristics directions:

$$\lambda^+ = \frac{dx}{dt} = V + a \tag{8}$$

$$\lambda^- = \frac{dx}{dt} = V - a \tag{9}$$

Using Equations (8) and (9), Equations (6) and (7) can be written as:

$$\left[\frac{\partial P}{\partial t} + \lambda^+\frac{\partial P^+}{\partial x}\right] + \rho a\left[\frac{\partial V}{\partial t} + \lambda^+\frac{\partial V^+}{\partial x}\right] + \rho a\frac{f}{2D}V|V| = 0 \tag{10}$$

and

$$\left[\frac{\partial P}{\partial t} + \lambda^+\frac{\partial P^+}{\partial x}\right] - \rho a\left[\frac{\partial V}{\partial t} + \lambda^+\frac{\partial V^+}{\partial x}\right] - \rho a\frac{f}{2D}V|V| = 0 \tag{11}$$

Adding Equations (10) and (11) and simplifying yields to:

$$\frac{\partial P}{\partial t} + 0.5\left[\lambda^+\frac{\partial P^+}{\partial x} + \lambda^-\frac{\partial P^-}{\partial x}\right] + 0.5\rho a\left[\lambda^+\frac{\partial V^+}{\partial x} - \lambda^-\frac{\partial V}{\partial x}\right] = 0 \tag{12}$$

Subtracting Equation (10) from Equation (11) and simplifying, yields to:

Application of an explicit lambda scheme for the above equation set needs the following transformations.

$$\frac{\partial V}{\partial t} + 0.5\left[\lambda^+\frac{\partial V^+}{\partial x} + \lambda^-\frac{\partial V^-}{\partial x}\right] + \frac{0.5}{\rho a}\left[\lambda^+\frac{\partial P^+}{\partial x} - \lambda^-\frac{\partial P^-}{\partial x}\right] + \frac{f}{2D}V|V| = 0 \tag{13}$$

Equations (12) and (13) are in so-called lambda form. Note that the spatial derivatives are marked with subscripts + and – to indicate the characteristic directions along which these derivatives are approximated. Predicted values of V_i^* and P_i^* can be obtained by substitution of finite-difference approximations for the time derivatives into Equations (12) and (13):

$$P_i^* = P_i^j - 0.5\Delta t\left[\lambda^+\frac{\partial P^+}{\partial x} + \lambda^-\frac{\partial P^-}{\partial x}\right] - 0.5\rho a\Delta t\left[\lambda^+\frac{\partial V^+}{\partial x} - \lambda^-\frac{\partial V^-}{\partial x}\right] \quad (14)$$

$$V_i^* = V_i^j - 0.5\Delta t\left[\lambda^+\frac{\partial V^+}{\partial x} - \lambda^-\frac{\partial V}{\partial x}\right] - \frac{0.5\Delta t}{\rho a}\left[\lambda^+\frac{\partial P^+}{\partial x} + \lambda^-\frac{\partial P^-}{\partial x}\right] - RV_i^j\left|V_i^j\right| \quad (15)$$

where: $R = \dfrac{f_g \Delta t}{2D}$

Then, the predictor–corrector scheme applied to our equation set yields to the following procedure (Gabutti, 1983). The spatial derivatives in Equations (14) and (15) are approximated as follow:

Predictor:

Part 1

$$\frac{\partial F^+}{\partial x} = \frac{F_i^j - F_{i-1}^j}{\Delta x} \quad (16)$$

$$\frac{\partial F^-}{\partial x} = \frac{F_{i+1}^j - F_i^j}{\Delta x} \quad (17)$$

Replacing the above finite-difference approximations in Equations (14) and (15) yields the predicted values of V_i^* and P_i^*.

Part 2

The predicted values of the time derivatives $\dfrac{\partial P^*}{\partial t}$ and $\dfrac{\partial V^*}{\partial t}$ of Equations (12) and (13) are calculated by using the following finite difference approximations:

$$\frac{\partial F^+}{\partial x} = \frac{2F_i^j - 3F_{i-1}^j + F_{i-2}^j}{\Delta x} \quad (18)$$

$$\frac{\partial F^-}{\partial x} = \frac{-2F_i^j + 3F_{i+1}^j - F_{i+2}^j}{\Delta x} \quad (19)$$

Corrector part

By considering the following finite difference approximations and using V^* and P^* instead of V and P in Equations (12) and (13), the corrected values of time derivatives $\dfrac{\partial P}{\partial t}$ and $\dfrac{\partial V}{\partial t}$ can be obtained.

$$\frac{\partial F^+}{\partial x} = \frac{F_i^* - F_{i-1}^*}{\Delta x} \quad (20)$$

$$\frac{\partial F^-}{\partial x} = \frac{F_{i+1}^* - F_i^*}{\Delta x} \quad (21)$$

Finally, the values of P and V at the unknown time level are determined from the following equations:

$$P_i^{j+1} = P_i^j + 0.5\Delta t\left(\frac{\partial P^*}{\partial t} + \frac{\partial P}{\partial t}\right) \quad (22)$$

$$V_i^{j+1} = V_i^j + 0.5\Delta t\left(\frac{\partial V^*}{\partial t} + \frac{\partial V}{\partial t}\right) \quad (23)$$

It can be noticed that above discretization is possible only at nodes 3, 4,..., N-1 as it is shown by part 2 of the predictor scheme. Then, a special treatment is needed at points near the boundaries. For this, a one sided finite-difference approximation can be used at nodes 2 and N-1.

Note that in Gabutti (1983) paper, two points finite-difference approximations was used at the nodes adjacent to the boundaries if three points were not available in the desired directions. In this study computational time interval was selected so that the Courant stability condition was satisfied at all nodes of the mesh (Streeter and Wylie, 1969). If necessary, the time interval can be reduced in some cases.

INITIAL AND BOUNDARY CONDITIONS

Initial and boundary conditions to the previous Equations (1) and (2) must be specified in order to obtain the appreciable solution for this differential equation set. Initial conditions of these systems are required to resolve initial pressure and velocity as a function of the position x along the pipeline. In this study they are given by the relation (8) for the pressure distribution. Boundary conditions must also be specified to obtain a unique solution. They depend on the considered cases. It is proposed in this work to treat numerically the two boundary conditions by the characteristics method. Then, integrating Equations (6) and (7) along the negative and the positive characteristics lines Equations (8) and (9) (Figure 1) yields to the following finite-difference equations:

$$V_i^{j+1} - V_{i-1}^j + (1/\rho a)_{i-1}^j (P_i^{j+1} - P_{i-1}^j) + RV_{i-1}^j\left|V_{i-1}^j\right| = 0 \quad (24)$$

$$V_i^{j+1} - V_{i+1}^j - (1/\rho a)_{i+1}^j (P_i^{j+1} - P_{i+1}^j) + RV_{i+1}^j\left|V_{i+1}^j\right| = 0 \quad (25)$$

Applying these equations to limit conditions yields (Figure 2):

$$(V_{N+1}^{j+1} - V_N^j) + (1/\rho a)_N^j (P_{N+1}^{j+1} - P_N^j) + RV_N^j\left|V_N^j\right| = 0 \quad (26)$$

$$(V_1^{j+1} - V_2^j) - (1/\rho a)_2^j (P_1^{j+1} - P_2^j) - RV_2^j\left|V_2^j\right| = 0 \quad (27)$$

A computational procedure to obtain P or V is necessary with the

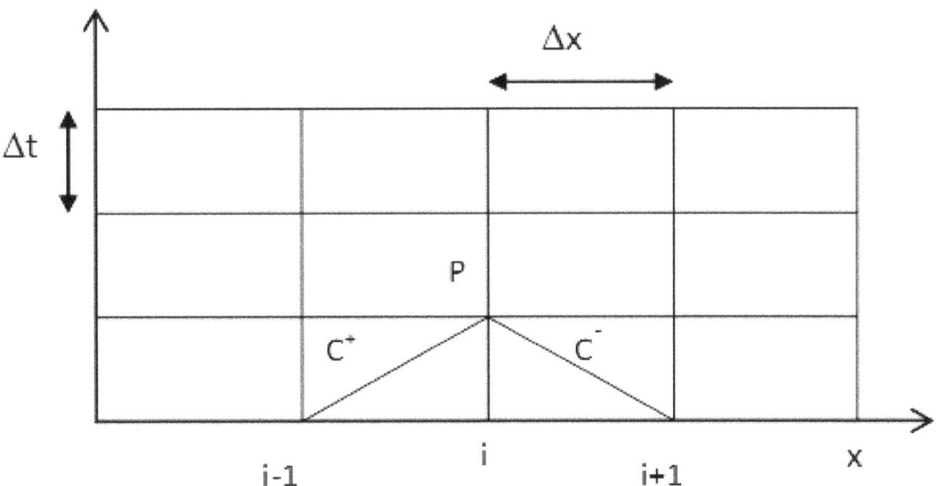

Figure 1. x, t Grid for method of characteristics.

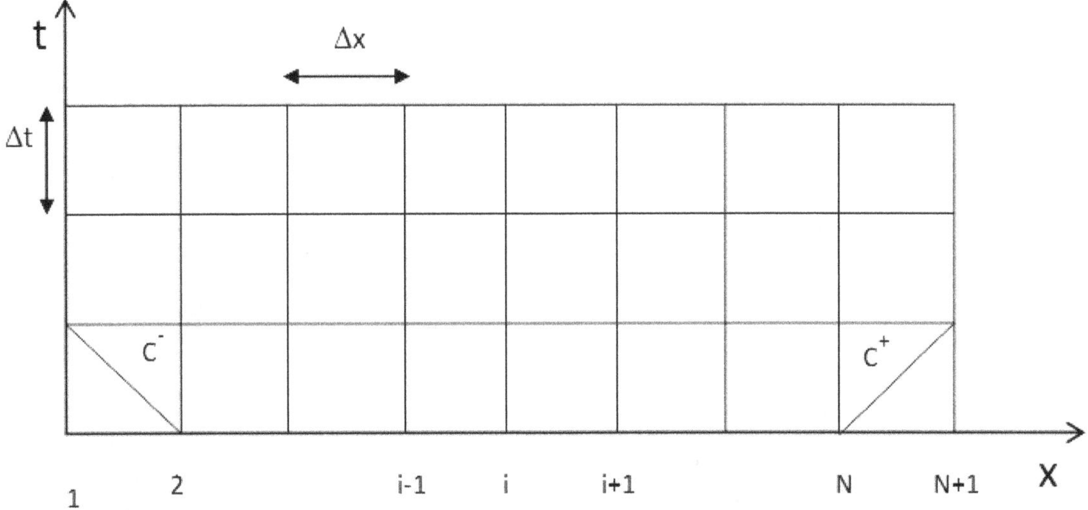

Figure 2. Characteristics at boundaries.

increments Δt , and equal space Δx. Values of fluid properties at the previous time are calculated flow from mesh i-1, i and i+1. The aforementioned stability criteria is required for:

$$\frac{\Delta t}{\Delta x} \leq \frac{1}{a + |V|} \tag{28}$$

RESULTS AND DISCUSSION

An example concerning a gas transient in a relatively short pipeline with an impulse supply of gas mass flux at the inlet, has been simulated using the previous predictor-corrector scheme. This example is taken from Zhou and Adewumi (1996) in which solutions have been obtained, respectively, by using the method of characteristics (MOC), and, a first-order three-point explicit Godunov scheme and a source free second-order five-point TVD scheme. A pipeline 91.44 m long, 0.609 m interior diameter and having initially a static pressure of 4136.8 kPa (with initial velocity V_i=0) with a shut downstream extremity. At time zero, upstream inflow begins to increase linearly and reaches 196 Kg/s at 0.145 s, then decreases linearly to zero again at 0.29 s, and then remains constant. The downstream end is closed. For the simulation of the above fast transient problem, the predictor-corrector lambda scheme adopts with the characteristics method, at the boundaries, the same Δt imposed by the stability Criteria (28).

For numerical simulation of this example, the previous predictor-corrector scheme adopts a grid size Δx=0.9144 m, Δt=0.811 x 10^{-3} s and C.F.L=0.312. These values can be compared with those used in the Godunov and TVD

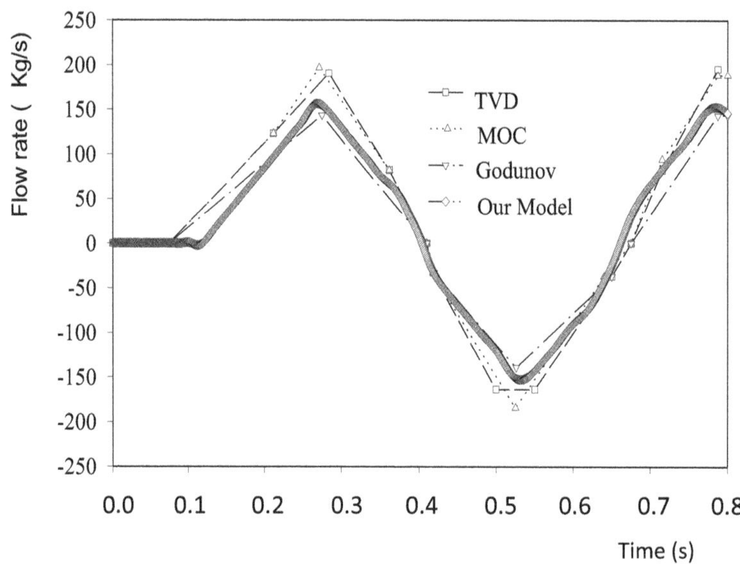

Figure 3. Flow rate variation at the midpoint of the pipeline.

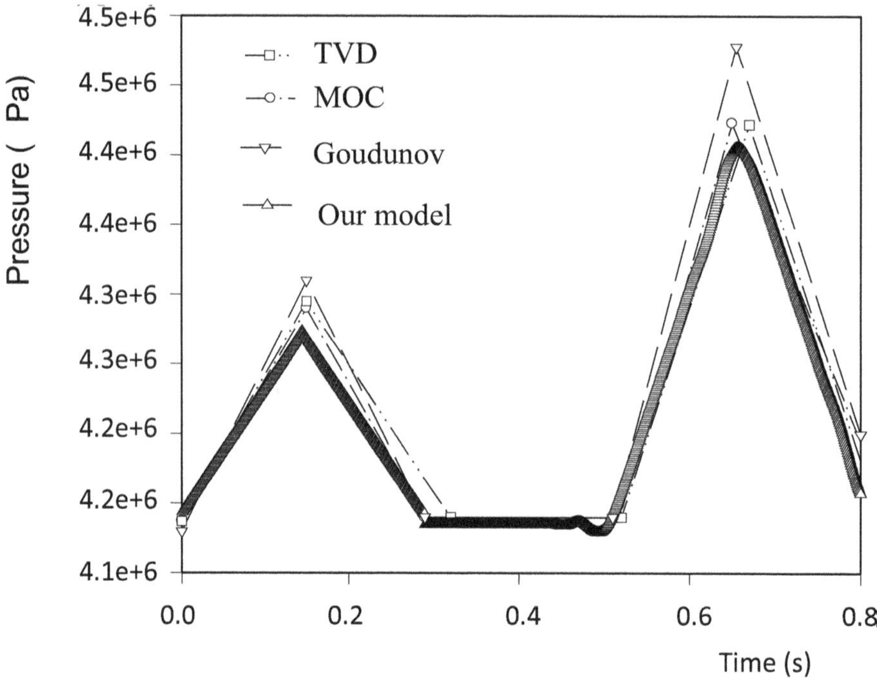

Figure 4. Pressure response at the inlet of the pipe.

Schemes (19) respectively: Δx=0.9144 m, Δt=0.09144 x 10^{-3} s and C.F.L=0.0348; Δx=0.9144 m, Δt=0.9 x 10^{-3} s and C.F.L=0.348. Some numerical results are illustrated by the following figures. It shown a time evolution rate of the gas (Figure 3) at midpoint of the gas line (x=0.5*L), where our results are compared with those obtained by Zhou and Adewumi (1996). Good agreement between them can be observed. The gas flow rate fronts are completely solved within the first 0.8 s. The behaviour of gas flow rate evolution at the midpoint of the pipeline is the result of reflected pressure impulse at the upstream end of the pipe.

A comparison of the predicted pressure (by the present model) at the inlet of the pipeline (x/L=0) with the reported data as shown in Figure 4. Again, relatively good agreement between the predicted results and the

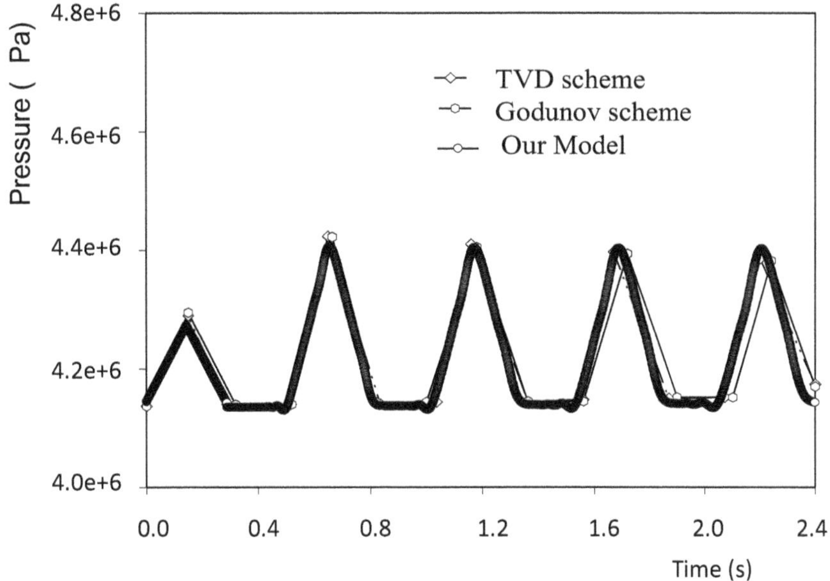

Figure 5. Pressure response at the inlet of the pipeline.

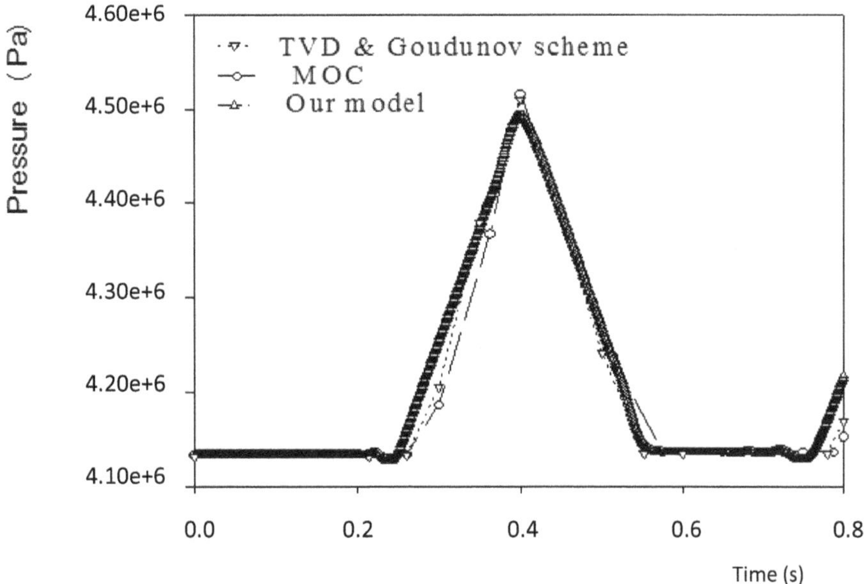

Figure 6. Pressure response at the outlet of the pipe.

reported data is obtained. It can be noticed that duration of the pressure pulse of the peak is the same. Pressure wave is maintained and captured during the first 0.80 s. The pressure wave fronts are reproduced during 2.4 s, without any loss of accuracy (Figure 5). Slight differences with an other methods can be due to some simplifications introduced by Zhou and Adewumi (1996), that is, the value of the friction losses coefficient. A comparison, as regards the pressure at the outlet point of the pipeline (x/L=1), between obtained numerical results and the reported data is shown in Figure 6. Good agreement can be noticed. At the outlet of the pipeline, the pressure wave fronts are completely resolved within the first 0.8 s.

In order to check the numerical method described herein, the pressure evolution has been calculated for time close to the end of the transient phenomenon. It is shown in Figure 7 that agreement is satisfactory. This indicates that the method described in the previous

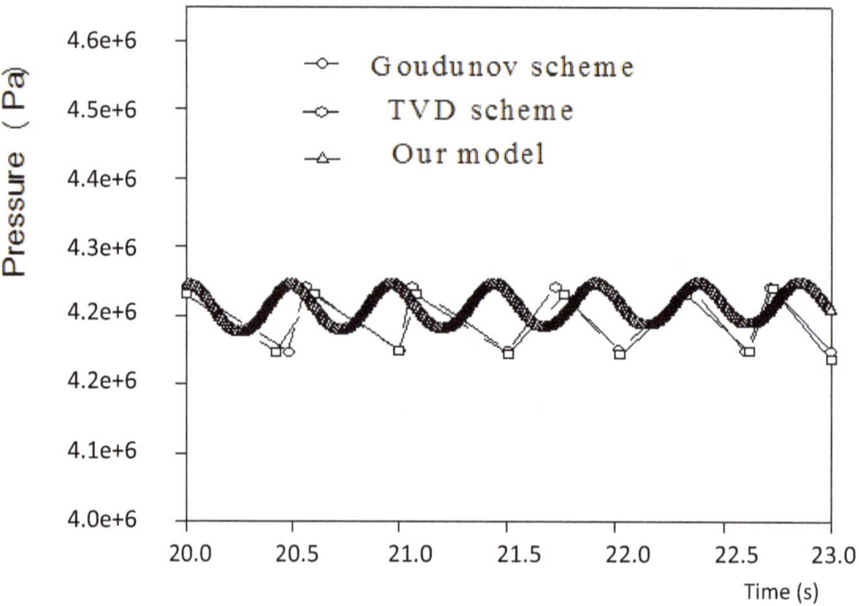

Figure 7. Pressure response at the inlet of the pipe.

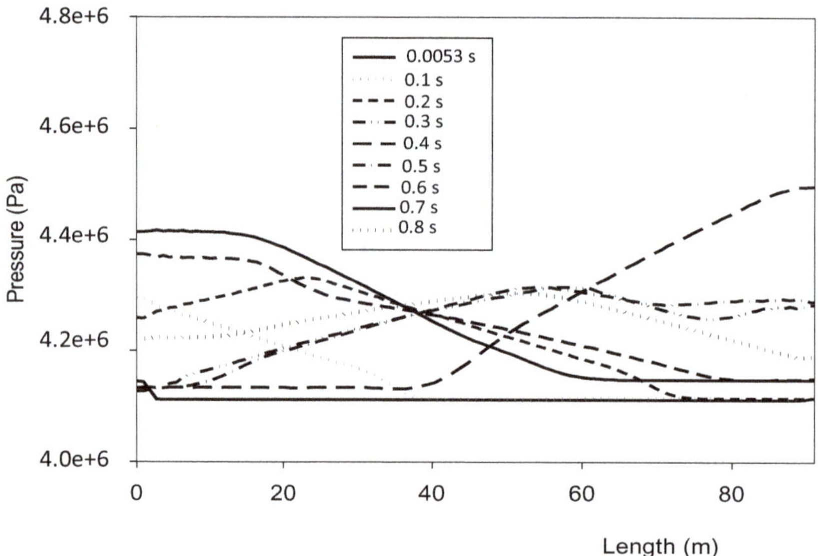

Figure 8. Axial transient pressure distribution.

sections is reliable. Also computed results show that the employed numerical simulation has not produced any undesirable effect.

The evolution of the pressure, corresponding to successive times relative to the pipeline filling, can be observed in Figure 8. Depending on the speed of propagation of sound in gas, then we note that the wave front (of pressure) is reflected from the downstream end.In Figure 9, we can observe the correspondence with Figures 2 and 4, namely the time change in pressure at the inlet and outlet of the pipeline. The incident and reflected pressure waves are explicitly shown in this figure. The longitudinal evolution of the flow rate from the inlet of the pipe is shown in Figure 10. Nevertheless, the speed reflected by the downstream end takes negative values to the upstream end, thereby causing a depression waves which propagate to the upstream.

Finally, the propagation phenomenon of the pressure and the flow velocity disturbances, initiated at upstream point that the pipeline, is well reproduced by the two

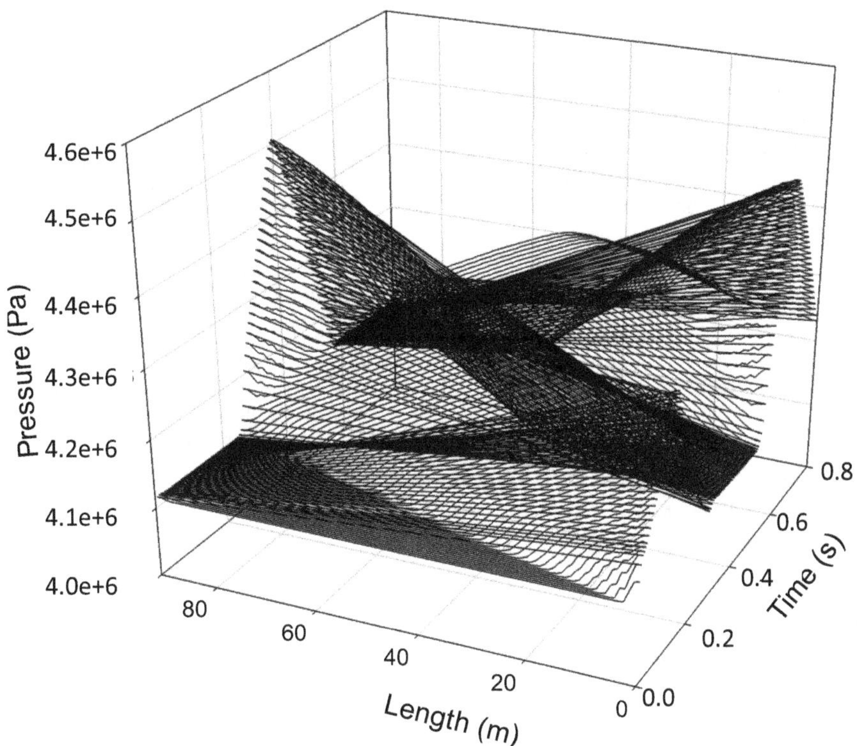

Figure 9. 3D Gas transients between the inlet and outlet.

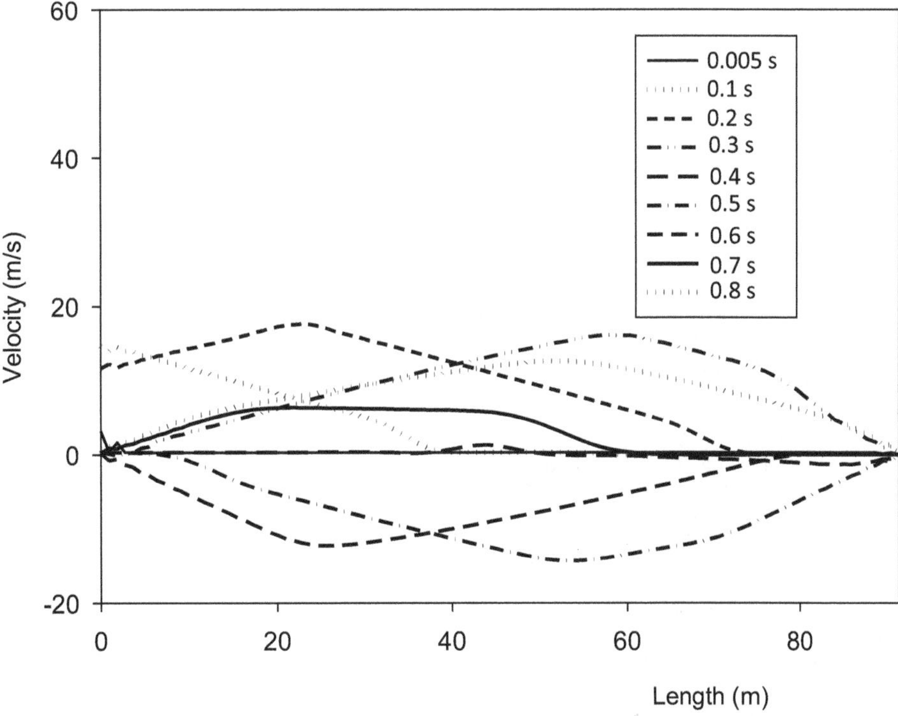

Figure 10. Axial transient velocity.

forms of difference schemes, that is, a second order scheme for the interior points and a method of characteristics for the extremities. The gas transients are contained in a very comprehensive and convenient.

CONCLUSION

The numerical methods presented in this study allow to calculate the propagation of pressure or velocity disturbances from one boundary of a gas pipeline. We have shown that the explicit lambda type of finite difference scheme presents an acceptable computational time. Also, the obtained results have shown that the accuracy of the previous scheme is comparable to the one of high resolution scheme for the considered examples.

Nomenclature

S: cross-sectional area of pipeline, vector; *a* :isothermal speed of sound; D: pipeline diameter; f_g: gas friction factor; g: gravitational acceleration; N: number of space intervals; μ: molecular gas weight; p: pressure; R: universal gas constant or $\mathbf{R} = \frac{\mathbf{f_g \Delta t}}{\mathbf{2D}}$; T: time; l, L: length; T: absolute gas temperature; V: gas velocity; X: axial co-ordinate , L; Z: compressibility factor or elevation; ρ: gas density;ρi: inlet gas density; Δt: uniform time step; Δx: uniform grid size.

Subscripts

g: gas; i: inlet, node.

Conflict of Interests

The author(s) have not declared any conflict of interests.

REFERENCES

Behbahani-Nejad M, Bagheri A (2010a). The Accuracy and Efficiency of a MATLABSimulink Library for Transient Flow Simulation of Gas Pipelines and Networks. J. Pet. Sci. Eng. 70:256–265. http://dx.doi.org/10.1016/j.petrol.2009.11.018

Behbahani-Nejad M, Shekari Y (2010b). The accuracy and efficiency of a reduced-order model for transient flow analysis in gas pipelines. J. Pet. Sci. Eng. 73:13-19. http://dx.doi.org/10.1016/j.petrol.2010.05.001

Dorao CA, Fernandino M (2011).Simulation of transients in natural gas pipelines. J. Nat. Gas Sci. Eng. 3(1):319-364. http://dx.doi.org/10.1016/j.jngse.2011.01.004

Edris E, Mahdi NS, Bahamin B (2012). Simulation of transient gas flow using the orthogonal collocation method. Chem. Eng. Res. Des. P. 90.

Gabutti B (1983). On two Upwind Finite-Difference Schemes for Hyperbolic Equations in Non-Conservative Form. Comput. Fluids. 11(3):207-230. http://dx.doi.org/10.1016/0045-7930(83)90031-2

Gato LMC, Henriques JCC (2005). Dynamic behavior of high pressure natural gas flow in pipelines. Int. J. Heat Mass Transf. 26:817-825.

Greyvenstein GP (2001). An implicit method for the analysis of transient flows in pipe networks. Int. J. Numer. Methods Eng. 53(5):1127-1143. http://dx.doi.org/10.1002/nme.323

Kameswara R, Eswaran K (1993). On the Analysis of Pressure Transients in Pipelines., Int. J. Pres. Ves. Piping. 56:107-129. http://dx.doi.org/10.1016/0308-0161(93)90120-I

Kessal M (2000). Simplified Numerical Simulation of Transients in Gas Networks, Chemical. Eng. Res. Design, Vol. 78, Part A. http://dx.doi.org/10.1205/026387600528003

Leveque RJ, Yee HC (1990). A study of Numerical Methods for Hyperbolic conservation laws with stiff source terms. J. Comp. Phys. 86:187-210. http://dx.doi.org/10.1016/0021-9991(90)90097-K

Lister M (1960). The Numerical Solution of Hyperbolic Partial Differential Equations By the Method of Characteristics, in Ralston, A., and Wilf, H.S., Eds, Mathematical Methods for Digital Computers, Wiley, New York, pp.165-179.

Moretti G (1979). The λ-Schemes, Computers & Fluids. 7(4):191-205. http://dx.doi.org/10.1016/0045-7930(79)90036-7

Streeter VL, Wylie EB (1969). Natural gas pipeline transients, SPEJ. pp. 357-364.

Zannetti L, Colasurdo G (1981). Unsteady Compressible Flow: A Computational Method Consistent with the physical Phenomenon. AIAA J. 19:951-956.

Zhou JJ, Adewumi MA (1996). Simulation of Transients in Natural Gas pipelines, SPEJ. pp. 202-208.

14

Similarities between photosynthesis and the principle of operation of dye-sensitized solar cell

Efurumibe E. L. and Asiegbu A. D.

Physics Department, College of Natural and Physical Sciences, Michael Okpara University of Agriculture, Umudike, Abia State, Nigeria.

Photosynthesis is the process of converting light energy into chemical energy and storing it in the bonds of sugar; while the operational principle of the dye-sensitized solar cell involve the absorption of light by a dye adsorbed by a photoanode. Electrons are released when the dye is illuminated. The comparison done here showed that there exist a lot of similarities between photosynthesis and the principle of operation of a dye-sensitized solar cell.

Key words: Photosynthesis, dye, solar, cell, similarities.

INTRODUCTION

Dye-sensitized solar cell is a new technology that generates electricity when exposed to sunlight. This technology has opened a new area of research interest for scientists. Currently researchers are working to improve on the photon-to-current conversion efficiency of the dye sensitized solar cell. Some researchers have varied the chemical composition of the dye (Ruthenium complex) with the aim of improving the photo-to-current conversion efficiency (Kuang et al., 2007). It is believed that there exist some similarities between photosynthesis and the operational principle of any dye-sensitized solar cell. Dye-sensitized solar cell, as mentioned above, requires sunlight before it can generate electricity. In the same way, plants require sunlight before they can produce sugar. Before we look into the similarities proper, we need to first look at the various concepts under investigation.

PHOTOSYTHESIS

Photosynthesis is the process of converting light energy to chemical energy and storing it in the bonds of sugar (Pessarakli, 2002). This process occurs in plants and some algae (Kingdom Protista). Plants need only light energy, CO_2, and H_2O to make sugar. The process of photosynthesis takes place in the chloroplasts, specifically using chlorophyll, the green pigment involved in photosynthesis (Farabee, 2007). Photosynthesis takes place primarily in plant leaves, and little to none occurs in stems, etc. The parts of a typical leaf include the upper and lower epidermis, the mesophyll, the vascular bundle(s) (veins), and the stomates. The upper and lower epidermal cells do not have chloroplasts, thus photosynthesis does not occur there. They serve primarily as protection for the rest of the leaf. The stomates are holes which occur primarily in the lower epidermis and are for air exchange: they let CO_2 in and O_2 out. The vascular bundles or veins in a leaf are part of the plant's transportation system, moving water and nutrients around the plant as needed. The mesophyll cells have chloroplasts and this is where photosynthesis occurs (Raghavendra, 2000).

Chlorophyll looks green because it absorbs red and blue light, thus making the red and blue colors invisible to our eyes. It is the green light which is not absorbed that finally reaches our eyes, making chlorophyll appear green. However, it is the energy from the red and blue light that are absorbed and invariably used for photosynthesis. The green light as can be seen is not absorbed by the plant, and thus cannot be used during photosynthesis. The overall chemical reaction involved in photosynthesis is:

$6CO_2 + 6H_2O$ (+ light energy) $\rightarrow C_6H_{12}O_6 + 6O_2$.

This is the source of the O_2 we breathe. It therefore becomes necessary for government at all level to check the act of deforestation going on in the country (Farabee, 2007).

There are two parts to photosynthesis:

1. The light reaction: It happens in the thylakoid membrane and converts light energy to chemical energy (Dahik, 2011). This chemical reaction must therefore need the presence of light before it can occur. Chlorophyll and several other pigments such as beta-carotene are organized in clusters in the thylakoid membrane and are involved in the light reaction. Each of these differently-colored pigments can absorb a slightly different color of light and pass its energy to the central chlorphyll molecule to do photosynthesis (Blankenship, 2002).

2. The dark reaction: It takes place in the stroma within the chloroplast, and converts CO_2 to sugar. This reaction does not need light directly for it to occur, but the products of the light reaction. The dark reaction involves a cycle called the Calvin cycle in which CO_2 and energy from adenosine triphosphate are used to form sugar (Blankenship, 2002).

DYE-SENSITIZED SOLAR CELL

Dye-sensitized solar cells are nanoparticulate photovoltaic cells that generate electricity when exposed to sunlight (Reijnders, 2009). Dye sensitized solar cells offer the prospect of very low-cost fabrication and present a range of attractive qualities that will facilitate market entry (Grätzel and Durrant, 2008). In most cases, the dye-sensitized solar cell is called in conjunction with the word: "mesoscopic" as: dye-sensitized mesoscopic solar cell. The word mesoscopic refers to a small scaled size (usually nanoscaled size). The standard dye-sensitized solar cell that uses Titanium oxide as its anode is often times called: Gratzel cell (Reijnders, 2009). This is because Michael Grätzel was the one who first developed a workable dye-sensitized cell based on Titanium oxide anode and a Rheutenium Dye. Grätzel received a millennium technology prize award for his work in 2010 (Hollister, 2010).

The progress realized recently in the fabrication and characterization of nanocrystalline materials has opened up vast new opportunities for the dye-sensitized solar cells. The Dye sensitized mesoscopic solar cells achieves optical absorption and charge separation processes by the association of a sensitizer (dye) as light-absorbing material with a wide-bandgap semiconductor (TiO_2) of nanocrystalline morphology (O'Regan and Grätzel, 1991). Dye-sensitized solar cells are said to be photo-electrochemical cells that make use of electrolytes in place of semiconductors. A sketch of the mesoscopic dye-sensitized solar cell is given in Figure 1.

In the working of the cell, when the sensitizer dye is illuminated, it releases electrons which are injected by a fast process into the conduction band of the titanium oxide (TiO_2) anode. As can be seen from Figure 2, the sensitizer dye is attached to the surface of the mesoporous titanium oxide nanocrystalline thin film. Photoexcitation of the sensitizer dye results in the injection of an electron into the conduction band of the TiO_2. The injected electron transports through the anode towards the external terminals where they could be utilized by a load. On the other hand, the sensitized dye becomes ionized after release of electrons. This ionized dye is regenerated by electron donation from the electrolyte (Grätzel and Durrant, 2008). The electrolyte on its own is regenerated by the transparent conducting oxide glass coated with platinum. The electrolyte in the Grätzel cell is an iodide (triodide) redox couple dissolved in a liquid organic solvent.

COMPARING PHOTOSYNTHESIS IN PLANT WITH THE OPERATIONAL PRINCIPLE OF THE DYE-SENSITIZED SOLAR CELL

The leaf of a plant represents the dye in the dye-sensitized solar cell. Under illumination, the dye in the dye-sensitized solar cell releases electrons which are transported through the anode to the external terminals for utilization (Figure 1). Similarly, under illumination by sunlight, the leaf of a plant releases sugar and oxygen. The oxygen goes into the atmosphere through the stomates while the sugar ($C_6H_{12}O_6$) is stored in the upper and lower epidermis through the vascular bundles. It can be observed that in both cases the products of the reactions (electrons, oxygen and sugar) are transported to areas of need. The electrons are transported through the anode to the external terminals, while the oxygen and the carbohydrate are transported through the stomates and the vascular bundles respectively. The chemical reaction involved in photosynthesis is given as:

$6CO_2 + 6H_2O$ (+ light energy) $\rightarrow C_6H_{12}O_6 + 6O_2$

while that involve in the operation of dye-sensitized solar cell is given as:

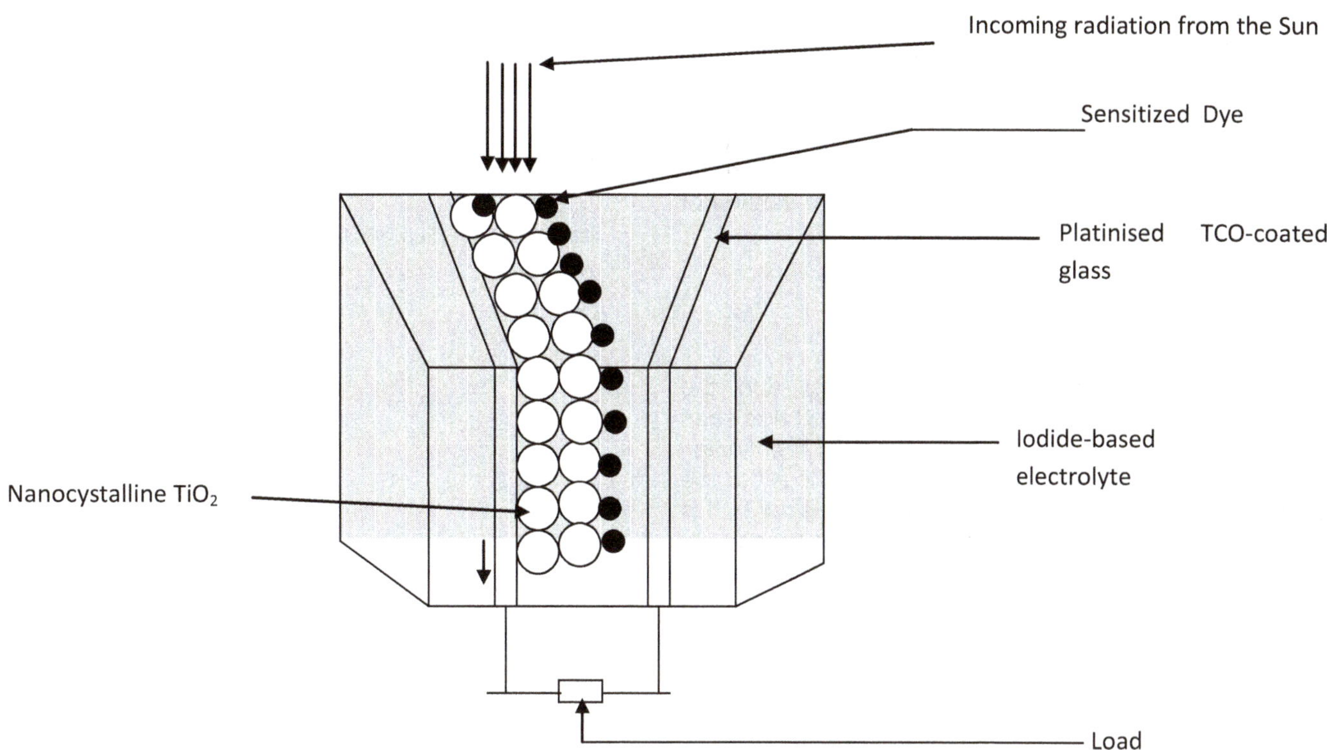

Figure 1. A sketch of the mesoscopic dye-sensitized solar cell.

$$D_c + \text{light} \rightarrow D_c^+ + e^-$$
$$D_c^+ + 2I \rightarrow D_g + I_2^-$$
$$2I_2^- \rightarrow I_3^- + I^-$$

SIMILARITIES

The aim of the study is to deduced the similarities between photosynthesis and the operational principle of dye-sensitized solar cell. Following the study, it was observed that both processes need light from the sun to fuction. In both cases, there is absorbtion of energy from the sun. In dye-sensitized solar cell, the absorbed energy is needed to release of electrons from the dye, whereas in photosynthesis, the absorbed energy is used to generate carbohydrate and oxygen from the leaf. In both cases there is transportation of the products down to areas of needs. In dye-sensitized solar cell, the electrons are transported through the anode to the external terminal where they are utilized, whereas the carbohydrate generated during photosynthesis is transported through the vascular bundles of the plant to the upper and lower epidermis where they are stored for use. The oxygen generated goes into the atmosphere through the stomates. In both cases there are certain chemical reactions that are involved. For photosynthesis, the chemical reaction occurs in the leaf; whereas the reaction for dye-sensitized solar cell occurs at the

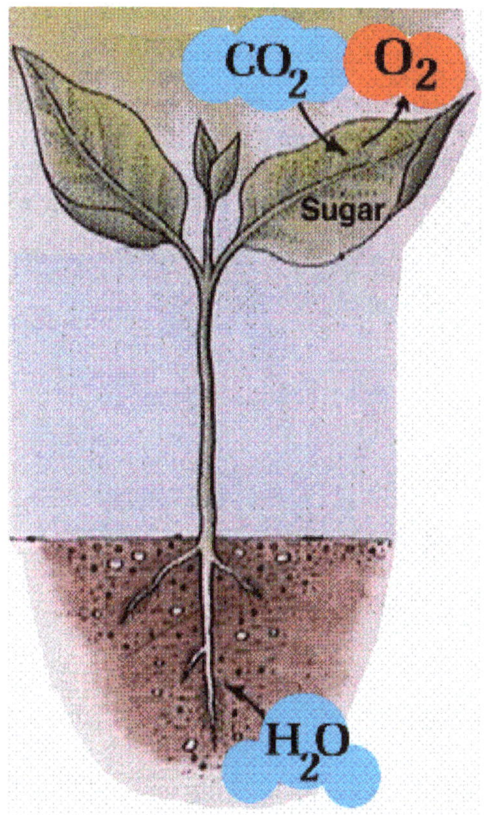

Figure 2. Photosynthetic process.

interface between the dye and the electrolyte. Both photosynthesis and the operational principle of dye sensitized solar cell are environmental friendly, since pollutants are not released in both processes. Lastly there is conversion of energy from light to electricity (in Dye-sensitized solar cell) and from light to chemical (in plants). In essence, in both cases there is conversion of light energy to other forms energy.

CONCLUSION

Having compared photosynthesis in plants and the operational principle of dye-sensitized solar cell and seen that they have similar feature, it is worthy of mentioning that dye-sensitized solar cell is a technology to be embraced. More researches should be carried out in this field of study that is mimicing photosynthetic process in its operation. For when dye-sensitized solar cell is fully adopted, the problem of 'carbon emmission' will be greatly reduced. And by this our environment would be protected.

REFERENCES

Blankenship RE (2002). Molecular Mechanism of Photosynthesis, Wiley-Blackwell, accessed online on 16/07/2011 from: www.books.google.com.

Dahik AC (2011). Investigating the effect of using artificial or direct light on the rate of photosynthesis, accessed online on 16/07/2011 from: www.free-ebookslink.com.

Grätzel M, Durrant RJ (2008). Dye-sensitized Mesoscopic solar cells, Giacomo Ciamician, New York. P. 217.

Farabee MJ (2007). Photosynthesis. Accessed from: http://www2.estrellamountain.edu/faculty/ farabee/ biobk/BioBookPS.html on 07/06/2011.

Hollister S (2010). Michael Grätzel, inventor of the dye-sensitized solar cell, wins 2010 Millennium Technology Prize, accessed on 10/07/2011 from: http://www.engadget.com.

Kuang D, Klein C, Snaith HJ, Humphry-Baker R, Zakeeruddin SM, Grätzel M (2008). A New ion-coordinating Ruthenium Sensitizer for Mesoscopic Dye-sensitized Solar Cells, Inorganica Chimica Acta. 361: 699-706.

O'Regan B, Grätzel M (1991). A low cost, high-efficiency solar cell based on dye-sensitized colloidal TiO_2 films, Nature. 353:737-739.

Pessarakli M (2002). Handbook of Plant and Crop Physiology, accessed online on 16/07/2011 from: www.books.google.com.

Raghavendra AS (2000). Photosynthesis: A Comprehensive Treatise, Cambridge University Press, accessed online on 16/07/2011 from: www.books.google.com.

Reijnders L (2009). Design Issues for Improved Environmental Performance of Dye-sensitized and Organic Nanoparticulate solar cells. J. Cleaner Prod. 18(2010):307-312, published by Elsevier.

15

Theoretical spin assignment and study of the A~100 – 140 superdeformed mass region by using ab formula

A. S. Shalaby[1,2]

[1]Applied Physics Department, Faculty of Applied Science, Taibah University, Saudi Arabia.
[2]Physics Department, Faculty of Science, Beni Suef University, Egypt.

Using an empirical formula of rotational spectra which consists of two parameters a and b (known as ab formula) and derived from the Bohr-Hamiltonian, we extract the spins of fifteen superdeformed bands observed in the A~ 100-140 mass region. The two parameters "a" and "b" were determined by using a search program to fit the proposed transition energies with their observed values. The calculated transition energies depend sensitively on the prescribed spins. The agreement between the calculated and observed transition energies is incredibly well when a correct spin assignment is made. The quality of the fit is generally good in the proposed nuclei, namely: ^{104}Pd(b1), ^{132}Ce(b1), ^{134}Nd(b1), ^{136}Nd(b2), ^{142}Sm(b1), ^{148}Eu(b1, b2), ^{131}Ce(b1, b2), ^{133}Ce(b1, b2), ^{133}Pr(b1, b3), ^{137}Sm(b1), ^{143}Eu(b1). Also, a good agreement is achieved between the extracted spin values in this paper with other theoretical models and also with the available experimental data.

Key words: Superdeformed nuclei, Even-A and Odd-A Nuclei of the superdeformed Mass Region 100-140, ab formula.

INTRODUCTION

The topic of Superdeformation has been at the forefront of nuclear structure physics since the observation of the first superdeformed band in ^{152}Dy (Twin et al., 1986; Bently et al., 1987). Experimental data on SD rotational bands are now available in different parts of the periodic table, namely; in the A~ 190, ~ 150, ~ 130, and ~ 80 mass regions (Hu and Zeng, 1997; Sevensson, 1999; Singh et al., 1996; Afanasjev et al., 1996).

Specially, the superdeformed mass region A~ 100-140 is of particular interest (Szymanski , 1996; Laird, 2002; Leoni et al., 2001; Khazov et al., 2005; Khazov et al., 2006) because of limited number of particles in these nuclei. Moreover, the poor study of these SD nuclei theoretically. The vast majority of the SD bands in this mass region show similar behavior of their dynamic moment of inertia, $\theta^{(2)}$, with the rotational frequency, $\hbar\omega$, in that, they exhibit a smooth decrease as $\hbar\omega$ increases.

For the SD bands, gamma ray energies are unfortunately the only spectroscopic information universally available. Because of the non-observation of the discrete linking transitions between the SD states and the low lying states at normal deformation (ND), the experimental data for the spin of the rotational bands is poor and the only way to obtain the value of the spin is doing theoretically. Several approaches for assigning spins to SD states have been proposed (Draper et al., 1990; Stephens, 1990; Becker et al., 1990; Wu et al., 1992; Hegazi et al., 1999; Khalaf et al., 2002; Shalaby, 2004). These approaches involve direct and indirect methods for assigning spin to the states in SD bands.

In the direct method like Bohr-Mottelson I (I+1) expansion (Bohr and Mottelson, 1975), ab formula (Wu et al., 1989; Holmberg and Lipas, 1968) and the abc formula (Wu et al., 1989), the energies of the states of rotational bands are expressed as a function of spin based on two or more parameters formula. Assuming various values for the spin I_0 of the lowest state in the SD bands, the two or more parameters can be adjusted to obtain a minimum root mean square deviation of the calculated and measured energies. On the other hand, the indirect methods rely mainly on the fitting of the experimental dynamical moment of inertia values with Harris formula (Stephens, 1990; Becker et al., 1990a; Hegazi et al., 1999; Shalaby, 2004; Becker et al., 1990b; Harris, 1964). The parameters obtained from the fit are then used to calculate the spin. In such a parameterization, the spin may be expressed as an expansion in the rotational frequency ħω. Such available approaches are usually referred to the best-fit method (BFM). There are two main problems within this method. On one hand, neither dynamic moment of inertia ($\theta^{(2)}$) nor rotational frequency (ħω) is quantities directly measured in experiment. One can estimate their values. On the other hand, this procedure contains an integration constant, which is in some way an additional parameter.

The method used in the present paper is a direct method based on a formula connecting directly spin and energy of a same level. To our opinion, this method has two advantages than the Harris three-parameter approach (Harris, 1964) done in our previous work (Shalaby, 2004) and others (Hegazi et al., 1999; Khalaf, 2002), where we study in our previous work (Shalaby, 2004) the SD mass region 60-90. First, it is rather general and is obtained within different approaches. Second, it is not necessary to introduce quantities, which are not directly measured in experiment. Spin assignment can be done by using only the transition energies. Unlike the rotational frequency employed in our previous work (Shalaby, 2004), the angular momentum is a directly measured quantity, hence the parameters exist in the present formula connecting the energy and spin, can be determined very easily from the observed rotational spectrum. A theory used to determine the spins of the considered superdeformed rotational bands in the A~ 100-140 mass region with the present ab approach was given. All the data on the fifteen SD bands observed in the A~ 100-140 region are analyzed by making use of this approach. The spins of all these SD bands are determined and the results seem reasonable. With the spin values thus assigned, the energy spectra of these SD bands were calculated and the results turned out unexpectedly well.

THEORY

The simple and effective method used to determine the spin values of the SD rotational bands is the ab-method (also called ab fitting) (Wu et al., 1992). In this method, the two parameter empirical expression for the rotational energy is given as:-

$$E(I) = a\left[\sqrt{1 + bI(I+1)} - 1\right] \qquad (1)$$

Which may be derived from the Bohr Hamiltonian for a well deformed nucleus with small axial symmetry. According to Equation (1), the transition energy from levels I to I-2 is:-

$$\begin{aligned} E_\gamma(I) &= E_\gamma(I \rightarrow I-2) \\ &= a\left[\sqrt{1 + bI(I+1)} - \sqrt{1 + b(I-2)(I-1)}\right] \end{aligned} \qquad (2)$$

For an SD cascade:

$$I_0 + 2n \rightarrow I_0 + 2n - 2 \rightarrow ...... \rightarrow I_0 + 4 \rightarrow I_0 + 2 \rightarrow I_0, \qquad (3)$$

The observed transition energies:

$$E_\gamma(I_0 + 2n), E_\gamma(I_0 + 2n - 2), ... E_\gamma(I_0 + 4) \text{ and } E_\gamma(I_0 + 2)$$

Can be least-squares fit by Equation (2) (ab fit) with fitting parameters a and b. It is found that, in general, the agreement between the calculated and observed transition energies depends sensitively on the prescribed level spins. When a correct "I_0" value is assigned, the calculated transition energies were found to coincide with the observed ones relatively well. However, if I_0 is shifted away from the correct ones even merely by ±1, the root mean square (rms) deviation:

$$\sigma = \left[\frac{1}{n}\sum_{i=1}^{n}\left|\frac{E_\gamma^{cal}\cdot(I_i) - E_\gamma^{exp}\cdot(I_i)}{E_\gamma^{exp}\cdot(I_i)}\right|^2\right]^{1/2} \qquad (4)$$

n which is the number of transitions involved in the fitting will increase drastically. Therefore, the spin value of the lowest (proposed) spin and hence all the spin values of the SD band levels can be determined. This quantity "σ" used in Wu et al. (1992) was taken to fit because the experimental error bars in determination of the level energies are used not reported. As the spin value of the lowest spin I_o is known, all the spin values of the SD band levels can be determined, it is clear that this fitting procedure is quite easy and straight forward and is much simpler than that using the Harris ω^2 - expansion (Hegazi et al., 1999; Khalaf, 2002; Shalaby, 2004).

We now remind the reader that a three-parameter Harris expansion of the cranking model (Harris, 1964) gives rise to the expression of energy:

$$E(\omega) = \alpha\omega^2 + \beta\omega^4 + \gamma\omega^6 \qquad (5)$$

Which leads to the dynamic moment of inertia (Shalaby, 2004)

$$\theta^{(2)}(\omega) = 2\alpha + 4\beta\omega^2 + 6\gamma\omega^4 \ (\hbar^2 \text{ MeV}^{-1}) \qquad (6)$$

Table 1. involves the three fitting parameters and the bandhead spin resulting from our present calculations, other works and the available experimental data for the assumed even-A and odd-A SD bands.

SD band	$E_\gamma(I_0+2\rightarrow I_0)$ (KeV)	Parameter			Band head spin or proposed spin (I_0)		
		ax10^5 (KeV)	b x 10^{-4}	$J_0 = \hbar^2/ab$ ($\hbar^2$/MeV^{-1})	Present work	Previous works [Ref.]: [1](Singh et al., 1996), [2](Singh et al., 2002), [3](Shalaby et al., 2012)	Exp.data[Ref.] (Singh et al., 1996)
^{104}Pd(b1)	1263.0	1.93	2.01	25.78	15	22[2], 21[3]	-
^{132}Ce(b1)	808.0	1.61	1.42	43.74	15	17[2], 18[1], 28[3]	-
^{134}Nd(b1)	663.9±0.5	0.72	3.21	43.57	13	15[2,3]	14
^{136}Nd(b2)	888.0±1.0	5.57	0.325	55.24	23	30[3]	23
^{142}Sm(b1)	680.0	1.14	1.75	50.13	15	23[2], 22[3]	-
^{148}Eu(b1)	748.0	1.24	1.40	57.60	20	30[2], 31[3]	-
^{148}Eu(b2)	844.0	1.27	1.38	57.06	23	26[2], 35[3]	-
^{131}Ce(b1)	591.0	1.61	1.37	45.34	11.5	12.5[2], 16.5[3]	-
^{131}Ce(b2)	847.0	1.43	1.54	45.41	17.5	21.5[2], 26.5[3]	-
^{133}Ce(b1)	748.0	1.44	1.50	46.30	15.5	19.5[2], 21.5[1], 24.5[3]	-
^{133}Ce(b2)	720.3	1.49	1.46	45.97	14.5	16.5[2], 18.5[1], 21.5[3]	-
^{133}Pr(b1)	871.0	1.81	1.16	47.62	19.5	25.5[1,2], 22.5[3]	-
^{133}Pr(b3)	821.0	1.51	1.51	43.86	16.5	25.5[2], 24.5[3]	-
^{137}Sm(b1)	379.0	1.99	0.98	51.43	9.5	8.5[1], 6.52, 8.5[3]	-
^{143}Eu(b1)	483.0	2.31	0.73	59.14	12.5	14.5[2,3]	15.5

This equation can be rewritten as: -

$$\theta^{(2)} = A + B\omega^2 + C\omega^4 \quad (\hbar^2 \text{ MeV}^{-1}) \qquad (7)$$

The spin can be predicted by integrating $\theta^{(2)}$ (Equation 7) with respect to ω

$$\hbar\hat{I} = A\omega + \left(B/3\right)\omega^3 + \left(C/5\right)\omega^5 + i_0 \ (\hbar) \qquad (8)$$

where the constant of integration (i_0) is called the aligned spin (resulting from the alignment of a pair of high-j particles) and it is always equal to zero.

RESULTS AND DISCUSSION

All the transition energies in the fifteen SD bands observed in the A~ 100-140 mass region have been least-squares fit by Equation (2) and the results are encouraging. We have obtained the values of the expansion coefficients a, b by using the Levenberg-Marquardt method (Flannery et al., 1992), to fit the proposed transition energies with their observed values. These fitting were done for the SD bands ^{104}Pd(b1), ^{132}Ce(b1), ^{134}Nd(b1), ^{136}Nd(b2), ^{142}Sm(b1), ^{148}Eu(b1, b2), ^{131}Ce(b1, b2), ^{133}Ce(b1, b2), ^{133}Pr(b1, b3), ^{137}Sm(b1), ^{143}Eu(b1) in the A~ 100-140 mass region, where b1, b2 and b3 refer to band 1, 2 and 3, respectively.

The spin assignments for each SD bands and also the corresponding fitting parameters are given in Table 1. For most of the fifteen SD bands, the spin assignments relatively coincide with the results of another theoretical results (Singh et al., 2002), our previous work (Shalaby et al., 2012), and also with the available experimental data (Singh et al., 1996).

We have two types of superdeformed nuclei

Even-A nuclei (7 nuclei)

As a first illustrative example, the results of the least-squares fitting of ^{104}Pd(b1) are listed in Table 2. It is seen that the E_2-transition energies can be reproduced relatively well when I_0 = 15, that is, E(17>15) = 1263.0 MeV. The deviations between the calculated and observed E(I) values are mostly less than 30 MeV. However, if I_0 is assumed to be 14 or 16, the rms deviation immediately increases (Figures 1 to 7). Therefore, the assignments of I_0 = 14 or 16 are completely unacceptable.

Similarly, we have obtained the results of the least-squares fitting of the other 6 even – A nuclei and these are listed in Tables 3 to 8.

Odd-A nuclei (8 nuclei)

As a first illustrative example, the results of the least-squares fitting of ^{131}Ce(b1) are listed in Table 9. It is seen

Table 2. Spin determination for the SD band ^{104}Pd(b1). I_0 is the spin value prescribed to the lowest level observed. $E_\gamma(I) = E_\gamma(I \rightarrow I-2)$ is the transition energy from level I to I-2. $\delta = E_\gamma^{exp.}(I) - E_\gamma^{cal.}(I)$. σ is the rms deviation defined by Equation 4.

Observed[a] E_γ(I) values (KeV)	Calculated E_γ(I) values (KeV)								
	I_0=14[b]			I_0=15[c]			I_0=16[d]		
	I	E_γ(I)	δ	I	E_γ(I)	δ	I	E_γ(I)	δ
2079.0	28	2034.6	44.4	29	2050.1	28.9	30	1897.2	181.8
1919.0	26	1914.3	4.7	27	1924.6	-5.6	28	1794.5	124.5
1763.0	24	1788.7	-25.7	25	1795.5	-32.5	26	1687.3	75.7
1638.0	22	1657.9	-19.9	23	1663.0	-25.0	24	1575.6	62.4
1511.0	20	1522.2	-11.2	21	1527.2	-16.2	22	1459.6	51.4
1381.0	18	1381.5	-0.51	19	1388.3	-7.3	20	1339.3	41.7
1263.0	16	1236.2	26.8	17	1246.4	16.6	18	1214.9	48.1
σ		1.38×10^{-2}			1.19×10^{-2}			5.16×10^{-2}	

[a]Reference (Singh et al., 2002): [b]a = 1.22×10^5 KeV, b = 3.40×10^{-4}; [c]a = 1.93×10^5 KeV, b = 2.01×10^{-4}; [d]a = 1.13×10^5 KeV, b = 3.22×10^{-4}.

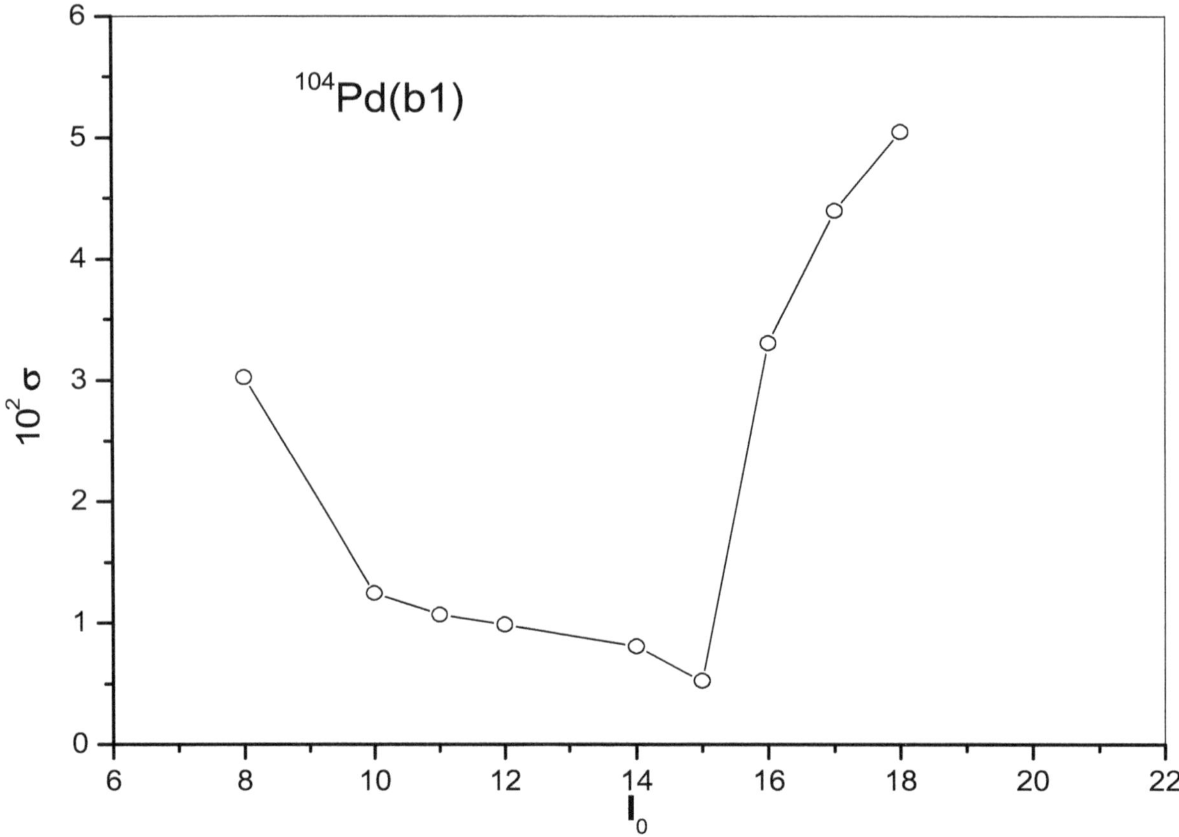

Figure 1. The rms deviation for various spin assignments in ^{104}Pd(b1). I_0 is the value prescribed to the lowest level observed [E $\gamma(I_0 + 2 \Longrightarrow I_0)$ = 1263 KeV].

that the E_2-transition energies can be reproduced relatively well when I_0 = 11.5, that is, E(13.5→11.5) = 591.0 MeV. The deviations between the calculated and observed E(I) values are mostly less than 81 MeV. However, if I_0 is assumed to be 10.5 or 12.5, the rms deviation immediately increases (Figures 8 to 15). Therefore, the assignments of I_0 = 10.5 or 12.5 are completely unacceptable.

Similarly, we have obtained the results of the least-squares fitting of the other 7 odd – nuclei and these are

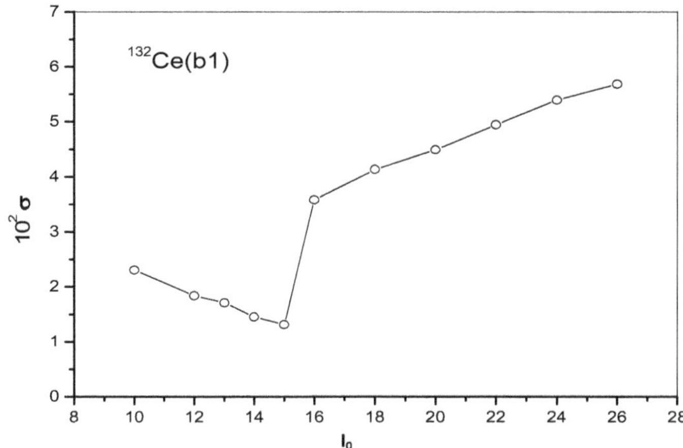

Figure 2. The rms deviation for various spin assignments in ^{132}Ce(b1). I_0 is the value prescribed to the lowest level observed [E $\gamma(I_0 + 2 ==> I_0)$ = 808 KeV].

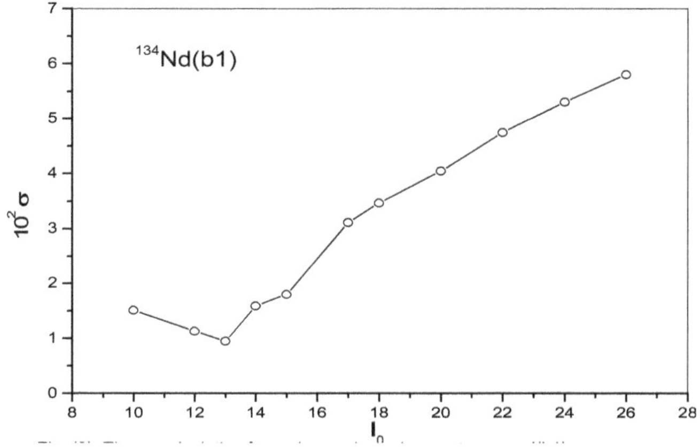

Figure 3. The rms deviation for various spin assignments in ^{134}Nd(b1). I_0 is the value prescribed to the lowest level observed [E $\gamma(I_0 + 2 ==> I_0)$ = 663.9 KeV].

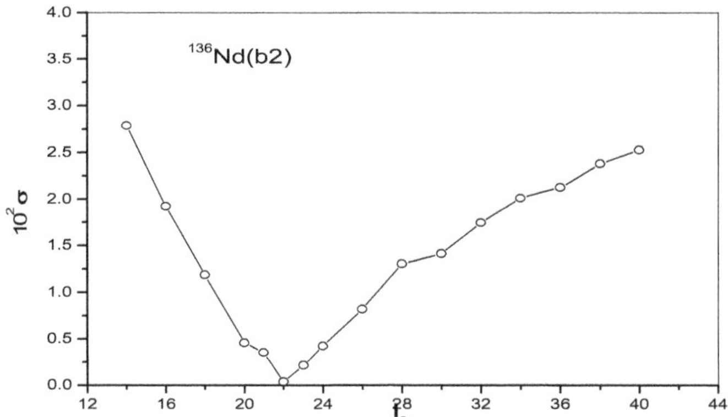

Figure 4. The rms deviation for various spin assignments in ^{136}Nd(b2). I_0 is the value prescribed to the lowest level observed [E $\gamma(I_0 + 2 ==> I_0)$ = 888 KeV].

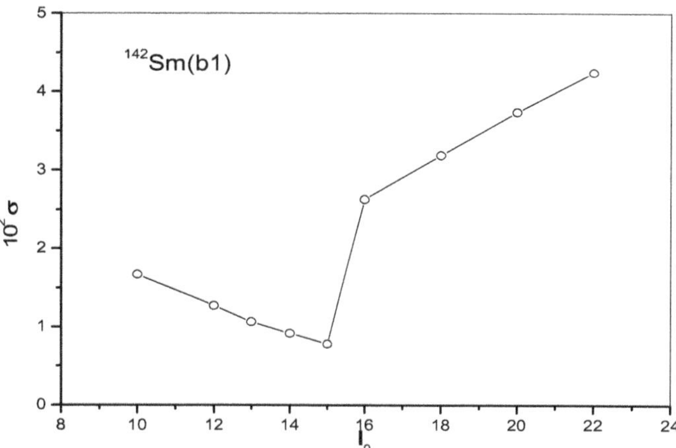

Figure 5. The rms deviation for various spin assignments in ^{142}Sm(b1). I_0 is the value prescribed to the lowest level observed [E $\gamma(I_0 + 2 ==> I_0)$ = 680 KeV].

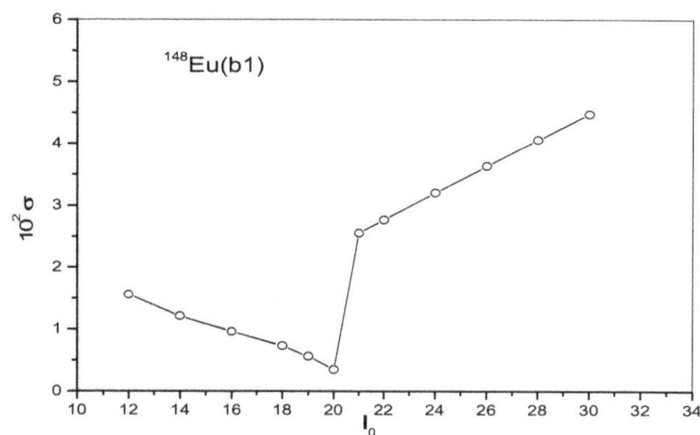

Figure 6. The rms deviation for various spin assignments in ^{148}Eu(b1). I_0 is the value prescribed to the lowest level observed [E $\gamma(I_0 + 2 ==> I_0)$ = 748 KeV].

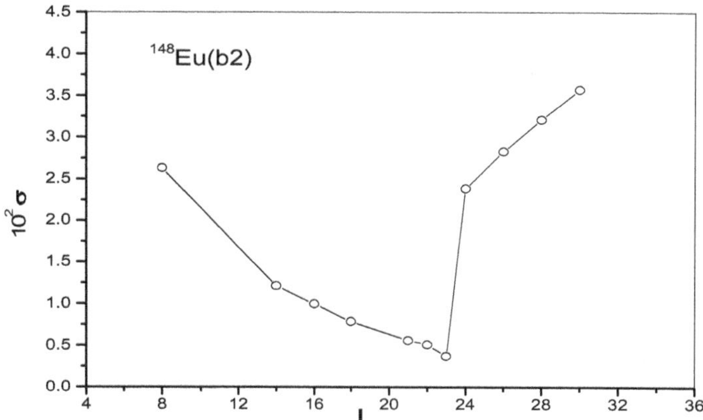

Figure 7. The rms deviation for various spin assignments in ^{148}Eu(b2). I_0 is the value prescribed to the lowest level observed [E $\gamma(I_0 + 2 ==> I_0)$ = 884 KeV].

Table 3. Spin determination for the SD band $^{132}Ce(b1)$. I_0 is the spin value prescribed to the lowest level observed. $E_\gamma(I)= E_\gamma(I\rightarrow I-2)$ is the transition energy from level I to I-2. $\delta = E_\gamma^{exp.}(I)- E_\gamma^{cal.}(I)$. σ is the rms deviation defined by Equation 4.

Observed[a] $E_\gamma(I)$ values (KeV)	Calculated $E_\gamma(I)$ values (KeV)								
	I_0=14[b]			I_0=15[c]			I_0=16[d]		
	I	$E_\gamma(I)$	δ	I	$E_\gamma(I)$	δ	I	$E_\gamma(I)$	δ
2030.0	48	1908.3	121.7	49	1920.0	110.0	50	1730.1	299.9
1930.0	46	1851.9	78.1	47	1859.7	70.3	48	1690.2	239.8
1836.0	44	1793.2	42.8	45	1797.6	38.4	46	1648.1	187.9
1739.0	42	1732.2	6.8	43	1733.6	5.4	44	1603.7	135.3
1650.0	40	1669.0	-19.0	41	1667.7	-17.7	42	1556.8	93.2
1565.0	38	1603.4	-38.4	39	1599.9	-35.0	40	1507.5	57.5
1484.0	36	1535.5	-51.5	37	1530.4	-46.4	38	1455.6	28.4
1406.0	34	1465.2	-59.2	35	1458.9	-52.9	36	1401.0	5.0
1333.0	32	1392.6	-59.6	33	1385.7	-52.7	34	1343.7	-10.7
1262.0	30	1317.7	-55.7	31	1310.6	-48.6	32	1283.5	-21.5
1193.0	28	1240.4	-47.4	29	1233.9	-40.9	30	1220.5	-27.5
1126.0	26	1160.9	-34.9	27	1155.4	-29.4	28	1154.6	-28.6
1059.0	24	1079.3	-20.3	25	1075.3	-16.3	26	1085.8	-26.8
993.0	22	995.5	-2.5	23	993.7	-0.65	24	1014.1	-21.1
928.0	20	909.8	18.2	21	910.5	17.5	22	939.5	-11.5
864.0	18	822.1	41.9	19	826.0	38.0	20	862.2	1.8
808.0	16	732.8	75.3	17	740.2	67.8	18	782.2	25.8
σ		4.04×10^{-2}			3.61×10^{-2}			6.06×10^{-2}	

[a]Reference (Singh et al., 2002): [b]$a = 1.21\times10^5$ KeV, $b = 2.00\times10^{-4}$; [c]$a = 1.61\times10^5$ KeV, $b = 1.42\times10^{-4}$; [d]$a = 7.19\times10^4$ KeV, $b = 3.26\times10^{-4}$.

Table 4. Spin determination for the SD band $^{134}Nd(b1)$. I_0 is the spin value prescribed to the lowest level observed. $E_\gamma(I)= E_\gamma(I\rightarrow I-2)$ is the transition energy from level I to I-2. $\delta = E_\gamma^{exp.}(I)- E_\gamma^{cal.}(I)$. σ is the rms deviation defined by Equation 4.

Observed[a] $E_\gamma(I)$ values (KeV)	Calculated $E_\gamma(I)$ values (KeV)								
	I_0=12[b]			I_0=13[c]			I_0=14[d]		
	I	$E_\gamma(I)$	δ	I	$E_\gamma(I)$	δ	I	$E_\gamma(I)$	δ
1450.0	36	1393.3	56.7	37	1402.2	47.8	38	1370.2	79.8
1367.3	34	1341.8	25.5	35	1347.0	20.3	36	1318.4	48.9
1289.9	32	1287.1	2.8	33	1289.1	0.76	34	1264.0	25.9
1216.0	30	1229.0	-13.0	31	1228.5	-12.5	32	1207.0	9.0
1143.8	28	1167.5	-23.7	29	1165.1	-21.3	30	1147.4	-3.6
1074.8	26	1102.5	-27.7	27	1098.8	-24.0	28	1085.1	-10.3
1007.4	24	1033.9	-26.5	25	1029.7	-22.3	26	1020.1	-12.7
942.2	22	961.7	-19.5	23	957.8	-15.6	24	952.4	-10.2
876.5	20	886.0	-9.4	21	883.2	-6.7	22	882.1	-5.6
807.8	18	806.7	1.1	19	806.0	1.8	20	809.3	-1.5
736.7	16	724.1	12.6	17	726.3	10.4	18	734.0	2.7
663.9	14	638.2	25.7	15	644.1	19.8	16	656.4	7.5
σ		2.26×10^{-2}			1.86×10^{-3}			2.11×10^{-2}	

[a]Reference (Singh et al., 2002): [b]$a = 5.43\times10^4$ KeV, $b = 4.53\times10^{-4}$; [c]$a = 7.15\times10^4$ KeV, $b = 3.21\times10^{-4}$; [d]$a = 6.93\times10^4$ KeV, $b = 3.17\times10^{-4}$

listed in Tables 10 to 16. The results for the transition energies in fifteen superdeformed bands observed in A~ 100-140 mass region are given in Tables 17 to 24, where the experimental data for the transition energies (labeled Exp[a]) are taken from Ref. (Singh et al., 2002) and the calculated transition energies (labeled Cal[b]) are done at

Table 6. Spin determination for the SD band ^{142}Sm(b1). I_0 is the spin value prescribed to the lowest level observed. $E_\gamma(I)= E_\gamma(I\rightarrow I\text{-}2)$ is the transition energy from level I to I-2. $\delta = E_\gamma^{exp.}(I)- E_\gamma^{cal.}(I)$. σ is the rms deviation defined by Equation 4.

Observed[a] E_γ(I) values (KeV)	Calculated E_γ(I) values (KeV)								
	I_0=14[b]			I_0=15[c]			I_0=16[d]		
	I	E_γ(I)	δ	I	E_γ(I)	δ	I	E_γ(I)	δ
1733.0	50	1665.5	67.5	51	1675.4	57.6	52	1539.8	193.2
1668.0	48	1621.7	46.3	49	1628.7	39.3	50	1506.1	161.9
1603.0	46	1576.0	27.0	47	1580.3	22.8	48	1470.6	132.4
1538.0	44	1528.2	9.8	45	1530.1	7.9	46	1433.2	104.8
1475.0	42	1478.4	-3.4	43	1478.1	-3.1	44	1393.8	81.2
1411.0	40	1426.5	-15.5	41	1424.3	-13.3	42	1352.4	58.6
1348.0	38	1372.5	-24.5	39	1368.8	-20.8	40	1308.8	39.2
1286.0	36	1316.2	-30.2	37	1311.4	-25.4	38	1263.0	23.0
1224.0	34	1257.8	-33.8	35	1252.2	-28.2	36	1215.0	9.0
1163.0	32	1197.2	-34.2	33	1191.3	-28.3	34	1164.6	-1.6
1102.0	30	1134.3	-32.3	31	1128.5	-26.5	32	1111.8	-9.8
1041.0	28	1069.3	-28.3	29	1064.0	-23.0	30	1056.6	-15.6
981.0	26	1002.1	-21.1	27	997.8	-16.8	28	998.9	-18.0
920.0	24	932.8	-12.8	25	929.9	-9.9	26	938.9	-18.9
860.0	22	861.5	-1.5	23	860.4	-0.40	24	876.4	-16.4
800.0	20	788.1	11.9	21	789.4	10.6	22	811.5	-11.5
739.0	18	713.0	26.0	19	716.9	22.1	20	744.3	-5.3
680.0	16	636.1	43.9	17	643.2	36.8	18	675.0	5.0
σ		2.70×10^{-2}			2.26×10^{-2}			4.77×10^{-2}	

[a]Reference (Singh et al., 2002): [b]a = 9.05×10^4 KeV, b = 2.33×10^{-4}; [c]a = 1.14×10^5 KeV, b = 1.75×10^{-4}; [d]a = 6.51×10^4 KeV, b = 3.10×10^{-4}.

Table 7. Spin determination for the SD band ^{148}Eu(b1). I_0 is the spin value prescribed to the lowest level observed. $E_\gamma(I) = E_\gamma(I\rightarrow I\text{-}2)$ is the transition energy from level I to I-2. $\delta = E_\gamma^{exp.}(I)- E_\gamma^{cal.}(I)$. σ is the rms deviation defined by Equation 4.

Observed[a] E_γ(I) values (KeV)	Calculated E_γ(I) values (KeV)								
	I_0=19[b]			I_0=20[c]			I_0=21[d]		
	I	E_γ(I)	δ	I	E_γ(I)	δ	I	E_γ(I)	δ
1555.0	51	1520.09	34.91	52	1526.89	28.11	53	1396.12	158.88
1499.0	49	1478.05	20.95	50	1482.96	16.04	51	1366.33	132.67
1443.0	47	1434.44	8.56	48	1437.65	5.35	49	1334.95	108.05
1388.0	45	1389.21	-1.21	46	1390.95	-2.95	47	1301.93	86.08
1331.0	43	1342.36	-11.36	44	1342.84	-11.84	45	1267.17	63.83
1276.0	41	1293.86	-17.86	42	1293.33	-17.33	43	1230.62	45.38
1220.0	39	1243.69	-23.69	40	1242.40	-22.40	41	1192.20	27.80
1165.0	37	1191.85	-26.85	38	1190.06	-25.06	39	1151.84	13.16
1111.0	35	1138.34	-27.34	36	1136.34	-25.34	37	1109.50	1.50
1057.0	33	1083.16	-26.16	34	1081.23	-24.23	35	1065.12	-8.12
1004.0	31	1026.33	-22.33	32	1024.77	-20.77	33	1018.64	-14.64
951.0	29	967.87	-16.87	30	966.99	-15.99	31	970.05	-19.05
900.0	27	907.82	-7.82	28	907.91	-7.91	29	919.31	-19.31
848.0	25	846.23	1.77	26	847.58	0.42	27	866.43	-18.43
798.0	23	783.15	14.85	24	786.06	11.94	25	811.41	-13.41
748.0	21	718.65	29.347	22	723.41	24.59	23	754.29	-6.29
σ		1.92×10^{-2}			1.71×10^{-2}			4.60×10^{-2}	

[a]Reference (Singh et al., 2002): [b]a = 1.01×10^5 KeV, b = 1.80×10^{-4}; [c]a = 1.24×10^5 KeV, b = 1.40×10^{-4}; [d]a = 5.96×10^4, KeV, b = 3.02×10^{-4}.

Table 8. Spin determination for the SD band ^{148}Eu(b2). I_0 is the spin value prescribed to the lowest level observed. $E_\gamma(I) = E_\gamma(I \rightarrow I-2)$ is the transition energy from level I to I-2. $\Delta = E_\gamma^{exp.}(I) - E_\gamma^{cal.}(I)$. σ is the rms deviation defined by Equation 4.

Observed[a] $E_\gamma(I)$ values (KeV)	Calculated $E_\gamma(I)$ values (KeV)								
	I_0=22[b]			I_0=23[c]			I_0=24[d]		
	I	$E_\gamma(I)$	δ	I	$E_\gamma(I)$	δ	I	$E_\gamma(I)$	δ
1544.0	50	1514.4	29.6	51	1522.4	21.6	52	1406.3	137.7
1489.0	48	1470.8	18.3	49	1477.1	11.9	50	1376.1	112.9
1434.0	46	1425.5	8.5	47	1430.4	3.6	48	1344.2	89.8
1378.0	44	1378.6	-0.60	45	1382.3	-4.3	46	1310.6	67.4
1322.0	42	1330.1	-8.1	43	1332.8	-10.8	44	1275.1	46.9
1269.0	40	1279.9	-11.0	41	1281.9	-12.9	42	1237.8	31.2
1212.0	38	1228.2	-16.1	39	1229.6	-17.6	40	1198.4	13.6
1158.0	36	1174.7	-16.7	37	1175.8	-17.8	38	1157.0	0.96
1104.0	34	1119.6	-15.6	35	1120.7	-16.7	36	1113.5	-9.5
1051.0	32	1062.9	-11.9	33	1064.2	-13.2	34	1067.8	-16.8
998.0	30	1004.5	-6.5	31	1006.4	-8.4	32	1019.9	-21.9
946.0	28	944.6	1.4	29	947.3	-1.3	30	969.7	-23.7
895.0	26	883.1	11.9	27	886.9	8.1	28	917.3	-22.3
844.0	24	820.1	23.9	25	825.2	18.8	26	862.5	-18.5
σ		1.31×10^{-2}			1.17×10^{-2}			4.19×10^{-2}	

[a]Reference (Singh et al., 2002): [b]a = 1.05×10^5 KeV, b = 1.74×10^{-4}; [c]a = 1.27×10^5 KeV, b = 1.38×10^{-4}; [d]a = 5.76×10^4 KeV, b = 3.23×10^{-4}.

Table 9. Spin determination for the SD band ^{131}Ce(b1). I_0 is the spin value prescribed to the lowest level observed. $E_\gamma(I) = E_\gamma(I \rightarrow I-2)$ is the transition energy from level I to I-2. $\delta = E_\gamma^{exp.}(I) - E_\gamma^{cal.}(I)$. σ is the rms deviation defined by Equation 4.

Observed[a] $E_\gamma(I)$ values (KeV)	Calculated $E_\gamma(I)$ values (KeV)								
	I_0=10.5[b]			I_0=11.5[c]			I_0=12.5[d]		
	I	$E_\gamma(I)$	δ	I	$E_\gamma(I)$	δ	I	$E_\gamma(I)$	δ
1822.0	44.5	1727.5	94.5	45.5	1741.0	81.0	46.5	1565.5	256.5
1732.0	42.5	1670.3	61.7	43.5	1678.6	53.4	44.5	1522.7	209.3
1640.0	40.5	1610.8	29.2	41.5	1614.4	25.6	42.5	1477.7	162.3
1550.0	38.5	1549.0	0.99	39.5	1548.4	1.6	40.5	1430.4	119.6
1464.0	36.5	1484.8	-20.8	37.5	1480.6	-16.6	38.5	1380.6	83.4
1381.0	34.5	1418.1	-37.1	35.5	1411.2	-30.2	36.5	1328.3	52.7
1301.0	32.5	1349.1	-48.1	33.5	1340.0	-39.0	34.5	1273.4	27.6
1225.0	30.5	1277.6	-52.6	31.5	1267.1	-42.1	32.5	1215.9	9.1
1151.0	28.5	1203.8	-52.8	29.5	1192.6	-41.6	30.5	1155.8	-4.8
1080.0	26.5	1127.6	-47.6	27.5	1116.5	-36.5	28.5	1092.9	-12.9
1011.0	24.5	1049.2	-38.2	25.5	1038.9	-27.9	26.5	1027.4	-16.4
943.0	22.5	968.5	-25.5	23.5	959.8	-16.8	24.5	959.1	-16.1
874.0	20.5	885.7	-11.7	21.5	879.3	-5.3	22.5	888.3	-14.3
804.0	18.5	800.9	3.1	19.5	797.6	6.4	20.5	814.9	-10.9
733.0	16.5	714.3	18.7	17.5	714.6	18.4	18.5	739.1	-6.1
662.0	14.5	625.9	36.1	15.5	630.6	31.4	16.5	660.9	1.1
591.0	12.5	558.7	32.3	13.5	566.9	24.1	14.5	600.8	-9.8
σ		3.55×10^{-2}			2.87×10^{-2}			5.78×10^{-2}	

[a]Reference (Singh et al., 2002): [b]a = 1.07×10^5 KeV, b = 2.21×10^{-4}; [c]a = 1.61×10^5 KeV, b = 1.37×10^{-4}; [d]a = 7.02×10^4 KeV, b = 3.15×10^{-4}

the two fitting parameters (a and b) given in Table 1. These tables give successively the calculations for even-mass nuclei and even spin (Tables 17 to 20), odd-mass nuclei with odd spin (Tables 21 to 24). The fact that the

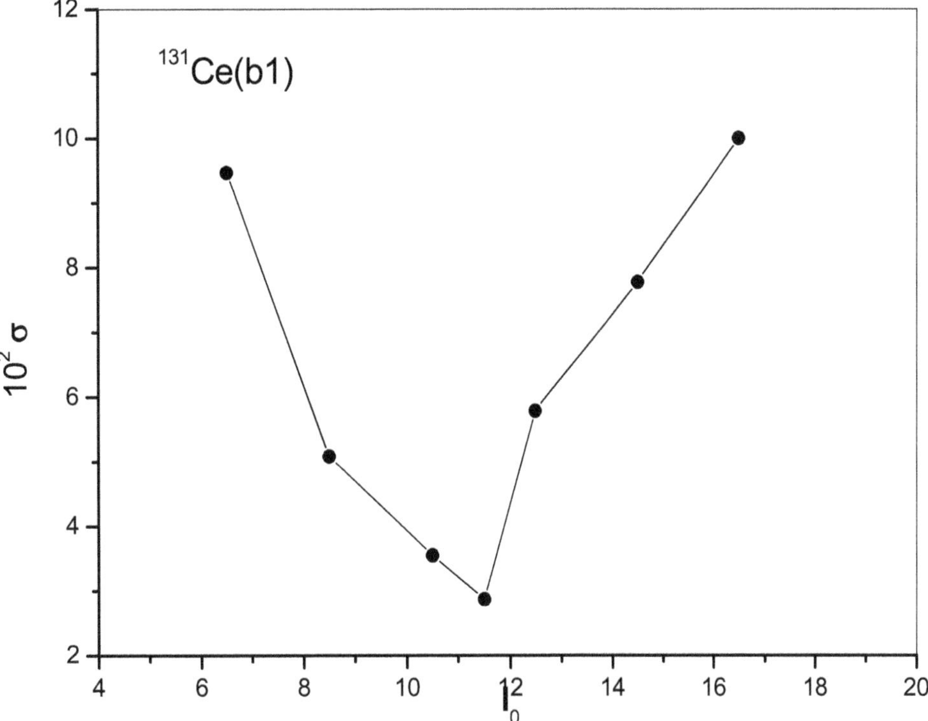

Figure 8. The rms deviation for various spin assignments in ^{131}Ce(b1). I_0 is the value prescribed to the lowest level observed [E $\gamma(I_0 + 2 ==> I_0)$ = 591 KeV].

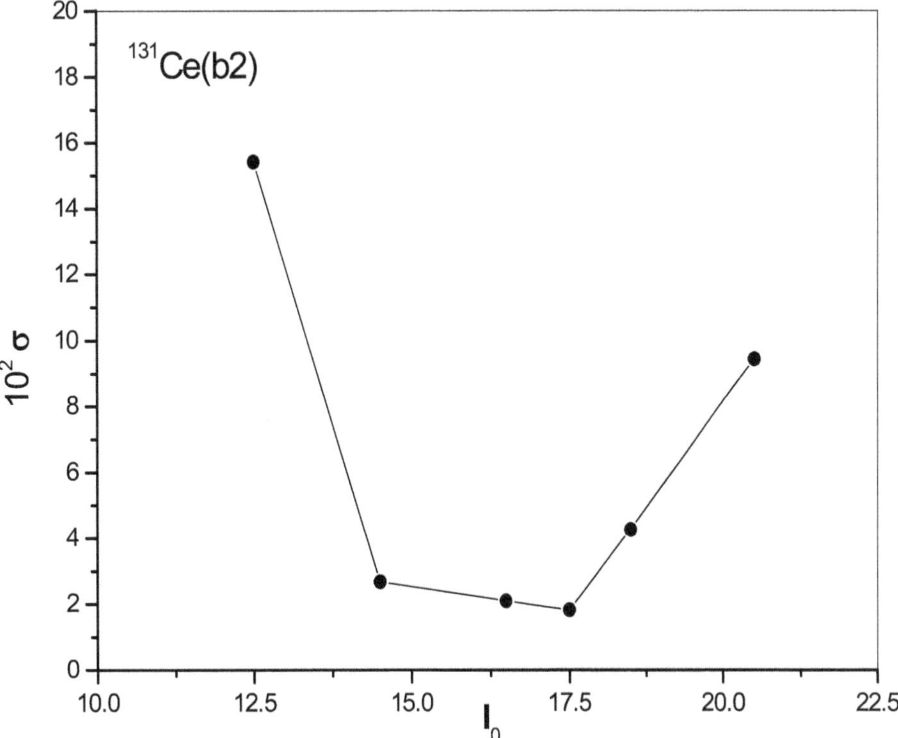

Figure 9. The rms deviation for various spin assignments in ^{131}Ce(b2). I_0 is the value prescribed to the lowest level observed [E $\gamma(I_0 + 2 ==> I_0)$ = 847 KeV].

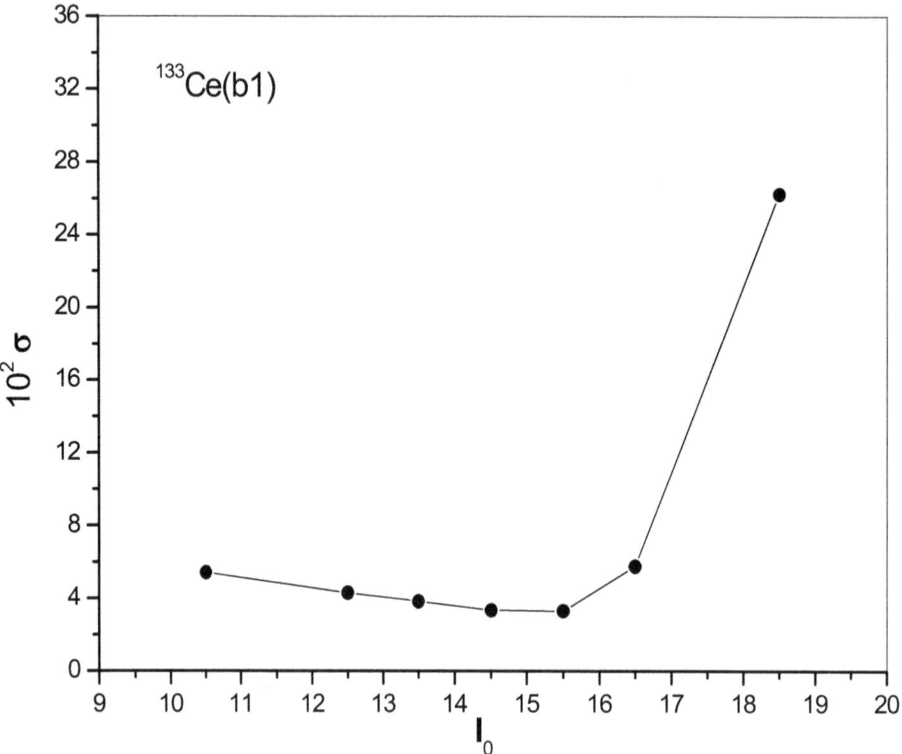

Figure 10. The rms deviation for various spin assignments in ^{133}Ce(b1). I_0 is the value prescribed to the lowest level observed [E $\gamma(I_0 + 2 \quad I_0)$ = 748 KeV].

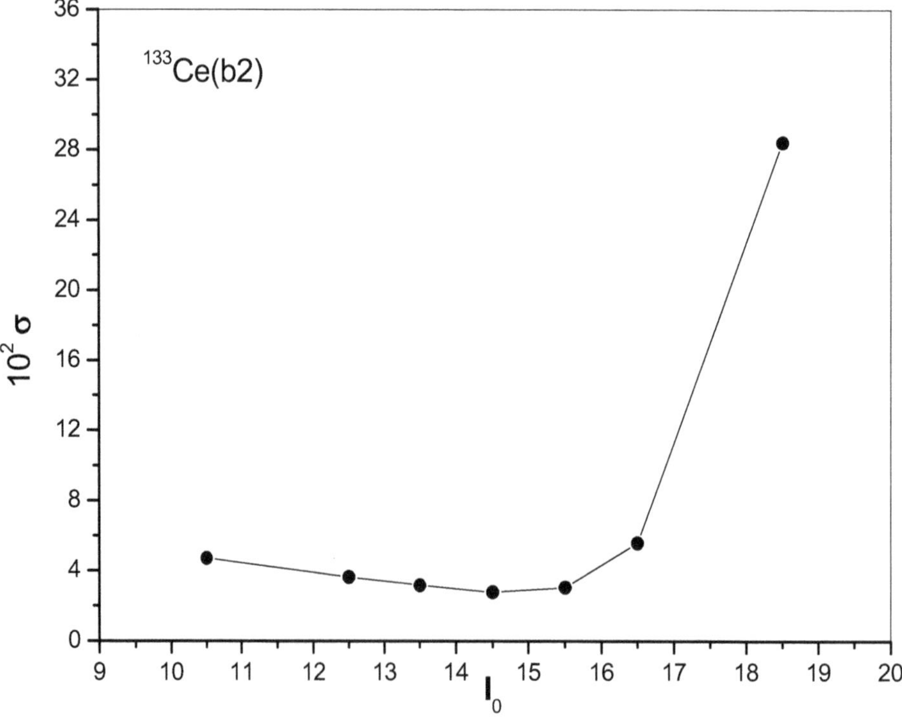

Figure 11. The rms deviation for various spin assignments in ^{133}Ce(b2). I_0 is the value prescribed to the lowest level observed [E $\gamma(I_0 + 2 \quad I_0)$ = 720.3 KeV].

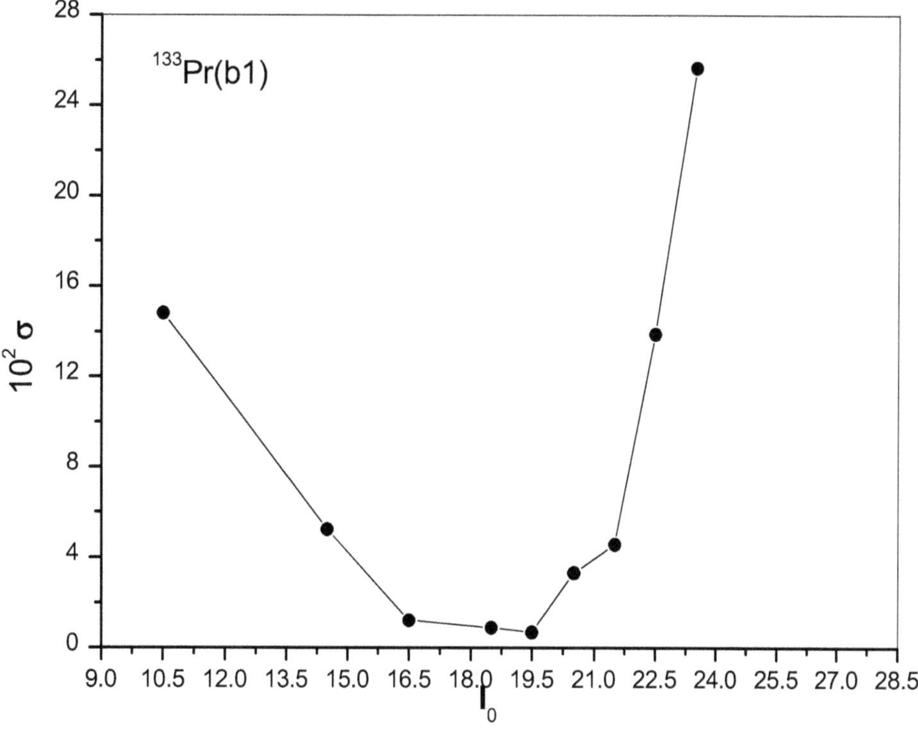

Figure 12. The rms deviation for various spin assignments in ^{133}Pr(b1). I_0 is the value prescribed to the lowest level observed [E $\gamma(I_0 + 2 \;\; I_0)$ = 871.0 KeV].

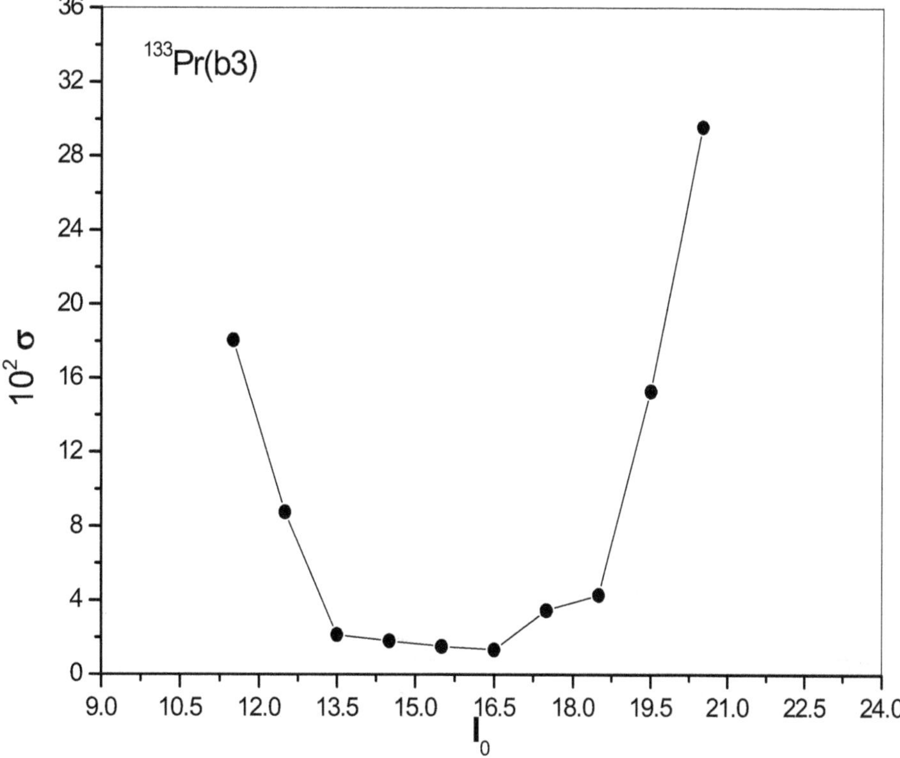

Figure 13. The rms deviation for various spin assignments in ^{133}Pr(b3). I_0 is the value prescribed to the lowest level observed [E $\gamma(I_0 + 2 \;\; I_0)$ = 821.0 KeV].

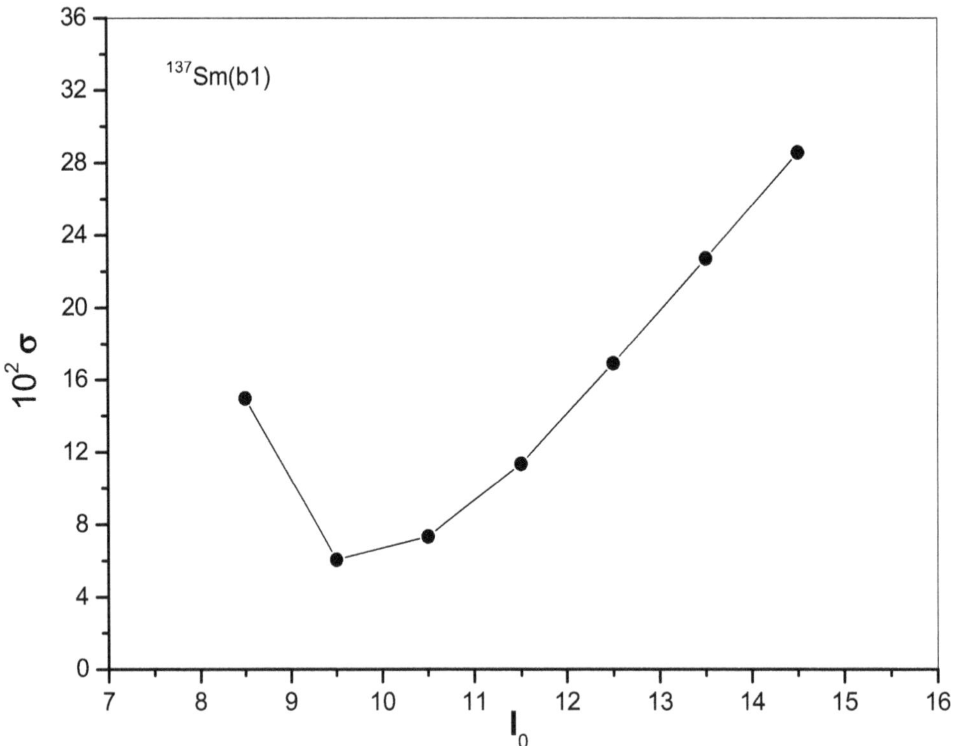

Figure 14. The rms deviation for various spin assignments in ^{137}Sm(b1). I_0 is the value prescribed to the lowest level observed [E $\gamma(I_0 + 2 \quad I_0)$ = 379.0 KeV].

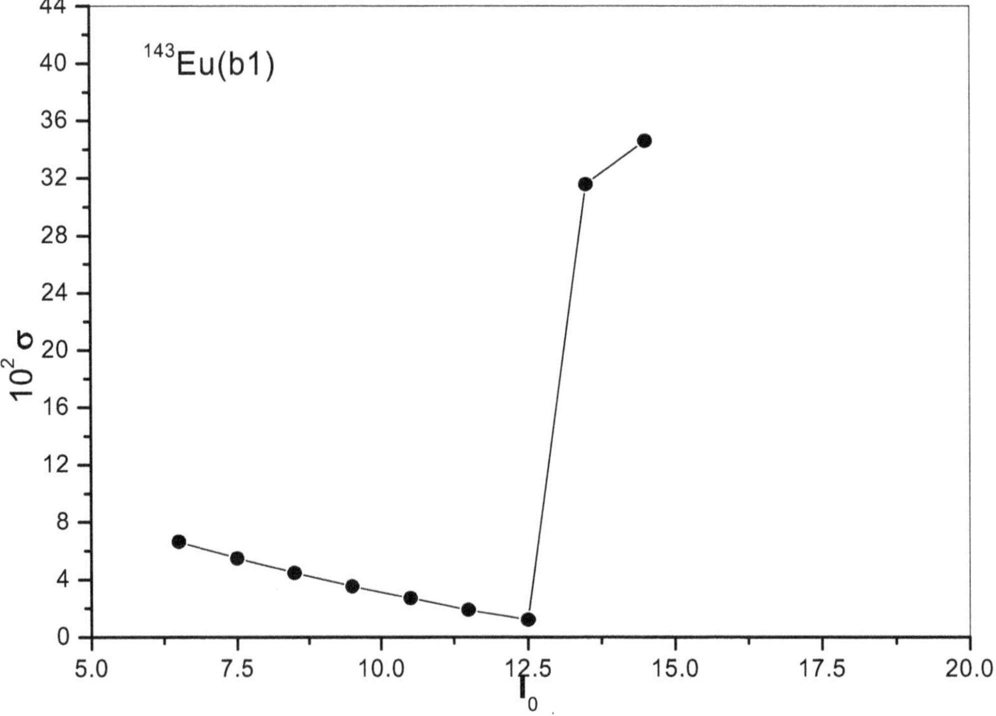

Figure 15. The rms deviation for various spin assignments in ^{143}Eu(b1). I_0 is the value prescribed to the lowest level observed [E $\gamma(I_0 + 2 \quad I_0)$ = 483.0 KeV].

Table 10. Spin determination for the SD band ^{131}Ce(b2). I_0 is the spin value prescribed to the lowest level observed. $E_\gamma(I)= E_\gamma(I\rightarrow I-2)$ is the transition energy from level I to I-2. $\delta= E_\gamma^{exp.}(I)- E_\gamma^{cal.}(I)$. σ is the rms deviation defined by Equation 4.

Observed[a] $E_\gamma(I)$ values (KeV)	Calculated $E_\gamma(I)$ values (KeV)								
	I_0=16.5[b]			I_0=17.5[c]			I_0=18.5[d]		
	I	$E_\gamma(I)$	δ	I	$E_\gamma(I)$	δ	I	$E_\gamma(I)$	δ
1723.0	42.5	1660.7	62.3	43.5	1672.1	50.9	44.5	1678.5	44.5
1635.0	40.5	1602.3	32.7	41.5	1611.9	23.1	42.5	1615.6	19.4
1552.0	38.5	1541.5	10.5	39.5	1549.6	2.4	40.5	1550.9	1.1
1471.0	36.5	1478.4	-7.4	37.5	1485.3	-14.3	38.5	1484.7	-13.7
1396.0	34.5	1413.0	-17.0	35.5	1418.8	-22.8	36.5	1416.7	-20.7
1322.0	32.5	1345.3	-23.3	33.5	1350.2	-28.2	34.5	1347.2	-25.2
1251.0	30.5	1275.2	-24.2	31.5	1279.5	-28.5	32.5	1276.1	-25.1
1181.0	28.5	1202.8	-21.8	29.5	1206.8	-25.8	30.5	1203.4	-22.4
1112.0	26.5	1128.2	-16.2	27.5	1132.1	-20.1	28.5	1129.3	-17.3
1043.0	24.5	1051.3	-9.3	25.5	1055.4	-12.4	26.5	1053.7	-10.7
976.0	22.5	972.3	3.7	23.5	976.8	-0.84	24.5	976.7	-0.72
908.0	20.5	891.2	16.8	21.5	896.5	11.5	22.5	898.4	9.6
847.0	18.5	808.2	38.8	19.5	814.4	32.6	20.5	818.9	28.1
σ		2.09×10^{-2}			1.82×10^{-2}			4.26×10^{-2}	

[a]Reference (Singh et al., 2002): [b]$a = 1.08\times10^5$ KeV, $b = 2.15\times10^{-4}$; [c]$a = 1.43\times10^5$ KeV, $b = 1.54\times10^{-4}$; [d]$a = 1.46\times10^5$ KeV, $b = 1.52\times10^{-4}$.

Table 11. Spin determination for the SD band ^{133}Ce(b1). I_0 is the spin value prescribed to the lowest level observed. $E_\gamma(I) = E_\gamma(I\rightarrow I-2)$ is the transition energy from level I to I-2. $\Delta = E_\gamma^{exp.}(I)- E_\gamma^{cal.}(I)$. σ is the rms deviation defined by Equation 4.

Observed[a] $E_\gamma(I)$ values (KeV)	Calculated $E_\gamma(I)$ values (KeV)								
	I_0=14.5[b]			I_0=15.5[c]			I_0=16.5[d]		
	I	$E_\gamma(I)$	δ	I	$E_\gamma(I)$	δ	I	$E_\gamma(I)$	δ
1928.0	48.5	1828.1	99.9	49.5	1815.0	113.0	50.5	1707.0	221.1
1833.0	46.5	1770.4	62.6	47.5	1759.6	73.4	48.5	1653.7	179.2
1743.0	44.5	1710.8	32.2	45.5	1702.5	40.5	46.5	1599.1	144.0
1655.0	42.5	1649.5	5.5	43.5	1643.6	11.4	44.5	1543.1	112.0
1570.0	40.5	1586.3	-16.3	41.5	1582.8	-12.8	42.5	1485.5	84.5
1489.0	38.5	1521.2	-32.2	39.5	1520.2	-31.2	40.5	1426.5	62.5
1411.0	36.5	1454.3	-43.3	37.5	1455.8	-44.8	38.5	1366.0	45.0
1337.0	34.5	1385.6	-48.6	35.5	1389.7	-52.7	36.5	1304.1	33.0
1267.0	32.5	1315.1	-48.1	33.5	1321.7	-54.7	34.5	1240.7	26.3
1198.0	30.5	1242.9	-44.9	31.5	1251.9	-53.9	32.5	1176.0	22.0
1132.0	28.5	1168.9	-36.9	29.5	1180.5	-48.5	30.5	1110.0	22.1
1068.0	26.5	1093.3	-25.3	27.5	1107.4	-39.4	28.5	1042.6	25.4
1003.0	24.5	1016.0	-13.0	25.5	1032.6	-29.6	26.5	974.0	29.1
937.0	22.5	937.3	-0.26	23.5	956.3	-19.3	24.5	904.1	33.0
873.0	20.5	857.0	16.0	21.5	878.6	-5.6	22.5	833.1	40.0
809.0	18.5	775.5	33.5	19.5	799.4	9.6	20.5	761.0	48.0
748.0	16.5	692.6	55.4	17.5	718.9	29.1	18.5	687.9	60.1
σ		3.36×10^{-2}			3.32×10^{-2}			5.75×10^{-2}	

[a]Reference (Singh et al., 2002), [b]$a = 1.49\times10^5$ KeV, $b = 1.48\times10^{-4}$; [c]$a = 1.44\times10^5$ KeV, $b = 1.50\times10^{-4}$; [d]$a = 1.61\times10^5$ KeV, $b = 1.21\times10^{-4}$.

calculated transition energies coincide with the observed results extremely well implies that the rotational spectrum in each SD band can be precisely described by Equation 1.

Conclusion

Fifteen Superdeformed bands observed in the nuclei of

Table 12. Spin determination for the SD band ^{133}Ce(b2). I_0 is the spin value prescribed to the lowest level observed. $E_\gamma(I) = E_\gamma(I \rightarrow I-2)$ is the transition energy from level I to I-2. $\delta = E_\gamma^{exp.}(I) - E_\gamma^{cal.}(I)$. σ is the rms deviation defined by Equation 4.

Observed[a] $E_\gamma(I)$ values (KeV)	Calculated $E_\gamma(I)$ values (KeV)								
	I_0=13.5[b]			I_0=14.5[c]			I_0=15.5[d]		
	I	$E_\gamma(I)$	δ	I	$E_\gamma(I)$	δ	I	$E_\gamma(I)$	δ
1731.3	45.5	1732.5	89.1	46.5	1749.3	72.3	47.5	1740.8	80.8
1731.3	43.5	1677.5	53.8	44.5	1690.3	41.0	45.5	1684.0	47.3
1643.0	41.5	1620.3	22.7	42.5	1629.5	13.5	43.5	1625.4	17.6
1557.3	39.5	1560.8	-3.5	40.5	1566.8	-9.5	41.5	1565.1	-7.8
1475.5	37.5	1499.1	-23.6	38.5	1502.4	-27.0	39.5	1503.0	-27.5
1397.9	35.5	1435.0	-37.1	36.5	1436.2	-38.3	37.5	1439.1	-41.2
1323.9	33.5	1368.7	-44.8	34.5	1368.3	-44.3	35.5	1373.4	-49.5
1253.2	31.5	1300.1	-47.0	32.5	1298.5	-45.3	33.5	1306.1	-52.9
1184.3	29.5	1229.3	-45.0	30.5	1227.1	-42.8	31.5	1237.0	-52.7
1118.2	27.5	1156.1	-38.0	28.5	1153.9	-35.7	29.5	1166.2	-48.0
1052.2	25.5	1080.8	-28.6	26.5	1079.2	-27.0	27.5	1093.8	-41.6
986.9	23.5	1003.4	-16.5	24.5	1002.8	-15.9	25.5	1019.9	-33.0
920.3	21.5	923.9	-3.6	22.5	925.0	-4.7	23.5	944.4	-24.1
854.2	19.5	842.4	11.8	20.5	845.8	8.4	21.5	867.5	-13.3
785.4	17.5	759.1	26.3	18.5	765.2	20.2	19.5	789.3	-3.9
720.3	15.5	674.1	46.2	16.5	683.4	36.9	17.5	709.7	10.6
σ		3.17×10^{-2}			2.78×10^{-2}			3.05×10^{-2}	

[a]Reference (Singh et al., 2002): [b]a = 1.09×10^5 KeV, b = 2.11×10^{-4}; [c]a = 1.49×10^5 KeV, b= 1.46×10^{-4}; [d]a = 1.45×10^5 KeV, b = 1.47×10^{-4}.

Table 13. Spin determination for the SD band ^{133}Pr(b1). I_0 is the spin value prescribed to the lowest level observed. $E_\gamma(I) = E_\gamma(I \rightarrow I-2)$ is the transition energy from level I to I-2. $\Delta = E_\gamma^{exp.}(I) - E_\gamma^{cal.}(I)$. σ is the rms deviation defined by Equation 4.

Observed[a] $E_\gamma(I)$ values (KeV)	Calculated $E_\gamma(I)$ values (KeV)								
	I_0=18.5[b]			I_0=19.5[c]			I_0=20.5[d]		
	I	$E_\gamma(I)$	δ	I	$E_\gamma(I)$	δ	I	$E_\gamma(I)$	δ
1530.0	38.5	1513.2	16.8	39.5	1513.2	16.8	40.5	1540.9	-10.9
1450.0	36.5	1448.5	1.5	37.5	1446.5	3.5	38.5	1476.2	-26.2
1380.0	34.5	1381.8	-1.8	35.5	1378.2	1.8	36.5	1409.9	-29.9
1299.0	32.5	1313.2	-14.2	33.5	1308.6	-9.6	34.5	1342.0	-43.0
1228.0	30.5	1242.5	-14.5	31.5	1237.4	-9.4	32.5	1272.5	-44.5
1156.0	28.5	1169.9	-14.0	29.5	1164.9	-8.9	30.5	1201.5	-45.5
1085.0	26.5	1095.4	-10.4	27.5	1091.0	-6.0	28.5	1129.0	-44.0
1013.0	24.5	1019.1	-6.1	25.5	1015.9	-2.9	26.5	1055.0	-42.0
943.0	22.5	941.1	1.9	23.5	939.5	3.5	24.5	979.7	-36.7
871.0	20.5	861.4	9.6	21.5	862.0	9.0	22.5	903.1	-32.1
σ		8.85×10^{-3}			6.79×10^{-3}			3.32×10^{-2}	

[a]Reference (Singh et al., 2002): [b]a = 1.28×10^5 KeV, b = 1.74×10^{-4}; [c]a = 1.81×10^5 KeV, b = 1.16×10^{-4}; [d]a = 1.64×10^5 KeV, b = 1.29×10^{-4}.

the mass region A~ 100-140 have been analyzed in terms of the expression for the γ-transition energies as a function of spin (ab formula). The use of direct expression of energy as a function of spin avoids necessary assumptions done in (Shalaby et al., 2012) about the values of quantities not directly measured in experiment like rotational frequency and dynamic moment of inertia. It is quite unexpected that so large a number of SD bands can be reproduced incredibly well by the simple two-parameter closed expression. It was found that the argument between the calculated and observed transition energies depends sensitively on the prescribed level

Table 14. Spin determination for the SD band ^{133}Pr(b2). I_0 is the spin value prescribed to the lowest level observed. $E_\gamma(I) = E_\gamma(I \to I-2)$ is the transition energy from level I to I-2. $\Delta = E_\gamma^{exp.}(I) - E_\gamma^{cal.}(I)$. σ is the rms deviation defined by Equation 4.

Observed[a] $E_\gamma(I)$ values (KeV)	Calculated $E_\gamma(I)$ values (KeV)								
	I_0=15.5[b]			I_0=16.5[c]			I_0=17.5[d]		
	I	$E_\gamma(I)$	δ	I	$E_\gamma(I)$	δ	I	$E_\gamma(I)$	δ
1656.0	39.5	1625.4	30.6	40.5	1637.0	19.0	41.5	1663.4	-7.4
1576.0	37.5	1561.8	14.2	38.5	1570.1	6.0	39.5	1597.1	-21.1
1500.0	35.5	1495.8	4.2	36.5	1501.3	-1.3	37.5	1529.0	-29.0
1420.0	33.5	1427.3	-7.3	34.5	1430.6	-10.6	35.5	1459.0	-39.0
1342.0	31.5	1356.3	-14.3	32.5	1358.0	-16.0	33.5	1387.3	-45.3
1264.0	29.5	1282.9	-18.9	30.5	1283.6	-19.6	31.5	1313.7	-49.7
1184.0	27.5	1207.1	-23.1	28.5	1207.3	-23.3	29.5	1238.4	-54.4
1110.0	25.5	1128.9	-19.0	26.5	1129.4	-19.4	27.5	1161.3	-51.3
1035.0	23.5	1048.4	-13.4	24.5	1049.7	-14.7	25.5	1082.7	-47.7
961.0	21.5	965.7	-4.7	22.5	968.4	-7.4	23.5	1002.4	-41.4
885.0	19.5	880.8	4.1	20.5	885.6	-0.64	21.5	920.7	-35.7
821.0	17.5	794.0	27.0	18.5	801.4	19.6	19.5	837.6	-16.6
σ		1.52×10^{-2}			1.33×10^{-2}			3.45×10^{-2}	

[a]Reference (Singh et al., 2002): [b]a = 1.09×10^5 KeV, b = 2.21×10^{-4}; [c]a = 1.51×10^5 KeV, b = 1.51×10^{-4}; [d]a = 1.57×10^5 KeV, b = 1.44×10^{-4}.

Table 15. Spin determination for the SD band ^{137}Sm(b1). I_0 is the spin value prescribed to the lowest level observed. $E_\gamma(I) = E_\gamma(I \to I-2)$ is the transition energy from level I to I-2. $\delta = E_\gamma^{exp.}(I) - E_\gamma^{cal.}(I)$. σ is the rms deviation defined by Equation (4).

Observed[a] $E_\gamma(I)$ values (KeV)	Calculated $E_\gamma(I)$ values (KeV)								
	I_0=8.5[b]			I_0=9.5[c]			I_0=10.5[d]		
	I	$E_\gamma(I)$	δ	I	$E_\gamma(I)$	δ	I	$E_\gamma(I)$	δ
1115.0	28.5	936.2	178.8	29.5	1084.0	31.0	30.5	1012.4	102.6
1045.0	26.5	870.4	174.6	27.5	1014.4	30.6	28.5	949.1	95.9
970.0	24.5	804.4	165.6	25.5	943.7	26.3	26.5	885.0	85.0
890.0	22.5	738.2	151.8	23.5	872.1	18.0	24.5	820.1	69.9
806.3	20.5	671.7	134.6	21.5	799.5	6.8	22.5	754.5	51.8
719.7	18.5	605.1	114.6	19.5	726.1	-6.4	20.5	688.2	31.5
629.1	16.5	538.3	90.8	17.5	651.9	-22.8	18.5	621.2	7.9
538.5	14.5	471.4	67.1	15.5	576.9	-38.4	16.5	553.7	-15.2
451.0	12.5	404.2	46.8	13.5	501.4	-50.4	14.5	485.7	-34.7
379.0	10.5	337.1	42.0	11.5	425.2	-46.2	12.5	417.2	-38.2
σ		1.50×10^{-1}			6.06×10^{-2}			7.33×10^{-2}	

[a]Reference (Singh et al., 2002): [b]a = 7.06×10^5 KeV, b = 2.39×10^{-5}; [c]a = 1.99×10^5 KeV, b = 9.77×10^{-5}; [d]a = 2.14×10^5 KeV, b = 8.17×10^{-5}.

Table 16. Spin determination for the SD band ^{143}Eu(b1). I_0 is the spin value prescribed to the lowest level observed. $E_\gamma(I) = E_\gamma(I \to I-2)$ is the transition energy from level I to I-2. $\delta = E_\gamma^{exp.}(I) - E_\gamma^{cal.}(I)$. σ is the rms deviation defined by Equation 4.

Observed[a] $E_\gamma(I)$ values (KeV)	Calculated $E_\gamma(I)$ values (KeV)								
	I_0=11.5[b]			I_0=12.5[c]			I_0=13.5[d]		
	I	$E_\gamma(I)$	δ	I	$E_\gamma(I)$	δ	I	$E_\gamma(I)$	δ
1743.0	55.5	1690.6	52.4	56.5	1707.9	35.1		2078.7	-335.7
1684.0	53.5	1644.5	39.5	54.5	1657.8	26.2		2030.9	-346.9
1623.5	51.5	1597.0	26.5	52.5	1606.7	16.8		1981.4	-357.8
1562.3	49.5	1548.2	14.1	50.5	1554.6	7.7		1929.9	-367.6
1502.2	47.5	1498.2	4.0	48.5	1501.6	0.06	49.5	1876.6	-374.4

Table 16. contd.

1442.7	45.5	1446.8	-4.1	46.5	1447.5	-4.8	47.5	1821.3	-378.6
1383.2	43.5	1394.0	-10.8	44.5	1392.6	-9.4	45.5	1764.0	-380.8
1324.1	41.5	1340.0	-15.9	42.5	1336.7	-12.6	43.5	1704.8	-380.7
1265.1	39.5	1284.6	-19.5	40.5	1279.8	-14.7	41.5	1643.5	-378.4
1206.5	37.5	1228.0	-21.5	38.5	1222.1	-15.6	39.5	1580.3	-373.8
1148.1	35.5	1170.1	-22.0	36.5	1163.5	-15.4	37.5	1514.9	-366.8
1089.9	33.5	1110.9	-21.0	34.5	1104.0	-14.1	35.5	1447.5	-357.6
1031.1	31.5	1050.5	-19.4	32.5	1043.8	-12.7	33.5	1378.1	-347.0
972.2	29.5	989.0	-16.8	30.5	982.7	-10.5	31.5	1306.7	-334.5
912.8	27.5	926.2	-13.4	28.5	920.9	-8.0	29.5	1233.3	-320.5
853.5	25.5	862.5	-9.0	26.5	858.3	-4.8	27.5	1158.0	-304.5
793.4	23.5	797.6	-4.2	24.5	795.0	-1.6	25.5	1080.8	-287.4
732.6	21.5	731.8	0.82	22.5	731.1	1.5	23.5	1001.8	-269.2
671.2	19.5	665.0	6.2	20.5	666.7	4.5	21.5	921.1	-249.9
609.0	17.5	597.4	11.6	18.5	601.6	7.4	19.5	838.7	-229.7
546.5	15.5	529.0	17.5	16.5	536.1	10.4	17.5	754.8	-208.3
483.0	13.5	460.0	23.0	14.5	470.1	12.9	15.5	669.5	-186.5
σ		1.91×10^{-2}			1.21×10^{-2}			3.16×10^{-1}	

[a]Reference (Singh et al., 2002): [b]$a = 1.54 \times 10^{5}$ KeV, $b = 1.16 \times 10^{-4}$; [c]$a = 2.31 \times 10^{5}$ KeV, $b = 7.32 \times 10^{-5}$; [d]$a = 1.33 \times 10^{5}$ KeV, $b = 1.71 \times 10^{-4}$.

Table 17. Calculation of the transition energies in Nd^{134}(b1) and Eu^{148}(b2).

Nd^{134}(b1)			Eu^{148}(b2)		
$E_γ(I)$ (KeV) Exp[a]	Cal[b]	Assigned I	$E_γ(I)$ (KeV) Exp[a]	Cal[b]	Assigned I
663.9	644.1	15	844.0	825.2	25
736.7	726.3	17	895.0	886.9	27
807.8	806.0	19	946.0	947.3	29
876.5	883.2	21	998.0	1006.4	31
942.2	957.8	23	1051.0	1064.2	33
1007.4	1029.7	25	1104.0	1120.7	35
1074.4	1098.8	27	1158.0	1175.8	37
1143.8	1165.1	29	1212.0	1229.6	39
1216.0	1228.5	31	1269.0	1281.9	41
1289.9	1289.1	33	1322.0	1332.8	43
1367.3	1347.0	35	1378.0	1382.3	45
1450.0	1402.2	37	1434.0	1430.4	47
			1489.0	1477.1	49
			1544.0	1522.4	51
	$σ = 1.86 \times 10^{-3}$			$σ = 1.17 \times 10^{-2}$	

spins. Therefore, the spins of each band levels can be determined.

For all fifteen SD bands, the spin values and also the corresponding parameters a and b are tabulated in Table 1. Of course, the spin assignments made in this paper remain to be confirmed from the future direct measurements.

It is seen from Table 1 that the values of the parameters a and b for all these bands are close to each other:

$a = (3.11 \pm 2.39) \times 10^{5}$ KeV,
$b = (1 : 3) \times 10^{-4}$

except for ^{136}Nd(b1) and ^{143}Eu(b1), where "b" is smaller than for the other SD bands. It is noted from these 15

Table 18. Calculation of the transition energies in Pd^{104}(b1) and Nd^{136}(b2).

Pd^{104}(b1)			Nd^{136}(b2)		
E_γ(I) (KeV) Exp[a]	Cal[b]	Assigned I	E_γ(I) (KeV) Exp[a]	Cal[b]	Assigned I
1263.0	1246.4	17	888.0	870.8	24
1381.0	1388.3	19	951.0	943.3	26
1511.0	1527.2	21	1017.0	1015.5	28
1638.0	1663.0	23	1081.0	1087.3	30
1763.0	1795.5	25	1151.0	1158.7	32
1919.0	1924.6	27	1221.0	1229.6	34
2079.0	2050.1	29	1290.0	1300.1	36
			1364.0	1370.1	38
			1438.0	1439.6	40
	$\sigma = 1.19 \times 10^{-2}$			$\sigma = 5.36 \times 10^{-3}$	

Table 19. Calculation of the transition energies in Ce^{132}(b1) and Sm^{142}(b1).

Ce^{132}(b1)			Sm^{142}(b1)		
E_γ(I) (KeV) Exp[a]	Cal[b]	Assigned I	E_γ(I) (KeV) Exp[a]	Cal[b]	Assigned I
808.0	740.2	17	680.0	643.2	17
864.0	826.0	19	739.0	716.9	19
928.0	910.5	21	800.0	789.4	21
993.0	993.7	23	860.0	860.4	23
1059.0	1075.3	25	920.0	929.9	25
1126.0	1155.4	27	981.0	997.8	27
1193.0	1233.9	29	1041.0	1074.0	29
1262.0	1310.3	31	1102.0	1128.5	31
1333.0	1385.7	33	1163.0	1191.3	34
1406.0	1458.9	35	1224.0	1252.2	35
1484.0	1530.4	37	1286.0	1311.4	37
1565.0	1599.9	39	1348.0	1368.8	39
1650.0	1667.6	41	1411.0	1424.3	41
1739.0	1733.6	43	1475.0	1478.1	43
1836.0	1797.6	45	1538.0	1530.1	45
1930.0	1859.7	47	1603.0	1580.3	47
2030.0	1920.0	49	1668.0	1628.7	49
			1733.0	1675.4	51
	$\sigma = 3.61 \times 10^{-2}$		$\sigma = 2.26 \times 10^{-2}$		

Table 20. Calculation of the transition energies in Eu^{148}(b1).

	Eu^{148}(b1)	
E_γ(I) (KeV) Exp[a]	Cal[b]	Assigned I
748.0	723.41	22
798.0	786.06	24
848.0	947.58	26
900.0	907.91	28
951.0	966.99	30
1004.0	1024.77	32
1057.0	1081.23	34

Table 20. Contd.

1111.0	1136.34	36
1165.0	1190.06	38
1220.0	1242.40	40
1276.0	1293.33	42
1331.0	1342.84	44
1388.0	1390.95	46
1443.0	1437.65	48
1499.0	1482.96	50
1555.0	1526.89	52
	$\sigma = 1.71\text{x}10^{-2}$	

Table 21. Calculation of the transition energies in Ce^{131}(b1) and Ce^{133}(b1).

	Ce^{131}(b1)			Ce^{133}(b1)	
E_γ(I) (KeV) Exp[a]	Cal[b]	Assigned I	E_γ(I) (KeV) Exp[a]	Cal[b]	Assigned I
591.0	566.9	13.5	748.0	692.6	16.5
662.0	630.6	15.5	809.0	775.5	18.5
733.0	714.6	17.5	873.0	857.0	20.5
804.0	797.6	19.5	937.0	937.3	22.5
874.0	879.3	21.5	1003.0	1016.0	24.5
943.0	959.8	23.5	1068.0	1093.3	26.5
1011.0	1038.9	25.5	1132.0	1168.9	28.5
1080.0	1116.5	27.5	1198.0	1242.9	30.5
1151.0	1192.6	29.5	1267.0	1315.1	32.5
1225.0	1267.1	31.5	1337.0	1385.6	34.5
1301.0	1340.0	33.5	1411.0	1454.3	36.5
1381.0	1411.2	35.5	1489.0	1521.2	38.5
1464.0	1480.6	37.5	1570.0	1586.3	40.5
1550.0	1548.4	39.5	1655.0	1649.5	42.5
1640.0	1614.4	39.5	1743.0	1710.8	44.5
1732.0	1678.6	41.5	1833.0	1770.4	46.5
1822.0	1741.0	43.5	1928.0	1828.1	48.5
	$\sigma = 2.87\text{x}10^{-2}$			$\sigma = 3.32\text{x}10^{-2}$	

Table 22. Calculation of the transition energies in Pr^{133}(b1) and Sm^{137}(b1).

	Pr^{133}(b1)			Sm^{137}(b1)	
E_γ(I) (KeV) Exp[a]	Cal[b]	Assigned I	E_γ(I) (KeV) Exp[a]	Cal[b]	Assigned I
871.0	903.1	22.5	379.0	425.2	11.5
943.0	979.7	24.5	451.0	501.4	13.5
1013.0	1055.0	26.5	538.5	576.9	15.5
1085.0	1129.0	28.5	629.1	651.1	17.5
1156.0	1201.5	30.5	719.7	726.1	19.5
1228.0	1272.5	32.5	806.3	799.5	21.5
1299.0	1342.0	34.5	890.0	872.1	23.5
1380.0	1409.9	36.5	970.0	943.7	25.5
1450.0	1476.2	38.5	1045.0	1014.4	27.5
1530.0	1540.9	40.5	1115.0	1084.0	29.5
σ = 6.79x10-3				$\sigma = 6.06\text{x}10^{-2}$	

Table 23. Calculation of the transition energies in Pr^{133}(b2) and Ce^{131}(b2).

	Pr^{133}(b2)			Ce^{131}(b2)	
E_γ(I) (KeV) Exp[a]	Cal[b]	Assigned I	E_γ(I) (KeV) Exp[a]	Cal[b]	Assigned I
821.0	837.6	19.5	847.0	818.9	20.5
885.0	920.7	21.5	908.0	898.4	22.5
961.0	1002.4	23.5	976.0	976.7	24.5
1035.0	1082.7	25.5	1043.0	1053.7	26.5
1110.0	1161.3	27.5	1112.0	1129.3	28.5
1184.0	1238.4	29.5	1181.0	1203.4	30.5
1264.0	1313.7	31.5	1251.0	1276.1	32.5
1342.0	1387.3	33.5	1322.0	1347.2	34.5
1420.0	1459.0	35.5	1396.0	1416.7	36.5
1500.0	1529.0	37.5	1471.0	1484.7	38.5
1576.0	1597.1	39.5	1552.0	1550.9	40.5
1656.0	1663.4	41.5	1635.0	1615.6	42.5
			1723.5	1678.5	44.5
	$\sigma = 1.33 \times 10^{-2}$			$\sigma = 1.82 \times 10^{-2}$	

Table 24. Calculation of the transition energies in Ce^{133}(b2) and Eu^{143}(b1).

	Ce^{133}(b2)			Eu^{143}(b1)	
E_γ(I) (KeV) Exp[a]	Cal[b]	Assigned I	E_γ(I) (KeV) Exp[a]	Cal[b]	Assigned I
720.3	683.4	16.5	483.0	470.1	14.5
785.4	765.2	18.5	546.5	536.1	16.5
854.2	845.8	20.5	609.0	601.6	18.5
920.3	925.0	22.5	671.2	666.7	20.5
986.9	1002.8	24.5	732.6	731.1	22.5
1052.2	1079.2	26.5	793.4	795.0	24.5
1118.2	1153.9	28.5	853.5	858.3	26.5
1184.3	1227.1	30.5	912.8	920.9	28.5
1253.2	1298.5	32.5	972.2	982.7	30.5
1323.9	1368.3	34.5	1031.1	1043.8	32.5
1397.9	1436.2	36.5	1089.9	1104.0	34.5
1475.5	1502.4	38.5	1148.1	1163.5	36.5
1557.3	1566.8	40.5	1206.5	1222.1	38.5
1643.0	1629.5	42.5	1265.5	1279.8	40.5
1731.3	1690.3	44.5	1324.1	1336.7	42.5
1821.6	1749.3	46.5	1383.2	1392.6	44.5
			1442.7	1447.5	46.5
			1502.2	1501.6	48.5
			1562.3	1554.6	50.5
			1623.5	1606.7	52.5
			1684.0	1657.8	54.5
			1743.0	1707.9	56.5
	$\sigma = 2.78 \times 10^{-2}$			$\sigma = 1.21 \times 10^{-2}$	

bands that the band head moment of inertia, $j_0 = \hbar^2/ab$, are close to each other, $j_0 \sim$ (25-60) $\hbar^2/MeV^{-1}$ and no evidence for odd-even difference is found. Moreover, the smallness of parameter b [~(1 : 3) x 10^{-4}, (Table 1)] implies the super-rigidity of these SD nuclei. The radius of convergence I_r for the I(I+1) expansion of Equation (1) can be estimated as

$I_r \approx 1/\sqrt{b} \approx 58\text{-}100.$

Conflict of Interests

The author(s) have not declared any conflict of interests.

REFERENCES

Afanasjev AV, Konig J, Ring P (1996). Superdeformed rotational bands in the A~ 140 – 150 mass region. Nucl. Phys. A. 608(2):107-175. http://dx.doi.org/10.1016/0375-9474(96)00272-2, http://dx.doi.org/10.1016/0375-9474(96)00257-6

Becker JA, Roy N, Henry EA, Yates SW, Kuhnert A, Draper JE, Korten W, Beausang CW, Deleplanque MA, Diamond RM, Stephens FS, Kelly WH, Azaiez F, Cizewski JA, Brinkman MJ (1990a). Level spin and moments of inertia in superdeformed nuclei near A~ 194. Nucl. Phys. A. 520:187c-194c. http://dx.doi.org/10.1016/0375-9474(90)91146-I

Becker JA, Roy N, Henry EA, Yates SW, Kuhnert A, Draper JE, Korten W, Beausang CW, Deleplanque MA, Diamond RM, Stephens FS, Kelly WH, Azaiez F, Cizewski JA, Brinkman MJ (1990b). Observation of superdeformation in 192Hg. Phys. Rev. C. 41(1):R9-R12. http://dx.doi.org/10.1103/PhysRevC.41.R9 PMid:9966368

Bently MA, Ball GC, Cranmer-Gordon HW, Forsyth PD, Howe D, Mokhtar AR, Morrison JD, Sharpey-Sahafer JF, Twin PJ, Fant P, Kalfas CA, Nelson H, Simpson J, Sletten G (1987). Intrinsic quadrupole moment of the superdeformed band in 152DY. Phys. Rev. Lett. 59:2141-2144. http://dx.doi.org/10.1103/PhysRevLett.59.2141.

Bohr A, Mottelson BR (1975. Nuclear Structure. (Benjamin, Reading, MA, II:4. Draper JE, Stephens FS, Deleplanque MA, Korten W, Diamond RM, Kelly WH, Azaiez F, Macchiavelli AO, Beausang CW, Rubel EC, Becker JA, Roy N, Henry EA, Brinkman MJ, Kuhnert A, Yates SW (1990). Spins in superdeformed bands in the mass 190 region. Phys. Rev. C42:R1791-R1795.

Flannery WH, Teukolsky BP, Vetterling WT (1992). Numerical Recipes in Fortran, (Cambridge University, Second edition, P. 683.

Hu ZX, Zeng JY (1997). Comparison of the Harris and ab expression for the description of nuclear superdeformed rotational bands. Phys. Rev. Lett. 56(5):2523-2527.

Harris SM (1964). Large-spin rotational states of deformed nuclei. Phys. Rev. Lett. 13(22):663-665. http://dx.doi.org/10.2172/4598915

Hegazi AM, Ghoniem MH, Khalaf AM (1999). Theoretical Spin Assignment for Superfeformed Rotational Bands in Mercury and Lead Nuclei. Egypt. J. Phys. 30(3):293-303.

Holmberg P, Lipas PO (1968). A new formula for rotational energies. Nucl. Phys. A117:552-560. http://dx.doi.org/10.1016/0375-9474(68)90830-0

Khalaf AM, El-Shal AO, Alam M, Sirag MM (2002). Spin prediction and Systematics of Moments of Inertia of Superdeformed Nuclear Rotational Bands in The Mass Region A~ 90. Egypt. J. Phys. 33(3):585-602.

Khazov Yu, Rodionov AA, Sakharov S, Singh B (2005). Nuclear Data sheets for A=132. J. Nuclear Data Sheets. 104(3):497-790. http://dx.doi.org/10.1016/j.nds.2005.03.001

Khazov Yu, Mitropolsky I, Rodionov A (2006). Nuclear data sheets for A=131. J. Nuclear Data Sheets 107(11):2715-2930. http://dx.doi.org/10.1016/j.nds.2006.10.001

Laird RW (2002). Quadrupole moments of highly deformed structures in the A~ 100 – 140 region: Probing the Single-Particle Motion in a Rotating Potential. Phys. Rev. Lett. 88:152501. http://dx.doi.org/10.1103/PhysRevLett.88.152501 PMid:11955192

Leoni S, Bracco O, Camera F, Million B, Algora A, Axeisson A, Bezoni G, Bergstrom N, Blasi N, Castoldi M, Frattini S, Gadea A, Herskend B, Kmiecik M, Biano GL, Maj A, Nyberg J, Pignanelli M, Styczen J, Vigezzi J, Zieblinski M, Zucchiatti A (2001). Quantum tunneling of the excited rotational bands in thr superdeformed nucleus 143Eu. Phys. Lett. B498:137-143. http://dx.doi.org/10.1016/S0370-2693(00)01367-8

Sevensson CE (1999). Decay out of the Doubly magic superdeformed band in the N=Z nucleus Z 60n. Phys. Rev. Lett. 82:3400-3403. http://dx.doi.org/10.1103/PhysRevLett.82.3400

Shalaby AS (2004). Spin determination for superdeformed rotational Bands in the A~ 60-90 Mass Region. Egypt. J. Phys. 35(2):221-235.

Shalaby AS, Al-Amry RW, Al-Full Z, Mahrous E (2012). Spin determination and study of the A~ 100-140 superdeformed mass region using three parameter empirical formula. J. Nonlinear Phenomena Complex Syst. 15(1):1-22.

Singh B, Firestone RB, Chu SYF (1996). Table of superdeformed nuclear bands and fission isomers. J. Nucl. Data Sheets 78:1. http://dx.doi.org/10.1006/ndsh.1996.0008

Singh B, Zywina R, Firestone RB (2002). Table of superdeformed nuclear bands and fission isomers. J. Nucl. Data Sheets. 97:241-592. http://dx.doi.org/10.1006/ndsh.2002.0018

Stephens FS (1990). Spin alignment in superdeformed rotational bands. Nucl. Phys. A520:91c-104c. http://dx.doi.org/10.1016/0375-9474(90)91136-F

Szymanski Z (1996). Identical rotational bands in the a~ 130 superdeformed region-analysed in terms of the pseudospin symmetry. Acta Phys. Pol. B27(4):1001-1010. Acta. Phys. Pol. B27:1001.

Twin PJ, Nyako BM, Nelson AH, Simpson J, Bentley MA, Cranmer-Gordon HW, Forsyth PD, Howe D, Mokhtar AR, Morrison JD, Sharpey-Sahafer JF, Sletten G (1986). Observation of a discrete-line superdeformed band up to 60ħ in 152Dy. Phys. Rev. Lett. 57:811-814. http://dx.doi.org/10.1103/PhysRevLett.57.811 PMid:10034167

Wu CS, Zeng JY, Xu FX (1989). Relations for the coefficients in the I(I+1) expansion for rotational spectra. Phys. Rev. C40(5):2337-2341. http://dx.doi.org/10.1103/PhysRevC.40.2337 PMid:9966232

Wu CS, Zeng JY, Xing Z, Chen XQ, Meng J (1992). Spin determination and calculation of nuclear superdeformed bands in A~ 190 region. Phys. Rev. C45(1):261-274. http://dx.doi.org/10.1103/PhysRevC.45.261 PMid:9967752

Design and evaluation of combined solar and biomass dryer for small and medium enterprises for developing countries

Okoroigwe E. C.[1,2] , Eke M. N.[1] and Ugwu H. U.[3]

[1]Department of Mechanical Engineering, University of Nigeria, Nsukka, Nigeria.
[2]National Centre for Energy Research and Development, University of Nigeria, Nsukka, Nigeria.
[3]Department of Mechanical Engineering, Michael Okpara University of Agriculture, Umudike, Abia State, Nigeria.

A small scale demonstration model consisting of a combined solar and biomass cabinet dryer with 3 equally spaced drying trays was designed, constructed and evaluated. The results, obtained using fresh yam chips as test material over a four day test period, were satisfactory and useful for optimization purposes. Maximum tray temperature of 53°C was obtained in combination with solar and biomass heating sources even though the ambient temperature for the test period was between 24 and 30°C. An optimal drying rate of 0.0142 kg/hr was achieved with the combined solar and biomass dryer, compared to the lower drying rate of 0.00732 kg/h for the solar drying and 0.0032 kg/h for the biomass drying. This study proved that the efficiency of agricultural dryers could be increased through the use of a combination of solar and biomass heating sources, compared to conventional dryers with only solar or only biomass heating sources. It implies that improvements in the design and construction of the various components of the system would lead to more efficient dryers for use in small and medium business enterprises for sustainable development of developing countries. Using combined solar and biomass dryers have the potential to increase the productivity and resultant economic viability of small and medium-scale enterprises producing and processing agricultural produce in developing countries. African countries, with large quantities of natural resources, like forests and solar radiation, could make the most use of these types of dryers.

Key words: Solar biomass dryer, drying rate, moisture loss, preservation, yam chips, solar radiation, small and medium-scale enterprises.

INTRODUCTION

Although most tropical regions of Africa experience high levels of solar radiation throughout the year, much of this radiation is absorbed by frequent rain and persistent cloud cover. Due to inadequate preservation techniques available to farmers in developing countries, large quantities of agricultural output with high moisture contents are lost annually due to decomposition by micro-organisms. This results in reduction of the net agricultural output and subsequent reduction in the gross domestic product (GDP) of the developing countries.

Dehydration is one of the oldest techniques employed in food or agricultural products storage and preservation (Montero et al., 2010; Montero, 2005; Mujumdar, 2000; Corvalan et al., 1995). The most common method of dehydration is by open air sun drying but this often results in food contamination and nutritional deterioration

(Ratti and Mujumdar, 1997). Food dehydration technology employs direct and/or indirect mixed mode systems with natural or forced distribution of heated air. According to Madhlopa and Ngwopa (2007), direct heating mode systems consist of direct heating of the items by direct sun radiation through transparent material enclosing the items while indirect heating mode systems consist of heating of air in a separate solar collector and circulating the same through the drying bed where it picks moisture from the crop. The mixed mode drying system combines the features of the two above mentioned systems. The study of solar drying and design of solar dryers are not new but the advent of the renewable energy industry has sparked renewed interest in these fields. There are many sources of literature on solar drying. Ekechukwu (1999) and Ekechukwu and Norton (1999 a, b) have presented overviews of drying principles, theories and solar drying technologies, with a description of low temperature air-heating solar collectors for crop drying applications.

Owing to intermittent solar radiation throughout the day, continuous drying of agricultural products can be accomplished through a combination of solar and non-solar heating sources in a mixed mode system. Thermal storage systems can also be used to increase the efficiency of solar dryers. These thermal storage system range from hybrid modes where, electrical resistances have been used to increase the heating of the air (Prasad and Vijay, 2005; Prasad et al., 2006) to the use of biomass as auxiliary heating source for the drying chamber. Auxiliary heating sources are used to provide uninterrupted supplies of thermal energy for the continuous operation of dryers, during periods of limited solar radiation like night time and cloudy The heated air from the auxiliary heating source passes through pipes or channels and then into the drying chamber via convection. Auxiliary heating source could be powered by electricity or biomass. However, biomass is the most widely used due to its availability and cost effectiveness in rural areas of developing countries. Many solar hybrid dryers have been designed, constructed and tested. Some of these have reached advanced and commercial stages, while some are still undergoing improvements. At the Asian Institute of Technology (AIT), Thailand (Elepaño et al., 2005) a hybrid dryer based on solar and biomass heat sources has been developed and commercialized. This dryer is constructed of bricks and mortar, and is more efficient than a conventional solar cabinet dryer made of steel or aluminum. The dryer has been tested with products such as banana and mushroom. Similarly in Nepal, the Research Centre for Applied Science and Technology (RECAST) developed a similar hybrid dryer based on combined biomass solar drying. These designs were constructed with concrete and bricks. It was capable of drying 10 kg of fresh agricultural products such as radish, carrot, ginger, mushroom, potato and pumpkin.

In Nigeria, some advances have been made in the development of solar hybrid dryers. Danshehu et al. (2008) evaluated a 150 kg kerosene-assisted solar cassava dryer which improved the dehydration process compared to open air solar drying. Similarly, Oparaku et al. (2003) evaluated a solar cabinet dryer with auxiliary heater where appreciable result was obtained. The problems facing the development of this technology in Nigeria include poor design and construction of the dryers, little or no mathematical modeling and poor choice of materials for construction. As a result of these reasons commercial solar dryers are yet to be realized in Nigeria. This has necessitated a search for suitable and efficient dryers based on local technology rather than importing dryers not suited for Nigerian conditions. Nigeria's geography, with tropical forests in the south and high solar radiation levels in the north 3.55 $kWh.m^{-2}.day^{-1}$ (Sambo, 2009), holds great potential for combined solar and biomass drying schemes. Utilising Nigeria's geography and natural resources for solar and biomass drying, could increase the durability and resultant availability and price stability of agricultural and aquatic products throughout the country. It is therefore the objective of this study to design, construct and evaluate the performance of a solar combined with biomass heating dryer for use in developing countries. This preliminary study of the authors utilizes fresh yam chips as test samples in a four day test period.

THEORY

Determination of the heat contribution in drying

The quantity of heat Q_w required, to evaporate moisture of mass m_w is estimated using the basic heat equation.

$$Q_w = m_w L_{vap} \tag{1}$$

where L_{vap} [kJ/kg] is the latent heat of evaporation of water, which can be calculated by the method of Youcef-Ali et al. (2001) as:

$$L_{vap} = 4.186\,(597 - 0.56T_{pr})\ \text{kJ/kg} \tag{2}$$

where T_{pr} [°C] is the product temperature, which can be assumed as the ambient temperature at the coldest weather condition.

$$Q_w = 4.186\, m_w\,(597 - 0.56T_{pr})\ \text{kJ} \tag{3}$$

This represents the heat energy required to dry the items in the dryer to appreciable moisture content status. The sketch below represents the various sources of heat to the drying chambers of the system. Total heat required, Q_w is the sum of the heat which enters the system from the collector, Q_c, the biomass stove, Q_s and the body of the drying chamber, Q_g (Figure 1).

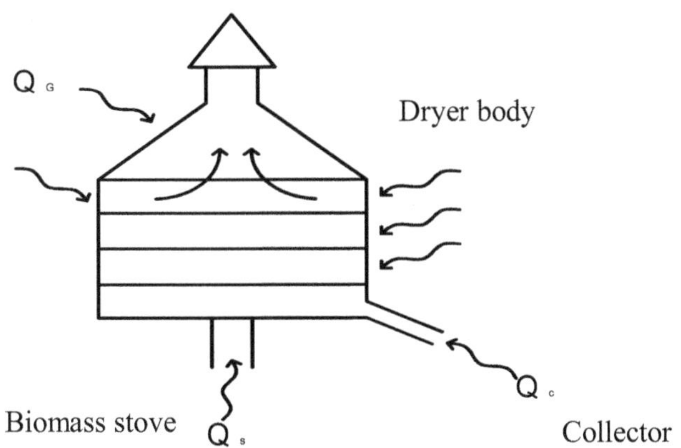

Figure 1. Sketch of the possible heat sources of the dryer.

$$Q_w = Q_c + Q_s + Q_g \quad (4)$$

where the quantity of the heat from the stove, Q_s is the product of the mass of biomass fuel used, m_c and the heating value of the biomass, H_v.

$$Q_s = \eta\, m_c H_v \quad \text{kJ} \quad (5)$$

According to Duffie and Beckman (1991), the useful energy output of a flat plat collector of area A_c is the difference between the absorbed solar radiation and the thermal loss. If I_T is the solar radiation incident on the collector, U_L the overall heat transfer coefficient of the collector and Q_u the heat output of the collector, then,

$$Q_u = A_c[I_T - U_L \Delta T] \quad (6)$$

Where ΔT = T - T_a the difference between the absorber plate temperature and the ambient temperature. Due to difficulty in estimating the absorber temperature because it depends on fluid flow characteristics, it has been suggested that the heat output of the collector be based on the heat removal factor F_R of the collector. But

$$F_R = \frac{m C_p (T_o - T_i)}{A_c[I_T - U_L(T_i - T_a)]} \quad (7)$$

The heat removal factor is a function of the collector efficiency factor F′ which can modify the heat removal factor as

$$F_R = \frac{m C_p}{A_c U_L}\left[1 - \exp\left(-\frac{A_c U_L F'}{m C_p}\right)\right] \quad (8)$$

A graphical representation of the ratio of these factors are provided in Duffie and Beckman (1991) which can be used to estimate the heat removal factor. In order to minimize the heat losses for maximum heat output, it is assumed that both the collector plate and the moving working fluid (air current) should be at the same temperature. Hence the maximum heat output of the absorber plate is

$$Q_u = A_c F_R\, [I_T - U_L(T_i - T_a)] \quad (9)$$

From above this is equivalent to the percentage of the total heat expected from the solar collector to remove m_w of moisture to dry the stuff. The solar radiation component of the Equation 9 can be estimated by the models suggested by Hottel and Wortz (1942) which were improved by Liu and Jordan (1963) as:

$$I_T = I_b R_b + I_d\left(\frac{1+\cos\beta}{2}\right) + I\rho_g\left(\frac{1-\cos\beta}{2}\right) \quad (10)$$

This is the total solar radiation on the tilted surface for 1 h. According to Ulgen (2006) maximum annual energy availability is obtained when the slope of the collector is equal to the angle of latitude of the location for low latitude countries ($\varphi \le 40°$). Hence, in Nsukka, solar collectors are tilted to the angle of latitude 7°. The heat developed by the direct solar radiation through the transparent body of the dryer can be estimated by the method of Harkness and Mehta (1978) and Okonkwo and Nwoke (2008). This method proposes that the heat developed through the transparent glass material can be estimated by

$$q_c = UA(T_o - T_i) \quad (11)$$

and

$$q_t = (I_D A_s \tau_1 + I_d A \tau_2) \quad (12)$$

Where q_c and q_t are heat transfer by conduction and solar radiation through the glass material respectively. The sum of these components gives the estimated total heat through the glass. From this the total area of the glass (A_s) can be calculated.

MATERIALS AND METHODS

The hybrid solar and biomass dryer (Figure 2) consists of a solar drying section and a biomass stove section. The dryer has the shape of a home cabinet with tilted transparent top, consisting primarily of a drying chamber, biomass stove, and solar collector. The dryer is provided with two heated air inlets: one at the top of the solar collector for the heated air leaving the collector, and the other at the base of the drying chamber for the heated air exiting the biomass stove. The chimney has a height of 180 cm from the ground and is located at the top of the drying chamber and serves as the air outlet. The drying chamber is 59.6 by 59.6 cm in

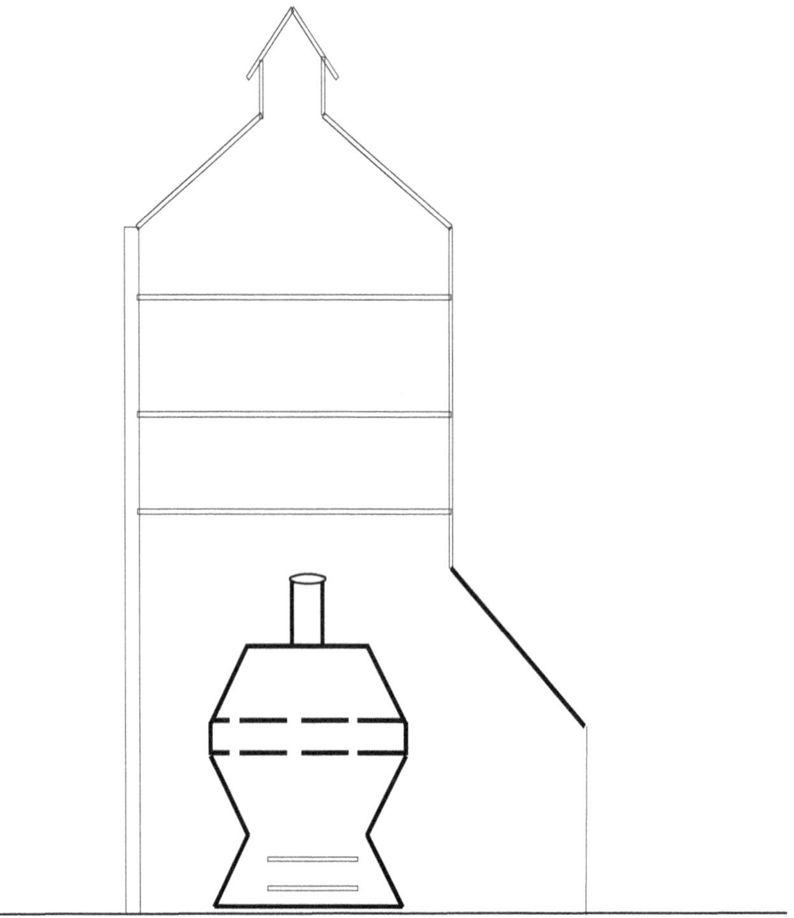

Figure 2. Dimensionless sketch of the combined solar and biomass dryer.

cross-section and has a height of approximately 104 cm. It has three tray levels, and is fitted with a piping system that channels the heat exiting the biomass stove to the different trays.

The biomass stove consists of three main components: the fuel chamber which doubles as the combustion chamber, the primary air inlet and frustum-shaped lid with a little pipe protrusion. The stove is approximately 43 cm in height. The flat plate solar collector of size 0.61× 0.64 m, consists of an absorber, insulation and cover plate. The movement of air from the inlets to the outlets, when the dryer is placed in the path of airflow, brings about a thermo-siphon effect which creates an updraft of solar heated air, which removes moisture from the drying chamber. Ambient air is used as source of air. The performance test of the dryer was carried over a period of four days. The first day consisted of measuring the temperature distribution across the trays of the dryer with no load. The second day consisted of measuring the moisture loss of yam chips on the dryer trays with only solar heating during the day, and then with only biomass heating during the night. The rest of the test period consisted of measuring the moisture loss of the yam chips on the dryer trays with a combination of solar and biomass heating. In all tests yam chips were placed in the open air to dry under direct sunlight as control. The yam chips were prepared from fresh yam purchased from the local market of Nsukka in Nigeria. The chips were washed to ensure that no impurities were involved in the experiment. Equal portions of yam chips of 120 g were weighed out using micro weighing balance and spread in the trays and open sun as control. Temperature changes were monitored using mercury thermometers fixed in the trays while ambient temperature was monitored with k-type thermocouples.

RESULTS AND DISCUSSION

Temperature distribution

The first test was conducted over a period of four hours from 14:00 to 18:00. Figure 3 compares the temperature distributions across the dryer trays over time with no load and with only direct solar radiation as heating source. Within the first hour the temperature of the three trays increased steadily from 38°C (tray 1) to the maximum of 44°C in tray 3. The collector and the transparent glass body performed well as this was reflected in the wide temperature difference between the ambient temperature and the temperature of the trays.

Moisture loss with solar drying

On the second day the system was evaluated with fresh

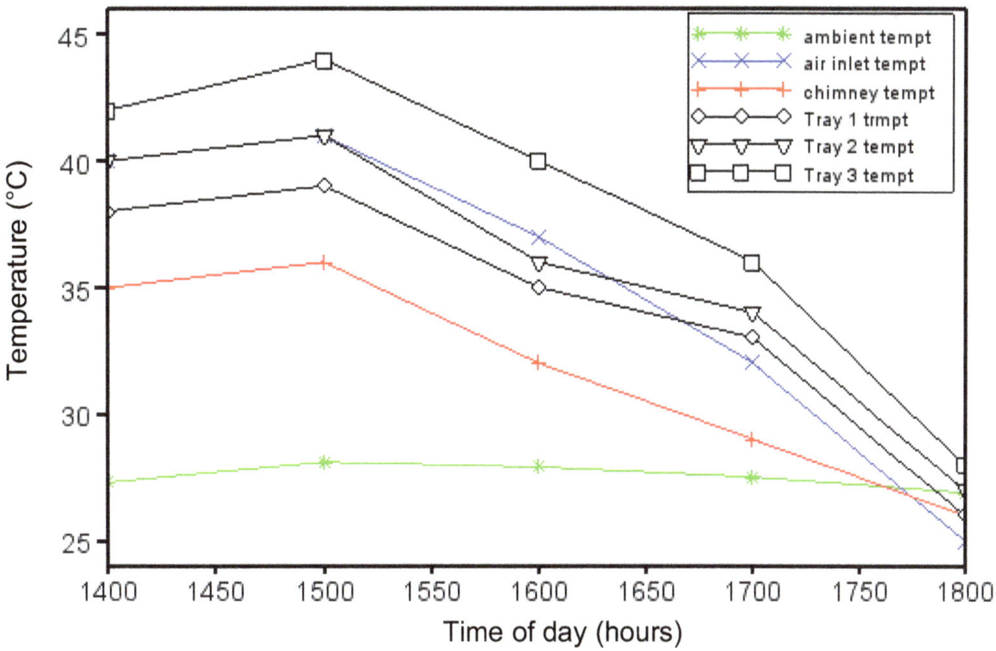

Figure 3. Temperature distribution across the dryer with no load and solar heating only on day one.

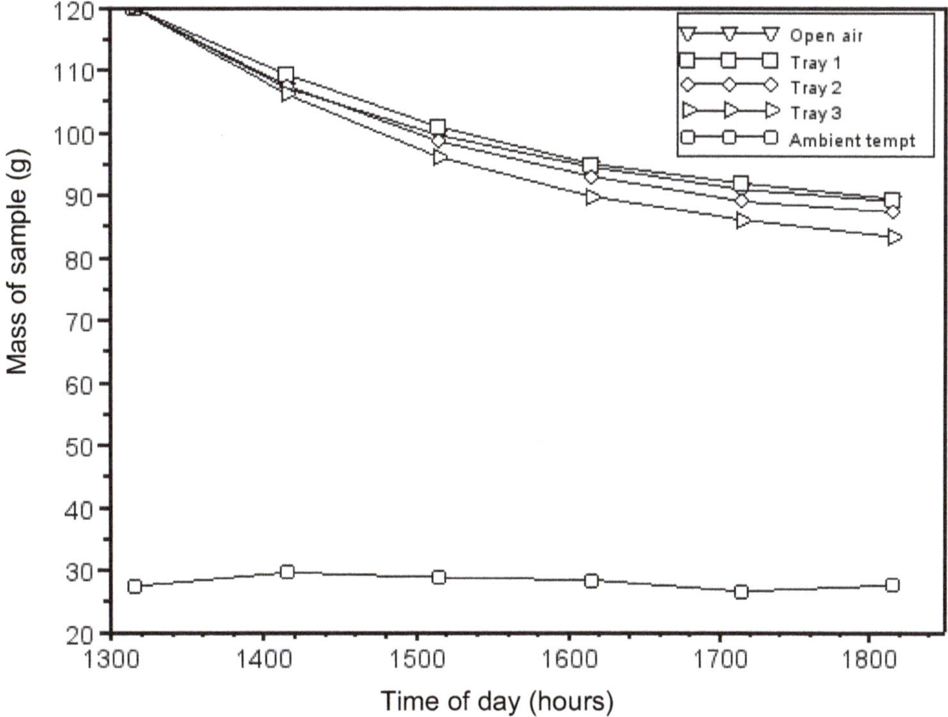

Figure 4. Mass loss of yam chips in the open air compared to yam chips in the dryer with only solar heating on day 2 starting from 13:15.

yam chips whose drying (mass loss) mechanism is shown in Figure 4. A high degree of moisture loss was achieved over the first hour. The mass loss in the open air was greater than in tray 1. This higher drying rate

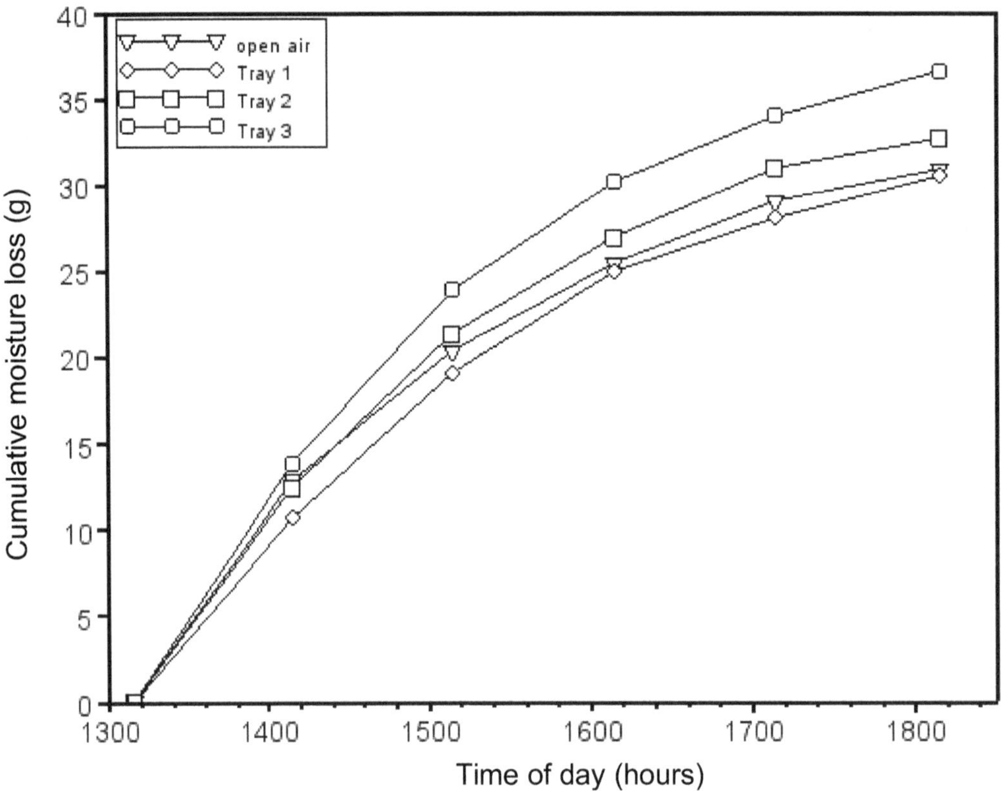

Figure 5. Cumulative moisture loss (removal) with solar drying only.

could have been caused by wind. Surface moisture from the yam chips was lost at a faster rate than its internal moisture, because it takes less energy to evaporate. The rate of total moisture loss slowly reduced after the surface moisture was lost and it took longer to evaporate the internal moisture of the yam chips. The drying effect in the dryer can be observed by the much slower drying rate of the yam chips in the open air compared to the drying rate of yam chips on the dryer trays, until dusk, when the tray temperatures dropped and returned to ambient conditions. The drying rate of 0.00732kg/hr obtained is comparable to 0.009 kg/h obtained by Ajao and Adedeji (2008) using a box type solar dryer for yam drying. The total (cumulative) moisture loss of the yam chips dried in the open air compared to those dried in the solar heated dryer is shown in Figure 5. The cumulative moisture loss in the open air was comparable to the first tray. This could be due to the fact that the first tray did not receive more solar radiation than those dried in the open air, because it was shaded by the upper two trays.

Moisture loss with biomass drying

Figure 6 shows the effect of adding an additional biomass heat source to the drying of the yam chips when solar radiation was not available. The marked moisture loss between 18:00 and 18:30 is expected due to the additional heat source which resulted in increase in temperature of the drying chamber. Within this period, the ambient temperature decreased and there was no sunshine. This implies that additional moisture removal was achieved through the biomass heat addition to the system.

Moisture loss with combined solar and biomass drying

Figures 7 – 9 present the results obtained when biomass was combined with solar drying in the day time. Figure 7 shows the rapid mass loss over the first 30 min while the maximum mass loss of 9.3 g of water occurred after one hour on tray 1. Combining solar heating with additional biomass heating improved the efficiency of the dryer compared to the drying modes where only solar or biomass heating was used. Combining solar and biomass drying reduced the moisture content of the yam chips to 70.83% in 2.5 hours (Figure 7), compared to the moisture content of 75% (Figure 4) of the yam chips with solar drying in 5 hours and 94.44% (Figure 6) in1.25 hours with biomass drying. Comparing this in terms of drying rate, using the tray with the largest cumulative moisture removal, we obtain 0.0142, 0.0032 and 0.00732 kg/h

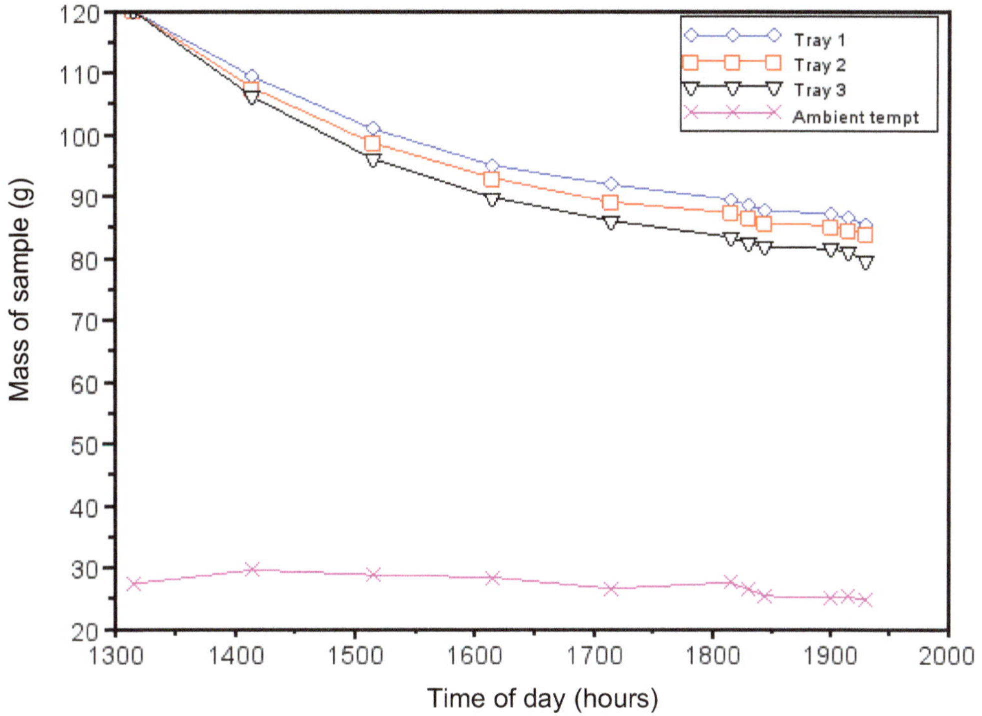

Figure 6. Mass reduction on solar drying in the day and biomass drying in the evening on day 2.

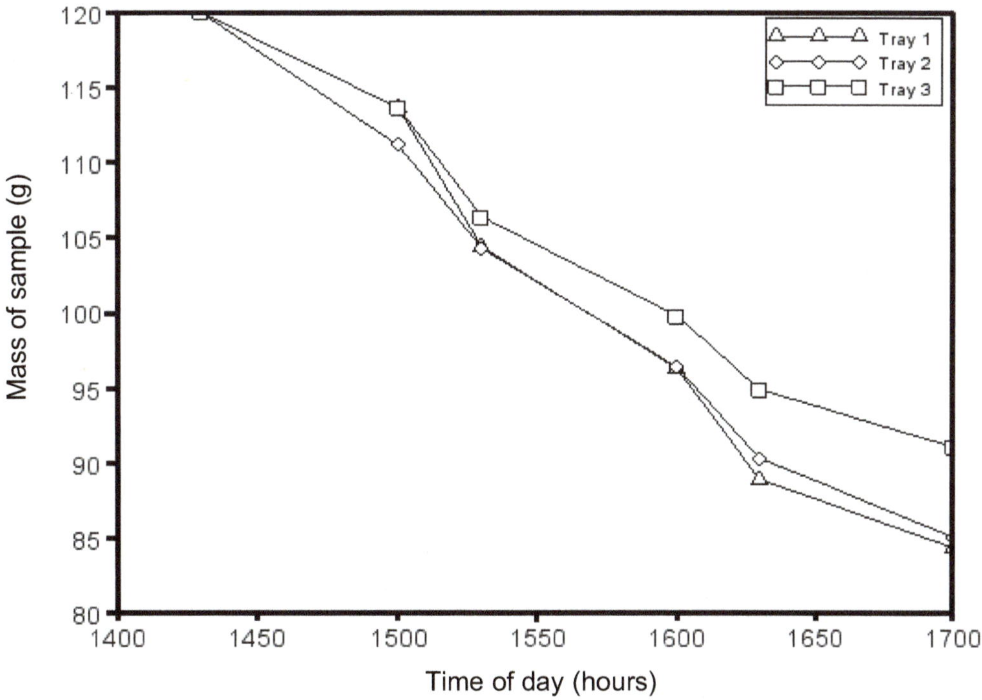

Figure 7. Mass loss with combined biomass and solar drying.

respectively for solar combined with biomass, biomass heating and solar heating only.

Figure 9 also shows that combined solar and biomass heating increased the maximum tray temperature to

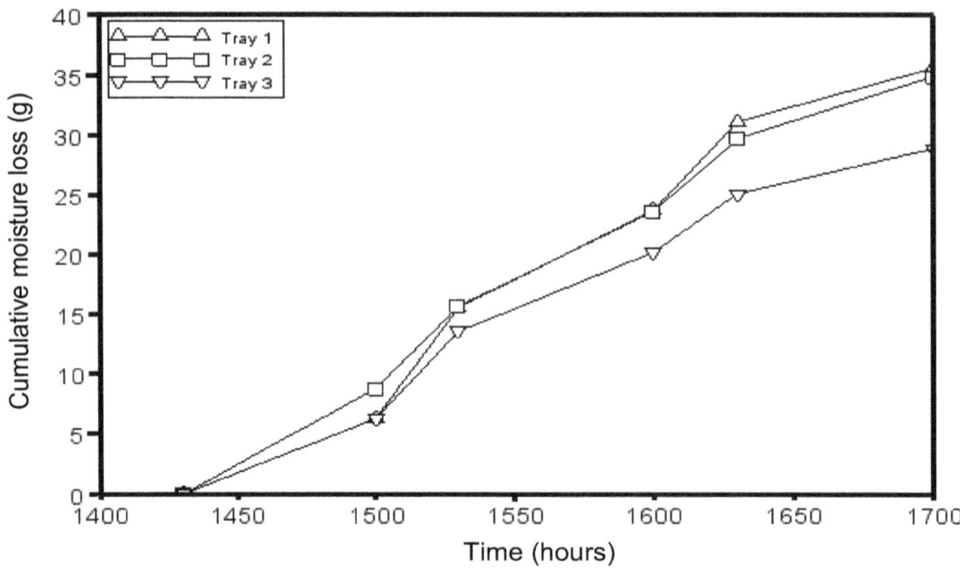

Figure 8. Cumulative moisture loss with combined biomass and solar drying.

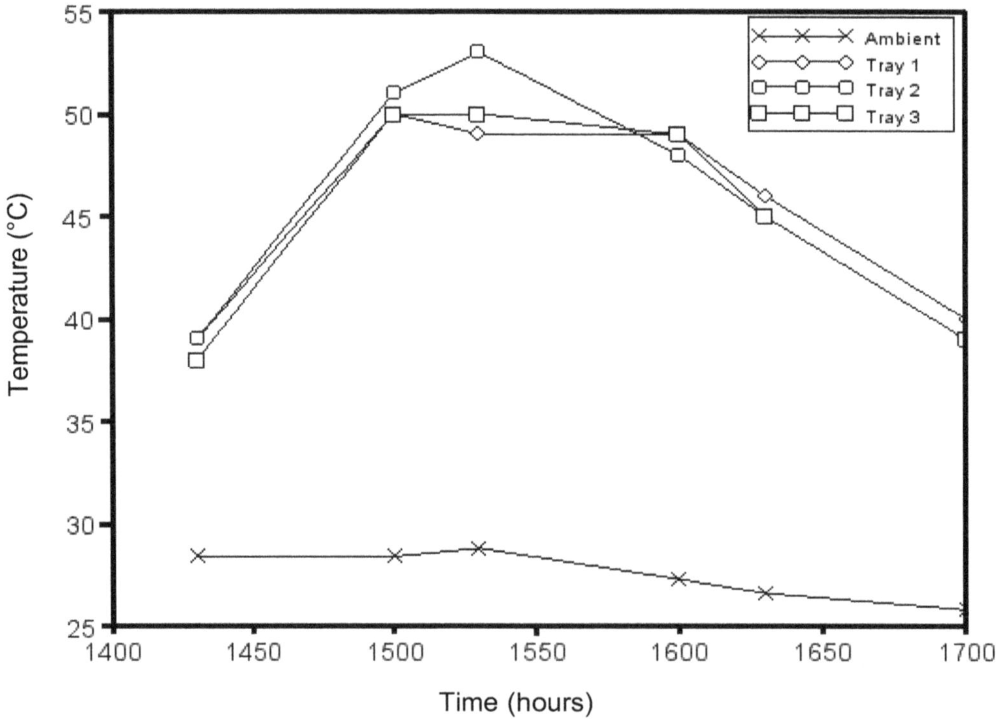

Figure 9. Temperature distribution in the trays with combined biomass and solar drying.

53°C, compared to 44°C with biomass heating and 32°C with solar heating (Figure 3).

Conclusion

This study proved that the efficiency of agricultural dryers could be increased through the use of a combination of solar and biomass heating sources, compared to conventional dryers with only solar or only biomass heating sources. Using combined solar and biomass dryers have the potential to increase the productivity and resultant economic viability of small and medium-scale enterprises producing and processing agricultural produce in developing countries. Countries, like Nigeria, with large quantities of natural resources, like forests and solar radiation, could make the most use of these types of dryers. It is believed that improvements in the construction of the various components of the system will improve the performance of the dryer for use in small and medium business enterprises.

ACKNOWLEGDMENT

The authors appreciate the suggestions of the reviewers which led to the improvement of different sections of the work

REFERENCES

Ajao KR, Adedeji AA (2008). Assessing some drying rates of some crops in Solar Dryer (Case study: Vegetables, tuber and grain crops). USEP J. Res. info. Civ. Eng. 5(1):1–12.

Corvalan R, Roman R, Saravia L (1995). Engineering of solar drying. CYTED-D, 1995 (in Spanish). IN [Monteroet al 2010].

Danshehu BG, Falayan CO, Chukwuka PC (2008). Performance Evaluation of a 150kg Kerosene Assisted Solar Cassava Dryer. Nig. J. Solar Energy. 19 (2):101-105.

Duffie JA, Bechman WA (1991). Solar Engineering of Thermal Processes. New York: John Wiley and Sons.

Ekechukwu OV, Norton B (1999a). Review of solar-energy drying systems. II: an overview of solar drying technology. Energy Convers. Manage. 40(6):615–655.

Ekechukwu OV, Norton B (1999b). Review of solar-energy drying systems. III: low temperature air-heating solar collectors for crop drying applications. Energy Convers. Manage. 40(6):657–667.

Ekechukwu OV (1999). Review of solar-energy drying systems. I: an overview of drying principles and theory. Energy Convers. Manage. 40(6):593–613.

Elepaño RA, Del Mundo RR, Gewali BM, Sackona P (2005). Technology Packages: Solar, Biomass and Hybrid Dryers. Regional Energy Resources Information Center, RERIC. pp. 2–18.

Harkness EL, Mehta ML (1978). Solar Radiation. Solar Radiation Control in Buildings. Applied Science Pubblishers. In [Okonkwo and Nwoke 2008] P. 102.

Hottel HC, Woertz BB (1942). Performance of Flat Plate Solar Heat Collectors. Transactions of ASME 64(91), 1942. In [Duffie and Bechman 1991].

Liu BY, Jordan RC (1963). The long Term Average Performance of Flat plate Solar Energy Collectors Solar Energy 7(53). In [Duffie and Bechman 1991].

Madhlopa A, Ngwalo G (2007). Solar dryer with thermal storage and biomass-backup heater. Solar Energy 81:449–462.

Montero I (2005). Modeling and construction of hybrid solar dryer for biomass byproducts. PhD thesis, Department of Chemical and Energetics Engineering, University of Extremadura, 2005. (in Spanish). In [Montero et al 2010] P. 286.

Montero I, Blanco J, Miranda T, Rojas S, Celma AR (2010). Design, construction and performance testing of a solar dryer for agro-industrial by-products. Energy Convers. Manage. 51:1510–1521.

Mujumdar AS (2000). Drying technology in agriculture and food sciences. Enfield – NH, USA: Science Publishers, Inc.; 2000. IN [Montero et al 2010].

Okonkwo WI, Nwoke OO (2008). Family Size Green House Solar Energy Crop Dryer. Nig. J. Solar Energy 19(2):6–10.

Oparaku NF, Unachukwu GO, Okeke CE (2003). Design, Construction and Performance evaluation of Solar Cabinet Dryer with auxiliary heater. Nig. J. Solar Energy. 14:41–50.

Prasad J, Vijay VK, Tiwari GN, Sorayan VPS (2006). Study on performance evaluation of hybrid drier for turmeric (Curcuma longa L.) drying at village scale. J. Food Eng. 75(4):497–502. In [Montero et al. 2010].

Prasad J, Vijay VK (2005). Experimental studies on drying of Zingiber officinale, Curcuma longa l and Tinospora cordifolia in solar–biomass hybrid drier. Renew Energy 30(14):2097–109. In [Montero et al. 2010].

Ratti C, Mujumdar AS (1997). Solar dryer of foods: modeling and numerical simulation. Solar Energy. 60:151–157.

Sambo AS (2009). Strategic Developments In Renewable Energy In Nigeria. International Association for Energy Economics Third Quarter 2009 pp.15–19. www.iaee.org/en/publications/newsletterdl.aspx?id=7 *assessed6/9/2012*

Ulgen K (2006). Optimum Tilt Angle for Solar Collectors. Energy Sources, Part A, 28:1171–1180.

Youcef-Ali S, Messaoudi H, Desmons JY, Abene A, Le Ray M (2001). Determination of the Average Coefficient of Internal Moisture Transfer during the Drying of a Thin Bed of Potato Slices. J. Food Eng. 48(2):95-101.

17

Financial planning for the preventive maintenance of the power distribution systems' critical components using the reliability-centered approach

Mansour Hosseini-Firouz and Noradin Ghadimi

Department of Engineering, Ardabil Branch, Islamic Azad University, Ardabil, Iran.

Among different maintenance strategies that exist for power distribution systems, the Reliability-Centered Maintenance (RCM) strategy attempts to introduce a structured framework for planning maintenance programs by relying on network reliability studies and cost/benefit considerations. For the implementation of the RCM strategy, the electricity distribution companies try to optimally utilize the existing financial resources in order to reduce the maintenance costs and improve the reliability of the network. The aim of this paper is to present a practical method for devising an appropriate maintenance strategy for network elements and for preventive maintenance budget planning, with the goal of improving the system reliability and reducing the maintenance costs. In the proposed method, the critical outage causes of the distribution system are determined on the basis of cost and reliability criteria, by the Technique for order preference by similarity to ideal solution (TOPSIS) method. Then, the optimum preventive maintenance budget is calculated by obtaining the cost functions of the critical elements and optimizing the overall cost function. In this investigation, the medium voltage distribution network of the "Haft Tir" district in Tehran has been chosen for the implementation of the proposed RCM strategy.

Key words: Reliability-centered maintenance (RCM), preventive maintenance (PM), technique for order preference by similarity to ideal solution (TOPSIS), power distribution system.

INTRODUCTION

One of the most important goals of power distribution companies is to provide uninterrupted and quality service to their customers (Brown, 2002). In this regard, the electricity distribution companies try to select an optimum maintenance strategy in order to reduce the failure rate of network equipment and increase the reliability of the system. This objective was not accomplished as desired in traditional maintenance methods in which the repair of network components was performed at specific time intervals. The huge maintenance expenditures and the inefficiency of these methods in reducing the outages in the system made it necessary to develop a more effective and comprehensive maintenance strategy; a strategy based on the ability to monitor network reliability and to consider the interrelationship between reliability and maintenance costs. These necessities gradually caused the maintenance strategies to be inclined towards the reliability-centered strategies and away from the time-based strategies. The Reliability-Centered Maintenance (RCM) strategy attempts to present an organized framework for the improvement of network reliability and the reduction of maintenance expenses by relying on cost/benefit studies and the reliability analysis of networks (Schneider et al., 2006; Ghadimi, 2012; Ahadi et al., 2014a). In the RCM strategy, the corrective and preventive maintenance strategies are subjected to cost/benefit analysis, and the optimum strategy is selected (Ahadi et al., 2014b).

The preventive maintenance strategy has many complexities relative to the corrective maintenance strategy. Knowing the priorities of elements for preventive measures and determining the proper time intervals between these activities are some of the challenges faced by the preventive maintenance engineers of power distribution companies (Ahadi et al., 2014b; Hagh et al., 2011). However, one of the most important problems that need to be addressed in preventive maintenance planning is the manner of allocating the preventive maintenance budget to the weak points of the network. The assessment of a preventive maintenance budget has always been difficult because of certain factors such as the outages due to environmental and human causes and the unknown nature of some causes of outages, especially the transient ones.

The maintenance budget limitation of the distribution companies, on one hand, and the complexity of assessing the PM budget, on the other hand, have made it necessary to conduct fundamental studies on the subject of maintenance budget planning. This issue is so important that in some cases, due to an incorrect assessment of the maintenance budget, network reliability may not improve much, even though vast maintenance resources are spent.

The establishments of an appropriate relationship between the preventive maintenance of an electricity distribution system and its reliability and the achievement of a RCM strategy have always been of interest to the researchers (Bertling et al., 2005; Ghadimi, 2012). However, due to the lack of proper network information, so far, RCM strategy has not been implemented adequately (Bertling et al., 2005). In Ahadi et al. (2014b) and Bertling et al. (2005), the implementation of the RCM strategy in the power distribution system of Stockholm, Sweden, has been discussed. In this method, optimal scenarios for dealing with the outage causes in the system are selected based on network reliability and cost-benefit analyses. The implementation of RCM in the distribution system has also been studied in Schneider et al. (2006) and Ghadimi (2012). In this method, by determining the failure rate of the critical failure modes, the strategies for dealing with these failure causes undergo cost-benefit analysis.

Also in some studies, the manner of implementing an optimal preventive maintenance strategy in electricity distribution systems for the purpose of network reliability improvement has been investigated. In Lie and Chun (1986) an algorithm is introduced for determining what type of preventive maintenance to consider for the components of a distribution network and when to apply this particular PM strategy. Sobhani and Ghadimi (2013) and Wallnerströmand et al. (2013) and deal with preventive maintenance planning based on risk assessment. As the distribution system components continue to operate, their failure probability increases and therefore the resulting risk goes up. An optimal maintenance model reduces the risk of component failure. In Hagh and Ghadimi (2014), the instant causes of failure in Tehran's electricity distribution system are classified and ranked, and the more important factors are selected for the implementation of preventive maintenance activities. In this reference, after finding the most prevalent causes of instantaneous failures, the variables with higher priorities are selected. In Mohammadi and Ghadimi (2014), the minimization of power outage cost and maintenance cost constitutes the objective function. Ultimately, the results of applying this method in the Birka system of Sweden are evaluated. In Teera-achariyakul et al. (2010), the duration of consumer power outage is minimized through the optimal allocation of a maintenance budget. In Maleh et al. (2013), the allocation of a maintenance budget for distribution feeders is investigated. In the proposed model, the failure rates of feeders are modeled as functions of PM budget, and the cost of PM is minimized by considering a suitable maintenance budget. Park et al. (2000), Canfield (1986) and Mohammadi and Ghadimi (2013) focus on the minimization of preventive maintenance cost. This cost includes the cost of power outage, cost of maintenance, cost of replacement and the annual cost of repairs. In Sittithumwat et al. (2004), the reliability indexes are expressed as probabilities, and by determining the maintenance level, it is attempted to reduce the system average interruption frequency index (SAIFI).

In large electricity distribution networks with numerous causes of power failure, the implementation of the RCM strategy based on determining the proper intervals between preventive activities and optimum maintenance scenarios faces many difficulties; because the assessment of different maintenance scenarios for network components and the reduction of component failures in these scenarios, and in brief, the cost/benefit analysis of maintenance scenarios is impossible without having a sufficient knowledge of the causes of network elements failures. Also, in planning for a maintenance budget, it is necessary to appropriately select a maintenance strategy for network components based on the costs incurred by component failures and the role of components in network reliability. Therefore, if an appropriate maintenance strategy is selected for the network components and the PM budget is optimally spent for the improvement of network reliability and the reduction of maintenance costs, a favorable RCM strategy can be achieved. This paper tries to present a practical method for selecting an optimal maintenance strategy and for the financial planning of a preventive maintenance budget for power distribution networks with the goals of improving network reliability and reducing maintenance costs. In this approach, after prioritizing the outage causes and recognizing the critical outage causes by the TOPSIS method, the maintenance cost functions of the outage causes are obtained with respect to the PM budget and network information, and then by minimizing the total maintenance cost, the optimum PM budget is calculated. To implement the proposed methods, the medium-voltage distribution network of the "Haft Tir" district in the city of Tehran has been selected as the sample network. The reliability information of this study is based on the outage information of the years 2005-2012 that has been extracted from the events logging software of the Greater Tehran Electricity Distribution Company (GTEDC) known as the ENOX Database.

Under proposed method of this paper, the process of PM budget allocation by the proposed method is described. This is followed by results of budget planning by this method for the sample distribution network. Effect of the proposed method on the improvement of network reliability and the reduction of maintenance cost in the studied network is further investigated, and the conclusion is presented.

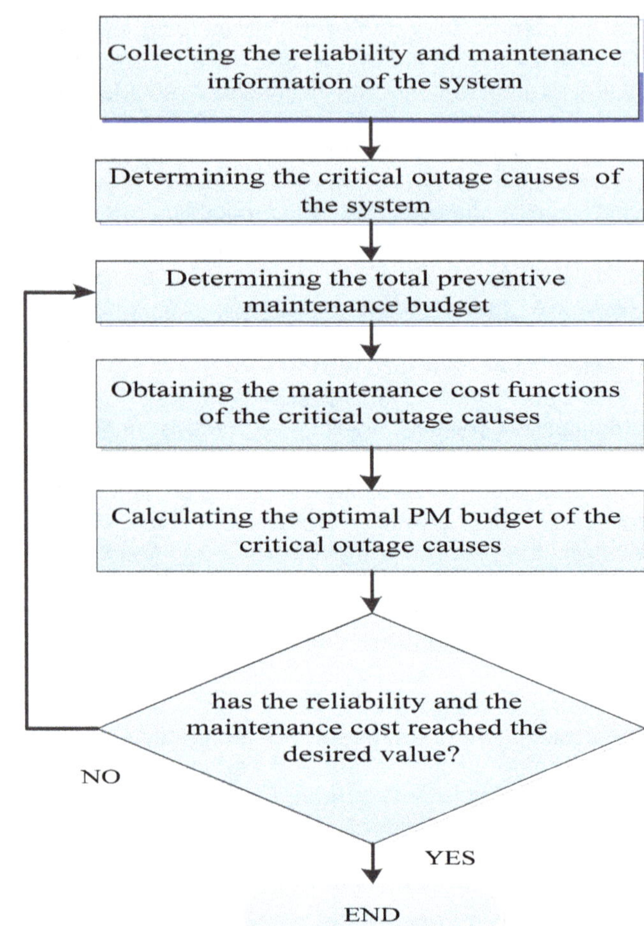

Figure 1. Flowchart for the allocation of preventive maintenance budget.

PROPOSED METHODS

For proper maintenance budget planning, first, it is necessary to devise a suitable maintenance strategy for the outage causes. An appropriate maintenance strategy is selected based on the role of different elements in network reliability and the costs imposed on the system.

After choosing an appropriate maintenance strategy and a PM strategy for the critical outage causes, it is necessary to establish the right relationship between the PM budget and network reliability and to determine the cost of maintaining network components, which mostly includes the cost of repairs, cost of energy not supply, cost of human resources and the cost of preventive maintenance. After obtaining the maintenance cost functions, the optimal PM budget of critical outage causes is calculated by optimizing the overall cost function. If the allocated budget does not lead to the reduction of maintenance cost and the improvement of network reliability as desired, the total PM budget will have to be increased. Figure 1 shows the flowchart of PM budget allocation procedure.

As is shown in the flowchart of Figure 1, the PM budget planning process comprises three major steps:

1) Prioritizing the outage causes and choosing the critical outage causes
2) Estimating the maintenance cost functions of the critical outage causes
3) Calculating the optimum budget of the critical outage causes

Prioritizing the outage causes and choosing the critical outage causes

Certainly, choosing an appropriate preventive maintenance strategy for sensitive and effective equipment in the distribution network rather than spending huge sums of money on the maintenance of all network elements, regardless of their role and importance in the system, will lead to more economical as well as optimal decisions. Those outage causes that have a higher influence on network reliability and the maintenance cost imposed on the network are called the "critical outage causes".

Since numerous factors such as the replacement cost of equipment, number of equipment, and the functions of elements in achieving network reliability must be considered in the selection of the critical outage causes, the multi-criteria decision-making methods (MCDMs) can be employed to prioritize the outage causes and to choose the right maintenance policy. In the proposed method, to prioritize the outage causes, these factors are compared with one another by the TOPSIS method by considering the indexes associated with network reliability, repair cost of components and the number of components. Weights are assigned to the main decision-making criteria based on the priorities of the electricity distribution companies and the opinions of these companies' maintenance engineers. After computing the priorities of the outage causes by the TOPSIS approach, the outage causes with higher priorities are selected for preventive maintenance strategy and those with lower priorities are selected for corrective maintenance strategy. Figure 2 shows the process of prioritizing the outage causes and selecting a maintenance strategy for them by the TOPSIS method.

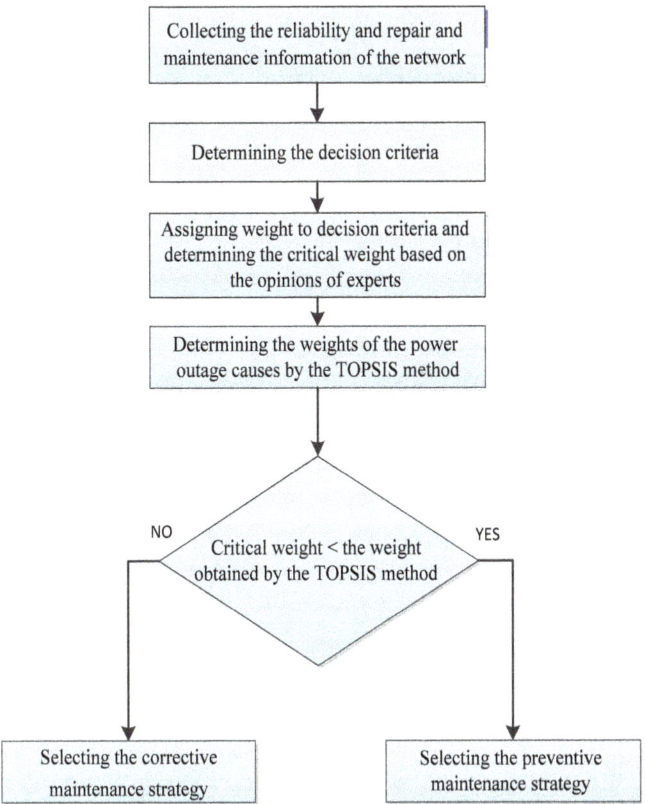

Figure 2. Flowchart for selecting a maintenance strategy for the outage causes by the TOPSIS method.

In the MCDMs of TOPSIS, which was presented in 1981 by Hwang and Yoon, the best solution is the one which is the closest to the positive-ideal solution and, at the same time, the farthest away from the negative-ideal solution (Ghadimi et al., 2014). The ideal solution represents a hypothetical choice which is the most favorable standardized weighted choice from each criterion among the considered choices; and the negative-ideal solution comprises the worst standardized weighted choice among the various choices.

The TOPSIS method evaluates a decision matrix that has M choices and N indices. A_i indicates the ith choice, X_j indicates the jth index and X_{ij} is the numerical value obtained from ith choice and jth index. In the following, the step-by-step prioritization of the choices by the TOPSIS method is described according to Ghadimi et al. (2014).

Step 1: Normalization of the decision matrix

This process tries to non-dimensionalize the existing quantities in the decision matrix. To do this, each value is divided by the magnitude of the vector corresponding to the same index. Every entry of the normalized decision matrix is obtained from Equation (1).

$$r_{ij} = \frac{X_{ij}}{\sqrt{\sum_{i=1}^{m} X_{ij}^{2}}} \tag{1}$$

Step 2: Assigning weight to the normalized matrix

The decision matrix is, in fact, parametric and it needs to become quantified. To this end, the decision-maker assigns a weight to each index and the sum of weights (W) is multiplied by the normalized matrix R.

$$W = (W_1, \dots, W_n) \tag{2}$$

$$\sum_{j=1}^{n} W_j = 1 \tag{3}$$

Step 3: Determining the ideal and negative-ideal solutions

The two virtual choices of A^* and A^- are defined as follows:

$$A^* = \{(\max_i V_{ij} \mid j\varepsilon J), (\min_i V_{ij} \mid j\varepsilon J') \mid i=1,2,\dots,m\} = \{V_1^*,\dots,V_n^*\} \tag{4}$$

$$A^- = \{(\min_i V_{ij} \mid j\varepsilon J), (\max_i V_{ij} \mid j\varepsilon J') \mid i=1,2,\dots,m\} = \{V_1^-,\dots,V_n^-\} \tag{5}$$

A^* is the positive-ideal and A^- is the negative-ideal solutions, J is the benefit criterion column and J' indicates the cost criterion column.

Step 4: Obtaining the separation measures

The separation of choice i from the positive- and negative-ideal solutions is estimated.

$$S_i^* = \sqrt{\sum_{j=1}^{n} (V_{ij} - V_j^*)^2} \quad i=1,2,..,m \tag{6}$$

$$S_i^- = \sqrt{\sum_{j=1}^{n} (V_{ij} - V_j^-)^2} \quad i=1,2,..,m \tag{7}$$

Step 5: Calculating the relative closeness to the ideal solution

This criterion is obtained from Equation (8).

$$C_i^* = \frac{S_i^-}{S_i^* + S_i^-} \quad 0 < C_i^* < 1 \tag{8}$$

Obviously, the less the separation of choice is from the ideal solution, the closer the relative closeness will be to 1.0.

Step 6: Ranking the choices

Finally, the choices are ranked in a descending order.

Table 1. Weights of selected criteria for the prioritization of the studied system's.

Decision criteria	Weights of the decision criteria
Number of outages	0.265
Duration of outages	0.24
Energy Not Supply	0.24
Equipment Replacement Cost	0.125
Number of Existing Elements	0.13

Estimating the maintenance cost functions of critical outage causes

To obtain the maintenance functions of the outage causes, it is necessary to determine their failure rate functions with respect to the PM budget. It is obvious that the bigger the PM budget is, the more the failure rate of components will be reduced. The function of failure rate versus PM budget is expressed as a function with exponential distribution according to Equation (9) (Schneider et al., 2006; Ghadimi, 2012; Maleh et al., 2013; Sarchiz et al., 2009)

$$\lambda(PM_i) = A_i + B_i e^{-C_i PM_i} \tag{9}$$

After determining the failure rate functions of the critical outage causes, the functions of repair cost, energy not supply cost and human resources cost are obtained by Equation (10) through Equation (12).

$$TC_i^{re} = \lambda(PM_i).C_i^{re} \tag{10}$$

$$TC_i^{ENS} = \lambda(PM_i).ENS_i^{avg}.C_{ENS} \tag{11}$$

$$TC_i^{hr} = \lambda(PM_i).hr_i^{avg}.C_{hr} \tag{12}$$

Calculating the optimum PM budget of the critical outage causes

The total maintenance cost is obtained by summing the repair cost, energy not supply cost, human resources cost and the preventive maintenance cost. By considering the total PM budget and the changes of the elements' failure rates, the optimum preventive maintenance budget is obtained by minimizing the total maintenance cost of the critical outage causes. The objective function and the governing constraints of the problem are according to Equation (13) through Equation (15).

$$Minimize: TCCM = \sum_{i=1}^{n} TC_i^{re} + TC_i^{ENS} + TC_i^{hr} \tag{13}$$

$$\sum_{i=1}^{n} PM_i = TCPM \tag{14}$$

$$\lambda_i^{\min} \le \lambda_i \le \lambda_i^{\max} \tag{15}$$

CASE STUDY RESULTS

To implement the proposed methods, the medium-voltage distribution network of the "Haft Tir" district in the city of Tehran has been selected as the sample system. The "Haft Tir" electricity distribution district, with an area encompassing about 140 km^2, is one of the largest regions in Tehran municipality. This investigation has been based on the outage information of the mentioned network for the years 2005-2012, which has been extracted from the incidents logging software (known as ENOX) of the Greater Tehran Electricity Distribution Company (GTEDC). The goal of this investigation is to plan the PM budget of the year 2013 for this system based on the system reliability and cost information of this system.

After collecting the reliability and maintenance cost information of the sample system, the process of PM budget allocation, according to the procedure described in proposed method, includes the steps of prioritizing the outage causes and determining a suitable maintenance strategy, obtaining the maintenance cost functions and calculating the optimal PM budget for the critical outage causes.

Prioritizing the outage causes and choosing the critical outage causes by the TOPSIS method

Identifying the critical outage causes is very important for the purpose of choosing an appropriate maintenance strategy. For certain outage causes such as operator error, equipment theft, disastrous climatic conditions and unanticipated events, proper preventive measures cannot be considered. After collecting the outage information of the sample system, separated by the cause of outage, the selected outage causes are classified into 13 groups for the selection of preventive and corrective maintenance policies. Table 2 shows the unplanned outage causes of the studied system.

The main selected criteria for the prioritization of outage causes include the number of outages, duration of outages, energy not supply, equipment replacement cost and the number of existing elements in the system. The weights of the decision criteria are determined based on the priorities of the Tehran power distribution company and the knowledge of the maintenance engineers of this company, according to Table 1.

Based on the information of the studied system, the columns and rows of the decision matrix are established

Table 2. Determining the priorities of the studied system's outage causes on the basis of decision criteria by the TOPSIS method.

Group	Failure cause	TOPSIS weight
1	Failure of capacitor bank	0.1174
2	Fault of the medium voltage cable	0.397
3	Failure of structure	0.298
4	Failure of transformer	0.575
5	Failure of lightning arrester	0.148
6	Failure of Insulator	0.125
7	Failure of disconnector switch	0.274
8	Failure of circuit breaker	0.437
9	Failure of cutout switches	0.134
10	Failure of cable terminations	0.204
11	Failure of recloser	0.1748
12	Fault in the main switch or the low voltage board	0.0439
13	Tree contact	0.259

Table 3. Critical outage causes of the studied system.

Group	Outage cause
A	Failure of transformer
B	Failure of cable terminations
C	Failure of circuit breaker
D	Failure of structure
E	Fault of the medium voltage cable
F	Failure of disconnector switch
G	Tree contact

based on the decision criteria and the information of outage causes, respectively; and the priorities of the outage causes are calculated by the TOPSIS method. The priorities of the outage causes of the studied system obtained by the TOPSIS method have been listed in Table 2. After determining the outage causes and considering a critical weight of 0.15 based on the opinions of the Greater Tehran Electricity Distribution Company's Engineers, seven groups of outage causes with priorities larger than the critical weight are selected for preventive maintenance strategy and the remaining outage causes are chosen for corrective maintenance strategy. Table 3 shows the outage causes of the studied system selected for PM budget planning.

Determining the maintenance cost functions and calculating the PM budget of critical outage causes

After selecting the critical outage causes of the system under study, based on the information of this system, the failure rate functions of the critical outage causes are determined as a function of PM budget, according to Equation (9). Table 4 shows the coefficients of the failure rate functions of the studied systems' critical outage causes.

The maintenance cost is obtained by adding up the repair cost, energy not supply cost, human resources cost and the preventive maintenance cost. The preventive maintenance budget of the critical outage causes is calculated according to Equation (13) through Equation (15) by optimizing the maintenance cost, considering the limitation of the total PM budget and taking into account the failure rate changes of the critical outage causes. Table 5 shows the information related to the studied system for determining the optimum PM budget, which includes the coefficients of the failure rate functions, average repair cost of elements, minimum and maximum failure rates of critical elements and the average amount of energy not supply and human resources for each time an element failures. The total PM budget of the investigated system for the year 2013 is 543,472 dollars, based on the planned budget of the Greater Tehran Electricity Distribution Company. Also, the cost of energy not supplied (C_{ENS}) has been considered as 8 dollars per kWh and the human resources

Table 4. coefficients of the failure rate functions of the sample systems's critical outage causes.

Group	A_λ (int/year)	B_λ (int/year)	$C_\lambda*10^5$
A	39	129.5	3.4
B	59	123.9	7.78
C	41	162.2	1.81
D	18	66.1	5.9
E	41	194.3	1.79
F	31	88.1	5.94
G	26	124.45	1.55

Table 5. Determining the priorities of the studied system's outage causes on the basis of decision criteria by the TOPSIS method.

Group	A_λ (Int/year)	B_λ (Int/year)	$C_\lambda*10^5$ (Int/year)	C^{rp} ($)	ENS_i^{avg} (KWh)	$\lambda_{\min}$ (Int/year)	$\lambda_{\max}$ (Int/year)	hr_i^{avg} (individual *hour)
A	39	129.5	3.4	285	1327	46	109	18
B	59	123.9	7.78	58	201	59	106	11
C	41	162.2	1.81	181	1021	58	156	14
D	18	66.1	5.9	170	936	18	28	21
E	41	194.3	1.79	123	1421	58	160	12
F	31	88.1	5.94	110	353	31	57	16
G	26	124.45	1.55	62	926	55	116	12

cost (C_{hr}) has been set as 11 dollars per individual per hour. The GAMS software and the nonlinear programming method have been employed to analyze the existing problem. Figure 3 shows the PM budget of the investigated system's critical outage causes estimated by the proposed method.

DISCUSSION

Here, the proposed method were implemented in the studied system. In addition to these method, it can be assumed that the share of each group of critical outage causes from the total PM budget of 2013 is similar to that of 2012 budget, and that no change is made in the budget planning method of 2013 relative to 2012. Thus, the PM budget of 2013 for the outage causes can be computed based on the 2012 budget for these causes. Figure 4 shows the PM budget planning for the studied system's outage causes by the proposed method along with the PM budget planning based on the 2012 budgeting procedure. In order to estimate the improvement of system reliability due to PM budget planning through the two said methods, the index of system reliability improvement is defined by Equation (16) (Ghadimi, 2012).

$$RII = \alpha\frac{\lambda-\lambda_{base}}{\lambda_{base}} + \beta\frac{U-U_{base}}{U_{base}} + \gamma\frac{ENS-ENS_{base}}{ENS_{base}} \quad (16)$$

Weighting coefficients α, β, γ are determined based on the priorities of the electricity distribution companies. To calculate the system reliability indexes of the two budget planning methods in the system under investigation, the functions of failure rate, outage duration and energy not supply versus PM budget are considered as exponential distribution functions according to Equation (17) through Equation (19).

$$\lambda(PM_i) = A_i^\lambda + B_i^\lambda e^{-C_i^\lambda PM_i} \quad (17)$$

$$U(PM_i) = A_i^U + B_i^U e^{-C_i^U PM_i} \quad (18)$$

$$ENS(PM_i) = A_i^{ENS} + B_i^{ENS} e^{-C_i^{ENS} PM_i} \quad (19)$$

The coefficients of functions (17) through (19) can be calculated with regards to the information of the studied system. Table 6 show the coefficients of failure rate, outage duration and energy not supply functions of the investigated network's critical outage causes.Thus, based on the obtained functions and the amount of PM budget,

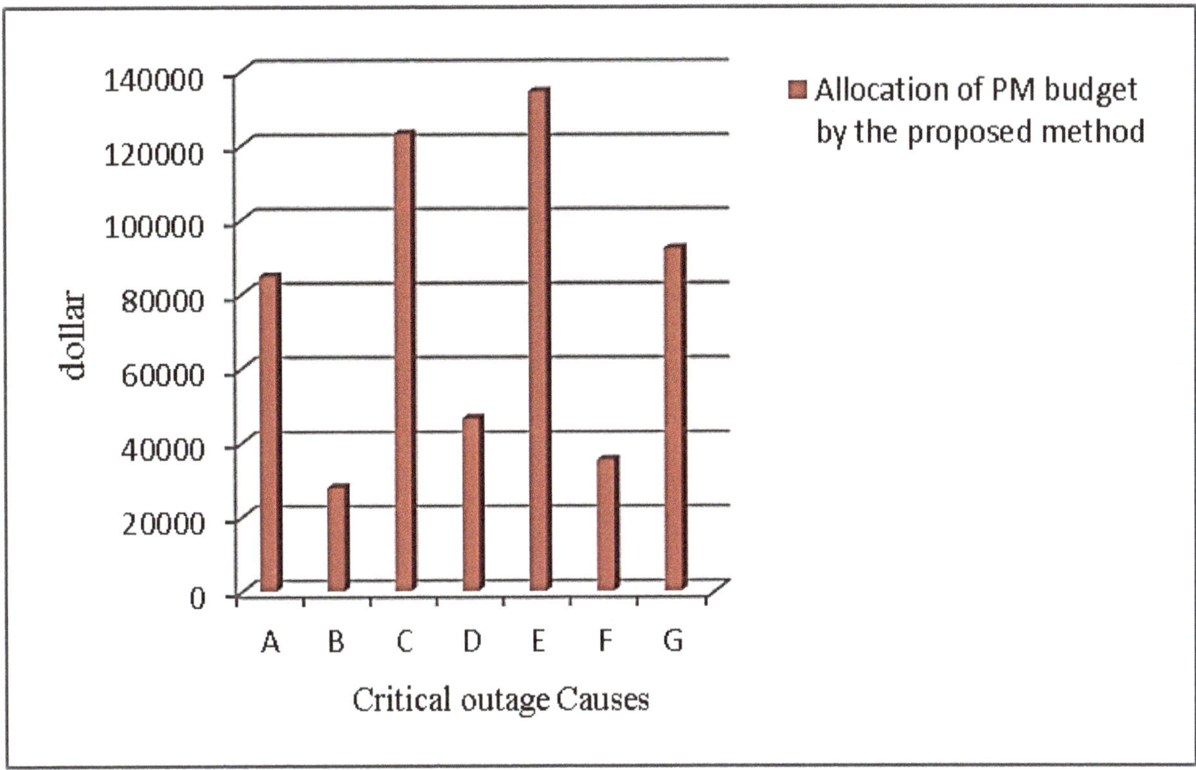

Figure 3. Allocation of PM budget to the studied system's critical outage causes by the cost optimization method.

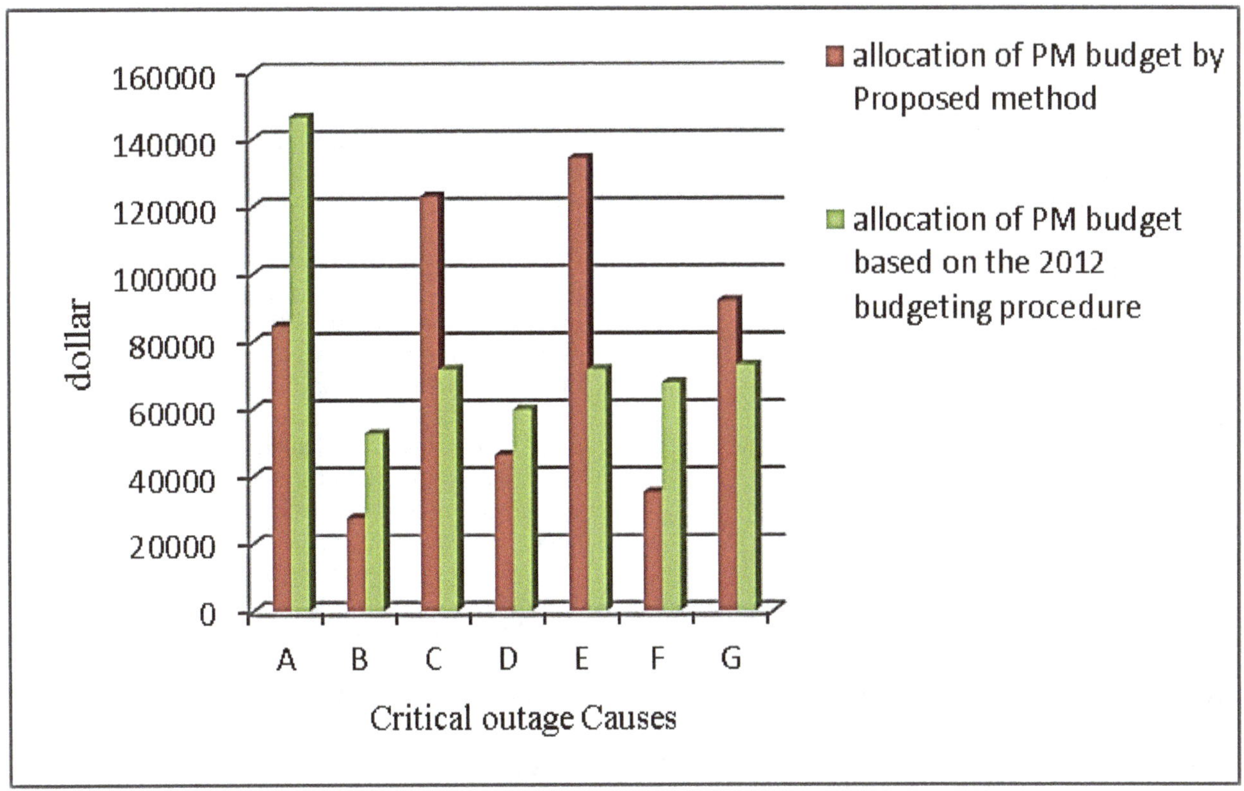

Figure 4. PM budgets of the studied network's critical outage causes obtained by the proposed method and by the budgeting procedure of the year 2012.

Table 6. Coefficients of failure rate, outage duration and energy not supply functions of the studied system's critical outage causes.

Group	A^{λ} (Int/year)	B^{λ} (Int/year)	$C^{\lambda} * 10^5$	A^U (Minute)	B^U (Minute)	$C^U * 10^5$	A^{ENS} (MWh)	B^{ENS} (MWh)	$C^{ENS} * 10^5$
A	39	129.5	3.4	1411	14280	0.637	19	114.39	0.691
B	59	123.9	7.78	1100	6234	0.221	6	12.04	0.241
C	41	162.2	1.81	1623	13650	0.690	26	108.37	0.351
D	18	66.1	5.9	1152	11260	0.487	12	62.22	0.256
E	41	194.3	1.79	1920	19680	0.392	41	125.4	0.314
F	31	88.1	5.94	905	6729	0.415	9	12.03	0.451
G	26	124.45	1.55	1423	13140	0.490	19	75.33	0.207

Table 7. Maintenance cost and reliability improvement index of the studied system obtained by proposed method and budget planning based on the year 2012 budget with considering $\alpha = 0.448$, $\beta = 0.338$, $\gamma = 0.214$.

Method	RII	Maintenance cost (dollar)
Proposed method	0.054	3,096,528
budget allocation of the year 2012 method	0.01	3,683,660

the reliability improvement index is determined according to Equation (16). Also, the maintenance costs of the studied system through the two budget planning approaches can be found by Equation (10) through Equation (13). Table 7 presents the indexes of reliability improvement and maintenance cost of the investigated system obtained by the proposed method and by budgeting procedure of the year 2012 with considering $\alpha = 0.448$, $\beta = 0.338$, $\gamma = 0.214$.

As is observed, by considering an identical total PM budget, the degree of network reliability improvement and the maintenance cost are higher in the proposed method compared to those obtained by the budget planning method of 2012. The establishment of an appropriate relationship between preventive maintenance, maintenance cost and network reliability in this method leads to the optimal expenditure of the PM budget for the improvement of network reliability and the reduction of maintenance costs.

Conclusion

Two major objectives are pursued in the implementation of preventive maintenance programs for an electricity distribution system: the improvement of network reliability and the reduction of maintenance costs. In this paper, a method has been presented for selecting the proper strategy for the maintenance of network components and for planning an appropriate preventive maintenance budget, with the goal of improving network reliability and reducing the maintenance cost. In this approach, after prioritizing the power outage causes and determining the critical power outage factors by the TOPSIS method, the maintenance cost function is obtained, and based on the reliability information of the network, the amount of budget that leads to a minimum maintenance cost is calculated.

The results of implementing this method in the medium voltage distribution network of the "Haft Tir" district in the city of Tehran indicate that by using this method in PM budget planning, network reliability improves and the maintenance cost is lowered. The establishment of an appropriate relationship between preventive maintenance, network reliability and maintenance cost in this method makes it possible to optimally spend the PM budget for the improvement of network reliability and the reduction of maintenance cost. Since the implementation of the RCM strategy based on the cost/benefit studies of different maintenance scenarios for network components runs into many difficulties in electricity distribution networks with numerous power outage causes, by applying this method, a favorable RCM strategy can be implemented.

Conflict of Interest

The authors have not declared any conflict of interest.

NOMENCLATURE; A^*, Positive ideal solution in the TOPSIS method; A^-, negative ideal solution in the TOPSIS method; $A_i^{ENS}, B_i^{ENS}, C_i^{ENS}$, coefficients of energy not supply functions with respect to the PM budget of the *i*th component; A_i^U, B_i^U, C_i^U, coefficients of outage duration functions with respect to the PM budget of the *i*th component; $A_i^\lambda, B_i^\lambda, C_i^\lambda$, coefficients of failure rate functions with respect to the PM budget of the *i*th component; C_{ENS}, cost of 1.0 kWh of energy not supply (in dollars); C_{hr}, average cost of one hour of human resources (in dollars); C_i^{re}, average repair cost of the *i*th component forevery failure (in dollars); ENS_i, annual energy not supply of the *i*th component (in MWh); ENS_i^{av}, average energy not supply in the failure of the *i*th component (inKwh); hr_i^{av}, average human resources needed in the failure of the *i*th component (in individual *hour); J, set of profit indexes in the TOPSIS method; J', set of cost indexes in the TOPSIS method ; PM_i, PM budget of the *i*th component (in dollars); RII, system reliability improvement index; r_{ij}, element of the *i*th row and *j*th column of the decision matrix normalized by the TOPSIS method; S_i^+, ideal separation in the TOPSIS method; S_i^-, negative ideal separation in the TOPSIS method; TC_i^{re}, annual repair cost of the *i*th component (in dollars) ; TC_i^{ENS}, annual energy not supply cost of the *i*th component (in dollars); TC_i^{hr}, annual human resources cost of the *i*th component (in dollars); $TCPM$, total cost of preventive maintenance (in dollars); U_i, annual outage duration of the *i*th component (in minutes);m TCM, total cost of maintenance (in dollars); X_{ij}, element of the *i*th row and *j*th column of the decision matrix in the TOPSIS method; α, β, γ, weight coefficients related to λ, U, ENS in the reliability improvement index; λ_i, annual failure rate of the *i*th component (in interruption per year); $\lambda_i^{base}, U_i^{base}, ENS_i^{base}$, λ, U, ENS of the *i*th component for allocating a PM budget equally to all the critical components; $\lambda_i^{\max}$, maximum annual failure rate of the *i*th component (in interruption per year); $\lambda_i^{\min}$, minimum annual failure rate of the *i*th component (in interruption per year);

REFERENCES

Ahadi A, Ghadimi N, Mirabbasi D (2014a). An analytical methodology for assessment of smart monitoring impact on future electric power distribution system reliability. Complexity. doi: 10.1002/cplx.21546.

Ahadi A, Noradin G, Davar M (2014b). "Reliability assessment for components of large scale photovoltaic systems." J. Power Sources 264:211-219.

Bertling L, Allan R, Eriksson R (2005). "A reliability-centered asset maintenance method for assessing the impact of maintenance in power distribution systems," Feb 2005, IEEE Trans. Power Syst. 20(1):75–82.

Brown RE (2002). "Electric Power Distribution Reliability,"New York: Marcel Dekker.

Canfield RV (1986). "Cost optimization of periodic preventive maintenance," IEEE Trans. Reliability. R-35(1):78-81.

Ghadimi N (2012). "Genetically tuning of lead-lag controller in order to control of fuel cell voltage." Sci. Res. Essays 7(43):3695-3701.

Ghadimi N, Afkousi-Paqaleh A, Ali E (2014). "A PSO-based fuzzy long-term multi-objective optimization approach for placement and parameter setting of UPFC." Arabian J. Sci. Eng. 39(4):2953-2963.

Hagh MT, Ghadimi N (2014). Multisignal histogram-based islanding detection using neuro-fuzzy algorithm. Complexity. doi: 10.1002/cplx.21556.

Hagh MT, Tarafdar M, Ghadimi N, Najafi S (2011). "Hybrid method to detect the anti-islanding mode protection for wind turbine with internally excited system." Int. Rev. Automatic Control 4.4 (2011).

Lie CH, Chun YH (1986). "An Algorithm for Preventive Maintenance Policy." IEEE Trans. Reliability. R-35(1):71-75.

Maleh MS, Ramtin RN, Noradin G (2013). "Placement of distributed generation units using multi objective function based on SA algorithm." Sci. Res. Essays 8(31):1471-1477.

Mohammadi M, Ghadimi N (2014). Optimal location and optimized parameters for robust power system stabilizer using honeybee mating optimization. Complexity. doi: 10.1002/cplx.21560.

Mohammadi M, Ghadimi N (2013). "Designing controller in order to control micro-turbine in island mode using EP algorithm." Sci. Res. Essays. 8(42):2100-2107.

Park DH, Jung GM, Yum JK (2000). "Cost minimization for periodic maintenance policy of a system subject to slow degradation" Reliability Eng. Syst. 68:105-112.

Sarchiz D, Bică D, Georgescu O (2009). "Mathematical model of Reliability Centered Maintenance (RCM). Power transmission and distribution networks applications", IEEE Bucharest Conference, Power Tech, bucharest, Romania.

Schneider D, Gaal A, Neumann C (2006). "Asset Management techniques," Electrical Power Energy Syst. 28:643–654.

Sittithumwat A, Soudi F, Tomsovic K (2004). "Optimal allocation of distribution maintenance resources with limited information" Electric Power Syst. Res. 68:208–220.

Sobhani B, Ghadimi N (2013). "Anti-islanding protection based on voltage and frequency analysis in wind turbines units." Int. J. Phys. Sci. 8(27):1408-1416.

Teera-achariyakul N, Chulakhum K, Rerkpreedapong D, Raphisak P (2010). "Optimal Allocation of Maintenance Budgets for Reliability target setting," Power and Energy Engineering Conference (APPEEC), Asia-Pacific, 2010.

Wallnerströmand C, Hilber P, Stenberg S (2013). "Asset management framework applied to power distribution for cost-effective resource allocation," Int. Trans. Electrical Energy Syst. DOI:10.1002/etep.1826, 2013.

Egbin power station generator availability and unit performance studies

F. E. Ogieva, A. S. Ike and C. A. Anyaeji

Department of Electrical Electronic Engineering, Faculty of Engineering, University of Benin, Benin City, Edo State, Nigeria.

This paper presents investigation on availability carried out on six steam unit generators in Egbin Thermal-Power Station in Nigeria. The availability investigation covers from 2005 to 2011 and was done through an exhaustive collection of data from samples of operating facilities in the power station. Data was collected from plant user maintenance log, operation records and manufacturers' data were also sources of information. This investigation used the IEEE std 762 generator performance indices amongst other calculated key operational availability indices in the evaluations and analysis of the collected data. A software program was developed, 'Function Outage Parameters (OP)', using the outage frameworks of data collected from the station. The program was implemented in MATLAB 11.5b which provided user-friendly Graphical User Interfaces (GUI) and corresponding output results in numerical values in tables of values and graphs. The data was used to evaluate all the six generating units available in the station. The result was used to appraise a periodic availability assessment of all the generating units. The study has demonstrated that availability has a very major impact on power generation and plant economy. The investigations ensured quantified (computed) comparative analysis for planned and unplanned outages by using results to estimate unit generators" performance capacity credibility. The availability results generated by stations values were: ST01 = 89%; ST02= 89.99%; ST03 =85.24%; ST04 = 87.45%; ST05= 86.50%; ST06 = 29.71% while the overall availability is 88.35%. Result shows reduction in plant availability is caused by increased number and duration of forced outages. The causes and durations of forced outages and unscheduled maintenances were identified through the study of outage causes. The use of a historic failure database to identify critical components for improvement of generating unit availability is demonstrated. While Nigeria is practically hungry for power supply availability to support economic growth and provide basic modern energy services to her people though the energy level is still abysmally low, the facts presented herein are sufficient to exhibit the importance of power availability and unit performance measurement in enhancing the country energy revolution and development.

Key words: Availability, performance, generators, steam turbines, maintenance, reliability.

INTRODUCTION

As power supply availability becomes the current catchphrase in business, industry, and society at large in

Nigeria, energy researches on availability is indispensible. The increasing competition in the electricity sector has had significant implications for plant operations; it requires thinking in strategic and economic rather than purely technical terms (André et al., 2007).

The overall power scene in Nigeria indicates heavy shortages almost in all states of the federation. The situation may be aggravated in coming years as the demand is increasing and if the power industry does not keep pace with the increasing energy demand.

In recent past, Nigeria has been referred to as a 'Nation that has Covenant with Darkness' by the Tell Magazine July 27, 2009. They were not far from the truth as a country with a population of over 140 million people had only 1500 MW of electricity to share at that time. This was put at 15.58 kW per individual per annum by the Central Intelligence Agency (CIA) Factbook (2007). That is about 1500 MW total generation. However, people have diverse view about the root cause of the electricity problems. Nigeria ranks abysmally low compared to other countries of Africa, as shown in the CIA Factbook (Tell Magazine July 27, 2009). The challenges of energy production vary from nations to nations. However, electric energy is produced and delivered practically on real time and there is no convenient method to readily store it, hence, it is said that electricity is simply ubiquitous.

While rapidly growing economies like Nigeria is hungry for practically any power to support economic growth and provide basic energy services to her people, the industrialized nations of the world are focusing on ensuring secured electricity supplies at competitive prices also in an environmentally acceptable way.

In order to achieve this goal, compulsory availability data documentation is crucial. The traditional measures used in reliability evaluation are probabilistic and consequently, they do not provide exact predictions (Richwine, 2004). They only state averages of past events and chances of future ones by means of most frequent values and long-run averages (Fernando, 1999). These measures that are mostly "factors" (Equivalent Availability Factor (EAF), the Forced Outage Factor (FOF) and Unit Capability Factor (UCF) use as their denominator the entire time period being considered (typically one year and above) without regard to whether or not the unit is required to generate (Richwine, 2004). Commercial availability is a proper availability evaluation used as a source of information that can be complemented with other economic and policy considerations for decision making in planning, design and operations in the power industry. Operational availability is the quantitative link between readiness objectives and supportability. The new "deregulated" (horizontal) structure in Nigeria is practically based on market principles, favouring competitions amongst private participants and consumer choice.

Under deregulation, a competitive power production becomes standard operation procedure. The quality of power a company produces becomes the measure of its success (Killich, 2006). Under the deregulation setting, energy particularly power generation should be decided by its quality. This supports the customer view point which is summed up into two concepts: technical and economical. Technical concept is all indicated in availability and reliability indices. The economical concept is integrated in electrical energy price which is required to be in the lowest possible range. While the managerial concepts which are figured in the performance indices are: availability, reliability and productivity (Mahmoud et al., 2000).

When deregulation is fully established, it will require the utility, Independent Power Producer (IPP), National Integrated Power Producers (IPP) and other Power Producers (perhaps Industrial Power Producers, IND) to bid power competitively at current market rates. In this case, the power producer that operates at the lowest cost per kilowatt-hour will thrive in this challenging environment. As we progress under deregulation thus, the traditional technical measures will become inadequate. This will thrust utilities to add specifics in terms of measurements that provides and help build on their traditional economics. This requires high importance to be placed on power plant performance and availability indices to form groundwork for performance and benchmarking (Richwine, 2011).

Turbine units more than 25 years in operation face serious threats in view of their remaining lifetime. Even in case of proper operation and maintenance talk less, absence of proper operation and maintenance (Stein and Cohen, 2003). The ageing of power plants leads to higher production cost which presently faces the Nigerian Electricity Generation Industry, mainly due to the following according to Stein and Cohen (2003):

i. Duration, occasioned by deterioration of original performance level (output and efficiency) and
ii. Decline in availability occasioned by increased number and duration of forced outages.

The availability of a complex system such as a steam turbine unit, is basically associated with its parts reliability and maintenance policy. This may be enhanced by proper recording of failure rates and maintenance frequencies and etc. Timely and appropriate recording of these data could help in product improvement by manufacturers (insight on design improvement) and to identify critical components for improvement to enhance system reliability, availability and maintainability evaluations based on a historical failure/outage database.

This question however highlights the need for systems that will consistently and rigorously seek to classify outage events using the performance indicators to justify their progress. Consequently, availability performance indicator amongst others is indispensable.

Background

The operation of a generating unit requires a coordinated operation of hundreds of individual components (Casazza and Delea, 2003). Each component has a different level of importance to the overall operation of the operating single unit. Failure of some pieces of equipment particularly the auxiliaries might cause little or no impairment in the operation of a generating unit.

Still, some might cause immediate or total shutdown of the unit if they fail. The failure rates of all the various components of a generating unit contribute to the overall unavailability of the unit. The unavailability of a generating unit due to component failure is known as its 'forced outage rate'. Generally, according to NERC/IEEE std 762, loss of generation have been distinguished to be caused by problems within and outside plant management control such as substation failure, transmission operating/repair errors, acts of terrorism or war, acts of nature (lightning) whether inside or outside the plant boundary (NERC/IEEE std 762 2006).

In a deregulated environment, competition is indispensible. Still, some might cause immediate or total shutdown of the unit if they fail. This has brought about the need for efficient allocation and use of available energy resources and power generation assets; effective scheduling of plant activities, such as outages and maintenance; greater use of analytical tools to conduct/benefit evaluation of proposed activities are changing the industry mindset (André et al., 2007). In another development, various components of a generating unit must be removed from service on a regular basis for preventive maintenance or to completely replace component(s) before forced outage results. This is called maintenance outage and major maintenance would include turbine overhauls, generator rewinds and boiler turbines, for which complete shutdowns are required. In summary, any condition requiring repairs which can be postponed to a weekend is referred to as 'maintenance outage'. If the unit must be removed from service during week days for a component problem, this is usually referred to as forced outage (NERC/IEEE std 762).

Meanwhile, forced outages are events whose specific occurrence cannot be predicted but can be described by using probabilistic measures. Maintenance outages are event which can be scheduled in advance. This difference is important in making analyses of total generator requirements for a system. The major area of judgment and discretion involved in classifying availability data is that they are usually influenced by economic and reliability considerations. For this reason, compilation and analyses of data requires extensive judgment and experience (Casazza and Delea, 2003).

With the traditional technical measure being considered inadequate in the now, supposedly competitive Nigeria Electricity Supply Industry NESI, there is need to place high importance on power plant availability measurement as font for performance measurement and benchmarking. Commercial availability accurately reflects more, the present-day market place. It therefore remains critical that the Nigerian power industry generate more meaningful metrics to evaluate commercial availability as the need to maximize utility from limited financial resource is equally important on both regulated and competitive environment. In a broader way, benchmarking with gap analysis offers a valuable input to the cost reduction and performance improvement in power generation management. The global liberalization of the electricity market is forcing utilities to deliver electrical energy with high efficiency and at a competitive price (Chirikutsi, 2007). The last sentence seems to be the 'catch-word' of the current deregulation exercise. Failure of some pieces of equipment particularly the auxiliaries might cause little or no impairment in the operation of a generating unit.

The combination of industry averages and the variability of distribution of data basing on technologies, size, age and mode of operation of the peer group plants are also of importance to performance improvements (Chirikutsi, 2007).

In this paper, performance measurements are considered to be based on statistical technical availability (Operational (commercial) Availability) of electric generating unit based on time and energy. The operational availability is considered appropriate for the following reason:

Availability measurements

Before you can begin to control anything, 'system' simple engineering methodology demands that, we must first measure it. The same applies to availability; even more so given the cost of implementing highly available systems can double for just a fraction of percentage of availability. The key is obviously to minimize downtime, since as downtime approaches zero, availability approaches 100%. Not all downtime results from unexpected system outages, since it also includes scheduled maintenance. Downtime consists of two categories: planned and unplanned, while unplanned downtime is the result of an unexpected system failure, planned downtime is that from planned system maintenance such as upgrades and patch installs.

This study is meant to improve procedures for estimating performances of generating units and systems of generating units from operational and technical angle. Hence, it is useful to discuss purposes and uses of some of the specific generating unit performance indices. For example, the forced outage rate (FOR) is used widely in generation system reliability and probabilities production cost studies. Indices including FOR, availability factor (AF), and unavailability factor (UF), are time based indices and depend strictly on the cumulative time in specific plant unit. But here, availability, reliability and productivity indices and parameters were evaluated to justify study objectives. The IEEE std 762 [IEEE Power Engineering Society, 2006] was used for the definitions and formulas.

Impact of downtime

Not all systems have the same level of dependency on availability. Downtime in some systems may be painful, like in the case of power generation supply, but the impact may be localized so that only a small group of users are affected (Islanding in transmission and distribution).

More than ever before, now availability has become a critical design criteria in energy industry–this is not to say that availability has not been important, but the impact of downtime and exposure has become much greater in considerations in repairable system design and implementations, particularly under deregulated market structure. More so, the desire to stand head-high above other competitors has also given this criterion a boost. The reason for this is that, we now provide systems that interact directly with customers, and there is no insulation between the system problems and those customers (Like the prepaid meter, and recharge cards etc.). There is a wide range of the cost of downtime, so it is useful to categorize the impact of downtime into different categories. Many applications can be classified into the following groups:

a. Mission critical: If the application is down, then critical production processes and/or customers are affected in a way that has massive impact on its profitability.
b. Business critical: Downtime that is often not visible to customers, but does have a significant cost associated with it.
c. Task critical: The outage affects only a few users, or the impact is limited and the cost is insignificant.

A close study of the above applications informs that the more mission critical oriented our application, the more the focus on availability efficiency should be. Unfortunately, increases in availability do not come for free. It is often tempting to try to increase system availability by first spending money on the system. Hence, precedence must be adhered to.

Availability performance

Availability performance is the ability of an item to be in a state to perform a required function under given conditions at a given instant of time or over a given time interval, assuming that the required external resources are provided. This ability depends on the combined aspects of reliability performance, maintainability performance and maintenance supportability (IEC 60050 (191-02-05)).

A power plant generator is an active component therefore in this case, everything is considered active. Such components will give an immediate feedback if there is a failure. Corrective maintenance is normally carried out shortly after a component has failed. The purpose is to bring the component back to a functional state as soon as possible. The component may be replaced or repaired. The calculation formulas assume that the repaired component will bring it to "as good as new" condition (Mahmoud et al., 2000).

All items are assumed operating unless failed. The exception would have been standby redundancy, but this scarcely exists in this power station because of high power supply demand.

The results in the analysis are based on two fundamental rules for combining probabilities:

1. If A and B are two independent events with probabilities P(A) and P(B) of occurring, then the probability P(AB) that both events will occur is the product:

$$P(AB) = P(A).P(B)$$

2. If two events A and B are mutually exclusive so that when one occurs the other cannot occur, the probability that either A or B will occur is:

$$P(AB) = P(A) + P(B)$$

This is used as a validation for fall calculations and computer simulations carried out.

In Javad (2005), like reliability, availability is considered a probability. If we considered a system which can be in one of two states, namely 'up (on)' and 'down (off)' as stated earlier. By 'up' it mean that the system is still functioning while by 'down' it mean that the system is not functioning; in this case it is being repaired or replaced, depending on whether the system is repairable or not.

Technically, availability performance is defined in four measures of: the availability function, limiting availability, the average availability function and limiting average availability. All of these measures are based on the function *X(t)*, which denotes the status of a repairable system at time *t*. The instant availability at time t (or point availability) is defined by (Javad, 2005):

$$A(t) = P(X(t) = 1) \quad (1)$$

This is the probability that the system is operational at time t. Because it is very difficult to obtain an explicit expression for A(t), other measures of availability have been proposed. One of these measures is the steady system availability (or steady-state availability or limiting availability) of a system, which is defined by:

$$A = \text{Limit}\ t\to\infty\ A(t) \quad (2)$$

This quantity is the probability that the system will be available after it has been run for a long time, and it is a very significant measure of the performance of a repairable system. Because it is very difficult to obtain an explicit expression for A(t), other measures of availability have also been proposed. For X(t) = 1, if the system is up and at time t = 0, system is down (Javad, 2005). The Equations (1) and (1) respectively, are used here only for the explanation of technical availability concept.

Any improvement in the unit's reliability and availability is associated with the requirement of additional effort through performance improvement. It is, therefore imperative to evolve techniques for reliability and availability allocation amongst various units of a system with minimum effort (Javad, 2005). However, some of these factors do not correctly describe the true state of the units.

For instance, if a peaking unit was required to generate 100 h/year but experienced forced outages during 25 of those demand hours (and no other outages over the 8760 h in the year), it would still have an EAF and UCF of: (8760-25)/8760 x 100 = 99.71% and a FOF and UCLF of (25)/8760 x 100 = 0.29% which are still relatively very high.

These numbers might look good on paper but the reality is that the unit could only produce 75% of the power required of it. So these factors do not correctly describe the unit's ability to produce its rated capacity when demanded.

Mathematically, Operational availability is defined:

$$\text{Mathematically, } A_o = \frac{\textit{Up Time}}{\textit{Operating Time}}$$

$$\text{Availability, } A_V = \frac{\textit{Available Hours}}{\textit{Period Hours}} \times \frac{100}{1}$$

Where, Available Hours = Period Hours – Forced Outage Hours – Scheduled Outage Hours.

It is the probability that an item will operate satisfactorily at a given point in time when used in an actual or realistic operating and support environment. It includes logistics time, ready time, and waiting or administrative downtime, and both preventive and corrective maintenance downtime. Other availability performance indices have been developed for accurate measures amongst which are equivalent availability etc.

The availability of a unit generator determines its performance credibility. The status of a generating unit is conveniently described as residing in one of several possible states. A hierarchical representation of these states is shown in Figure 1.

In any good electricity supply environment, power generation for an area must be simple (matrix) mix of three types of generations:

i. Based–Load Generation: This runs continuously to supply the minimum requirements of the area. This type has shock absorbing capabilities.
ii. Intermediate Generation: This runs to upgrade day time loads.
iii. Peaking Generation: This is started rapidly to meet the few peak hours on a peak day, or to provide immediate support for an area in the event of a contingency on the power system.

The last two fall within the range of frequency generators which are used for grid optimization. The two technical reasons for these categories are the ability of the generator to maneuver and the other, is its efficiency. A generator can maneuver if it can run at a wide range of output power levels, and change output power levels quickly.

Energy quality and availability

In a deregulated power structure, energy particularly power generation should be decided by its quality. This

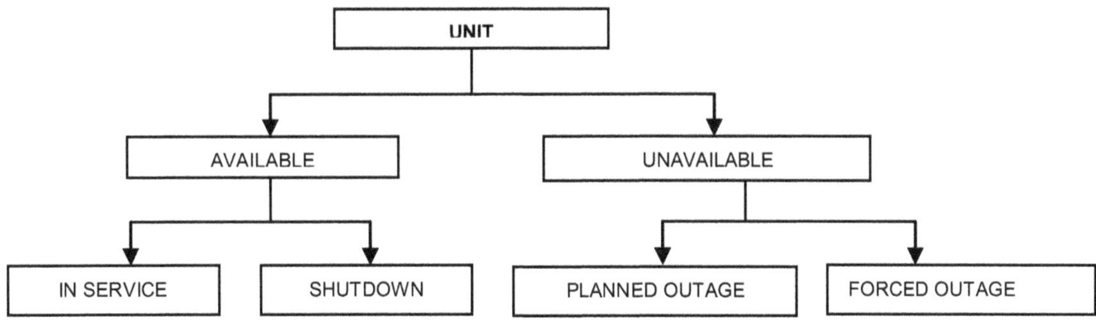

Figure 1. Simple generation unit states.

supports the customer view point which is summed up into two concepts: technical and economical. Technical concept is all indicated in availability and reliability indices. The economical concept is integrated in electrical energy price which is required to be in the lowest possible range. While the managerial concepts-which are figured in the performance indices-are: Availability, reliability and productivity (Mahmoud et al., 2000).

Generator performance measurement gains

A properly planned generator unit availability improvement program can go a long way to optimize overhaul intervals and many more. The cost advantage is immense and more so, there will be:

a. Long – term availability increase as a result of fewer overhauls on the generators,
b. Decrease in post–overhaul failures due to fewer overhauls performed on the system and subsequent overall improvement in availability,
c. Increased availability as result of specific repairs that will be made without overhaul required. Data monitoring helps to track increase in forced or maintenance outages and identifies components responsible.

Operational availability is the quantitative link between readiness objectives and supportability. Availability is a performance criterion for repairable systems that accounts for both the reliability and maintainability properties of a component or unit system.

It is defined as "a percentage measure of the degree to which machinery and equipment is in an operable and committable state at the point in time when it is needed". It is the degree (expressed as a decimal between 0 and 1, or the per-unit) to which one can expect a piece of equipment system to work properly when it is required. Technical considerations also classify the characteristic non-maintained and maintained systems. The non-maintained systems either fulfill their missions (by surviving beyond expected time) or fail it (by perishing before the expected time is completed). In contrast, maintained systems can be repaired (maintained) e.g. a unit generator, and put back into operation (Romeu, 2010). Ultimately, the contractual parties to deregulation in the entire energy sector that is, generation transmission and distribution are focusing on unilateral objectives, which normally are different from each other, and trying to reach them separately (Killich, 2006). In view of the forgoing, the operating requirements largely depend on reliability, maintainability and availability of the operating units of generators.

Maintenance cost advantage gains

According to GADS (2007), when performance improvement is properly planned, it is estimated that the cost of a turbine overhaul for one unit will be $3 million, making the annual cost of an overhaul done on a three-year interval $1 million. Extending the interval to seven years ($60,000 equivalent hours), the cost is about $400,000 a year. Total annual savings will be $600,000 a year per unit (Kopman et al., 1995).

Fuel savings

According to GADS, the fuel savings that results from repairs or modifications accomplished during an overhaul of a plant investigated was $1 million in a year when compared with the time the company started its investigation on optimization of overhaul intervals. This means that, extensive upgrade of old generators particularly through the life extension programs can almost assumes new units status. This in effect increases availability due to fewer overhauls. Post–overhaul failures decreases because of fewer overhauls performed and consequently, leads to overall improvement in availability.

Table 1. Egbin power plant commissioning dates. (Source: Egbin Thermal Business Unit Power Station).

Unit code name	Unit commissioning dates/year	Order of commissioning
ST-1	11/5/1986	3rd
ST-2	11/11/1985	2nd
ST-3	11/5/1985	1st
ST-4	11/11/1986	4th
ST-5	11/5/1987	5th
ST-6	11/11/1987	6th

Plant equipment availability will also increase because specific repairs could be made without requiring overhauls (Kopman et al., 1995).

To be able to manage this process, the availability engineer can handle this by using six standard review processes which include reason for improvement; definition of problem; careful analysis; solution projection; results and process improvement (Kopman et al., 1995). All steps must be supported by facts. We can establish the need for improvements by stratifying the areas of concerns with respect to impact to generation loss. We can study the description of events to define problems. Root cause analysis is performed to identify all possible causes of events.

BRIEF DESCRIPTION OF EGBIN ELECTRIC THERMAL POWER BUSINESS UNIT

The decision to site a thermal power station in Lagos metropolis came up in 1982 by the Federal Government of Nigeria under President Shehu Shagari. The Egbin power plant is located at Igede, near Egbin Town of Ikorodu Local Government of Lagos State, Nigeria. The power station is located about 40 km North East of the City of Lagos. It is situated by the Lagoon around Igede village. Its situation by the Lagoon satisfies the logistic need as well as the water supply requirements of the steam power plant. The Egbin power station is a thermal (steam) power plant. It also utilizes chemical energy of natural gas fuel or LPFO/HPFO (Low pour fuel oil/ High pour fuel oil) through combustion processes in the boiler to generate high pressure and temperature steam to run a three stage steam turbine. This is directly coupled to the generator motor at rated speed of 3000 rpm capable of generating maximum power of 220MW.

The Egbin power station consist of 6 (six) installed units each having a capacity of generating 220MW at maximum continuous rating (MCR). The station has a total installed capacity of 1320MW, the boiler at a capacity of 705t/h are designed for dual firing of natural gas and low/high pour fuel oil (LPFO/HPFO) (Table 1).

MAJOR CAUSES OF OUTAGES UNAVAILABILITY SUMMARISED FROM FIELD OUTAGE DATA RECORDS

This section summarizes the major interpretations for the various graphical presentations which includes description and causes of various major outages (As per Planned outage, Maintenance Outage, Forced outage) of the six (6) generating units within the period of investigation of the power station.

For every increase, it is either steady rise, sharp rise, an upward, trend, or a boom (a dramatic rise) and for every decrease either a decline, steady fall, sharp drop, a lump (a dramatic fall), or a reduction. Plateau normally levels out, does not change (steady), remained stable or stayed constant (maintained the same level).

Some of the reasons for the pattern exhibited by the different units' graphs are summarized. Some of these events are yearly repetitive and were summarized. The events (generated from the outage report and operators' log) which brought about the unavailability of the Egbin plant Units as reflected in the output graphs are: Industrial action, inspections and annual routine maintenance (RAM), annual overhauls, low gas head pressure making all BFP's trip, under frequency/ABC power failure, shutdown on ATS/Governor problem, SH output safety valve, tube leakage of secondary super heater, high main steam temperature, 330KV Switchyard inter-trip alarm, shattered current transformer in the Switch Yard, ground relay trouble, serious steam leakages, burners valve closed trip, condenser cleaning problems, de-mineralized water crisis, boiler tube leakage, natural gas header trip, fire outbreak due to frequency disturbance, bearings problems, heater bypass load runback failure, extreme low instrument air pressure, partial loss of flame, loss of excitation, generator hydrogen level, exploded furnace, system surge, unit service transformer fault, stage negative phase sequence, lifting of drum safety valve, very low main tank oil level, ATS failure, broken carbon brushes holders, loss of burner B1, generator main seal oil pump failure, pigging exercise at National Gas Company (NGC), superheater attemprator nozzle problem, generator rotor ground fault, Unit 6 was on forced outage due to furnace explosion and boiler tube leakage throughout the entire year 2006. Unit 6 was on forced outage due to furnace explosion and boiler tube leakage 2007 to 2011 in the years under review.

EGBIN DATA GENERATED FROM RAW FIELD DATA ARRANGED IN MATRIX FORM FOR ALL THE PARAMETERS ANALYSIS USING MATLAB SOFTWARE

The data in from the outage report from Egbin Power station rearranged, yielded the data used for MATLAB analysis as presented:

Some of the formulas amongst others inputted into the model program are listed as follow (IEEE Power Engineering Society, 2006)

Availability Factor – AF:
AF = (AH/PH) x 100 (%)
Forced Outage Rate – FOR:
FOR = (FOH/(FOH + SH)) x 100 (%)
Forced Outage Factor – FOF:
FOF = (FOH/PH) x 100 (%)
Service Factor – SF:
SF = (SH/PH) x 100 (%)

Planned Outage Factor – POF:
POH = (POH/PH) x 100 (%)
Where,
AH = Available Hours
PH = Period Hours
FOH = Forced Outage Hours
SH = Service Hours
POH = Planned Outage Hours

Egbine input data (from 2004 - 2011)

```
ESDH_Egbin = (ESDH2004;ESDH2005;ESDH2006;ESDH2007;ESDH2008;ESDH2009;ESDH2010;ESDH2011);
%========================================================
%FOH = input('Enter Forced Outage Hours = ');
FOH2004 = (215.67 127.25 480.32 238.52 308.43 204.23);
FOH2005 = (410.23 273.6 252.37 183.47 216.3 125.93);
FOH2006 = (384.07 207.63 1218.68 121.98 1194.77 0.00);
FOH2007 = (1501.02 1153.12 8230.56 757.43 1576.3 8760.00);
FOH2008 = (568.2 478.27 237.32 926.55 1137.82 8760.00);
FOH2009 = (460.57 474.49 1350.68 561.78 493.19 8760);
FOH2010 = (8.11 764.38 724 442.03 733.36 8760.00);
FOH2011 = (189.04 211.43 809.05 525.32 412.9 8760.00);
FOH_Egbin = (FOH2004;FOH2005;FOH2006;FOH2007;FOH2008;FOH2009;FOH2010;FOH2011);
%========================================================
%SH = input('Enter Service Hours = ');
SH2004 = (6535.39 6552.34 6789.34 7345.42 6360.75 7029.51);
SH2005 = (6654.76 7647.14 6947.38 7363.54 7270.58 7292.31);
SH2006 = (5845.43 6131.01 4712.19 4673.40 4960.67 941.52);
SH2007 = (3818.70 5636.87 436.32 4446.15 4747.05 0.00);
SH2008 = (5368.18 4669.41 1708.68 4971.34 5969.25 0.00);
SH2009 = (3450.37 4557.81 2563.45 3243.12 3662.06 0.00);
SH2010 = (330.79 6772.50 6709.79 7012.98 6022.14 0.00);
SH2011 = (7889.08 6479.23 6182.52 6694.26 6239.24 0.00);
SH_Egbin = (SH2004;SH2005;SH2006;SH2007;SH2008;SH2009;SH2010;SH2011);
%========================================================
%SS = input('Enter Starting Successes = ');
SS2004 = (25 28 35 32 22 19);
SS2005 = (44 44 41 24 32 30);
SS2006 = (33 35 32 21 37 2);
SS2007 = (41 41 0 30 41 0);
SS2008 = (32 33 13 34 38 0);
SS2009 = (35 30 38 36 28 0);
SS2010 = (4 45 49 63 42 0);
SS2011 = (24 19 31 36 29 0);
SS_Egbin = (SS2004;SS2005;SS2006;SS2007;SS2008;SS2009;SS2010;SS2011);
%========================================================
%SA = input('Enter Start Attempts = ');
SA2004 = (28 31 38 32 22 20);
SA2005 = (46 44 47 33 33 33);
SA2006 = (35 35 33 22 39 4);
SA2007 = (42 42 1 39 41 0);
SA2008 = (40 30 14 39 40 0);
SA2009 = (37 30 38 39 33 0);
SA2010 = (4 46 54 66 45 0);
SA2011 = (25 21 32 46 31 0);
SA_Egbin = (SA2004;SA2005;SA2006;SA2007;SA2008;SA2009;SA2010;SA2011);
%========================================================
%POH = input('Enter Planned Outage Hours = ');
POH2004 = (670.00 874.47 526.87 498.12 659.48 106.98);
POH2005 = (475.22 35.43 24.77 0.50 0.00 4.75);
POH2006 = (285.25 336.17 0 1006.07 185.68 0);
POH2007 = (304.85 191.8 0.00 845.73 38.02 0);
POH2008 = (148.08 273.25 0 501.83 141.62 0);
POH2009 = (253.63 0 397.72 294.55 155.18 0);
POH2010 = (0.00 208.35 0 243 378.69 0.00);
POH2011 = (1.93 171.5 701.88 192.21 166.67 0);
POH_Egbin = (POH2004;POH2005;POH2006;POH2007;POH2008;POH2009;POH2010;POH2011);
%========================================================
%MWH = input('Enter Megawatt Hour Produced = ');
MWH2004 = (1339755 1310468 1412183 1432356 1202182 1265311);
MWH2005 = (1364226 1529428 1458950 1435890 1381410 1422001);
MWH2006 = (1052177 919652 918877 925333 992133 195836);
MWH2007 = (706460 1014636 85083 880338 949410 0);
MWH2008 = (1052164 887188 324649 994267 1128188 0);
MWH2009 = (690074 865983 487056 648624 692130 0);
MWH2010 = (67811 1408680 1341958 1332467 1234539 0);
MWH2011 = (1617262 1347680 1236504 1271909 1279044 0);
MWH_Egbin = (MWH2004;MWH2005;MWH2006;MWH2007;MWH2008;MWH2009;MWH2010;MWH2011);
%========================================================
%NPC = input('Enter Nameplate Capacity = ');
NPC2004 = (220 220 220 220 220 220);
NPC2005 = (220 220 220 220 220 220);
NPC2006 = (220 220 220 220 220 220);
NPC2007 = (220 220 220 220 220 220);
NPC2008 = (220 220 220 220 220 220);
NPC2009 = (220 220 220 220 220 220);
NPC2010 = (220 220 220 220 220 220);
NPC2011 = (220 220 220 220 220 220);
NPC_Egbin = (NPC2004;NPC2005;NPC2006;NPC2007;NPC2008;NPC2009;NPC2010;NPC2011);
%========================================================
```

```
%RSH = input('Enter Reserve Shotdown Hours = ');
RSH2004 = (0 0 0 0 0 0);
RSH2005 = (0 0 0 0 0 0);
RSH2006 = (0 0 0 0 0 0);
RSH2007 = (0 0 0 0 0 0);
RSH2008 = (0 0 0 0 0 0);
RSH2009 = (0 0 0 0 0 0);
RSH2010 = (0 0 0 0 0 0);
RSH2011 = (0 0 0 0 0 0);
RSH_Egbin = (RSH2004;RSH2005;RSH2006;RSH2007;RSH2008;RSH2009;RSH2010;RSH2011);
%=========================================================
%FON = input('Enter Forced Outage Number = ');
FON2004 = (34 27 42 36 18 30);
FON2005 = (45 44 39 22 31 23);
FON2006 = (32 20 34 16 38 2);
FON2007 = (38 39 1 34 39 0);
FON2008 = (35 25 12 34 39 0);
FON2009 = (33 29 34 37 30 0);
FON2010 = (5 45 54 65 40 0);
FON2011 = (25 21 32 46 31 0);
FON_Egbin = (FON2004;FON2005;FON2006;FON2007;FON2008;FON2009;FON2010;FON2011);
%=========================================================
%FOH = input('Enter Full Forces Outage Hours = ');
FOH2004 = (215.67 127.25 480.32 238.52 308.43 204.23);
FOH2005 = (410.23 273.6 252.37 183.47 216.3 125.93);
FOH2006 = (384.07 207.63 1218.68 121.98 1194.77 0.00);
FOH2007 = (1501.02 1153.12 8230.56 757.43 1576.3 8760.00);
FOH2008 = (568.2 478.27 237.32 926.55 1137.82 8760.00);
FOH2009 = (460.57 474.49 1350.68 561.78 493.19 8760);
FOH2010 = (8.11 764.38 724 442.03 733.36 8760.00);
FOH2011 = (189.04 211.43 809.05 525.32 412.9 8760.00);
FOH_Egbin = (FOH2004;FOH2005;FOH2006;FOH2007;FOH2008;FOH2009;FOH2010;FOH2011);
%=========================================================
%EPOH = input('Enter Equivalent Planned Outage Hours, EPOH = ');
EPOH2004 = (670.00 874.47 526.87 498.12 659.48 106.98);
EPOH2005 = (475.22 35.43 24.77 0.50 0.00 4.75);
EPOH2006 = (285.25 336.17 0 1006.07 185.68 0);
EPOH2007 = (304.85 191.8 0.00 845.73 38.02 0);
EPOH2008 = (148.08 273.25 0 501.83 141.62 0);
EPOH2009 = (253.63 0 397.72 294.55 155.18 0);
EPOH2010 = (0.00 208.35 0 243 378.69 0.00);
EPOH2011 = (1.93 171.5 701.88 192.21 166.67 0);
EPOH_Egbin = (EPOH2004;EPOH2005;EPOH2006;EPOH2007;EPOH2008;EPOH2009;EPOH2010;EPOH2011);
%=========================================================
%MOH = input('Enter Full maintenance Outage Hours = ');
MOH2004 = [0.00 174.67 0.00 0.00 136.00 38.54];
MOH2005 = [201.93 480.85 490.93 24.00 398.00 0.00];
MOH2006 = [0 0 0 0 0 7960.52];
MOH2007 = [784.18 200 0 0 0 0];
MOH2008 = [0 254.8 6717.43 320.83 200.00 0];
MOH2009 = [1003.19 4.68 602.71 831.49 924.88 0];
MOH2010 = [316.29 100.39 0 0 0 0];
MOH2011 = [71.2 279.37 0 221.36 461.46 0];

MOH_Egbin = [MOH2004;MOH2005;MOH2006;MOH2007;MOH2008;MOH2009;MOH2010;MOH2011];(souce: NCC, 2011)
%=========================================================
%RUNCAP = input('Enter Running Capacity = ');
RC2004 = (205 200 208 195 189 180);
RC2005 = (205 200 210 195 190 195);
RC2006 = (180 150 195 198 200 208);
RC2007 = (185 180 195 198 200 0.0);
RC2008 = (196 190 190 200 189 0.0);
RC2009 = (200 190 190 200 189 0.0);
RC2010 = (205 208 200 190 205 0.0);
RC2011 = (205 208 200 190 205 0.0);
RC_Egbin = (RC2004;RC2005;RC2006;RC2007;RC2008;RC2009;RC2010;RC2011);
%=========================================================
%FAILEDSTART = input('Enter Failed Starts = ');
FS2004 = (3 3 3 0 0 1);
FS2005 = (2 0 6 9 1 3);
FS2006 = (2 0 1 1 2 2);
FS2007 = (1 1 1 9 0 0);
FS2008 = (2 3 1 5 2 0);
FS2009 = (2 0 0 3 5 0);
FS2010 = (0 1 5 3 3 0);
FS2011 = (1 2 1 10 2 0);
FS_Egbin = (FS2004;FS2005;FS2006;FS2007;FS2008;FS2009;FS2010;FS2011);
%=========================================================
%NOOFSTOP = input('Enter No Of Stops = ');
NOS2004 = (28 31 38 32 22 20);
NOS2005 = (46 44 47 33 33 33);
NOS2006 = (35 35 33 22 39 4);
NOS2007 = (42 42 1 39 41 0);
NOS2008 = (40 30 14 39 40 0);
NOS2009 = (37 30 38 39 33 0);
NOS2010 = (5 46 53 65 45 0);
NOS2011 = (24 20 31 45 31 0);
NOS_Egbin = (NOS2004;NOS2005;NOS2006;NOS2007;NOS2008;NOS2009;NOS2010;NOS2011);
%=========================================================
```

However, the analysis modeling output result shows some abnormally high availability values for some units which would not reflect the real situation. This may have been caused by frequent shutdowns and data manipulations; recording patterns which does not align with the IEEE std 762 reporting standards.

Parameters analysis

Data were generated for a total number of 22 parameters and indices from the data inputs entered. A corresponding numbers of graphs were also plotted after analysis by using MATLAB software. But only few out of the 22 parameters and indices are presented here. Some of the input data are also presented. The reasons for the graphical patterns are also presented as deduced from the outage report with reasons for major outages experienced within this period of seven years. They are presented before the final summary (Figure 2a-e).

Egbin output result data from MATLAB

Using the above data as inputs for the software program written, the following output data results and graphs are generated as presented in this article.

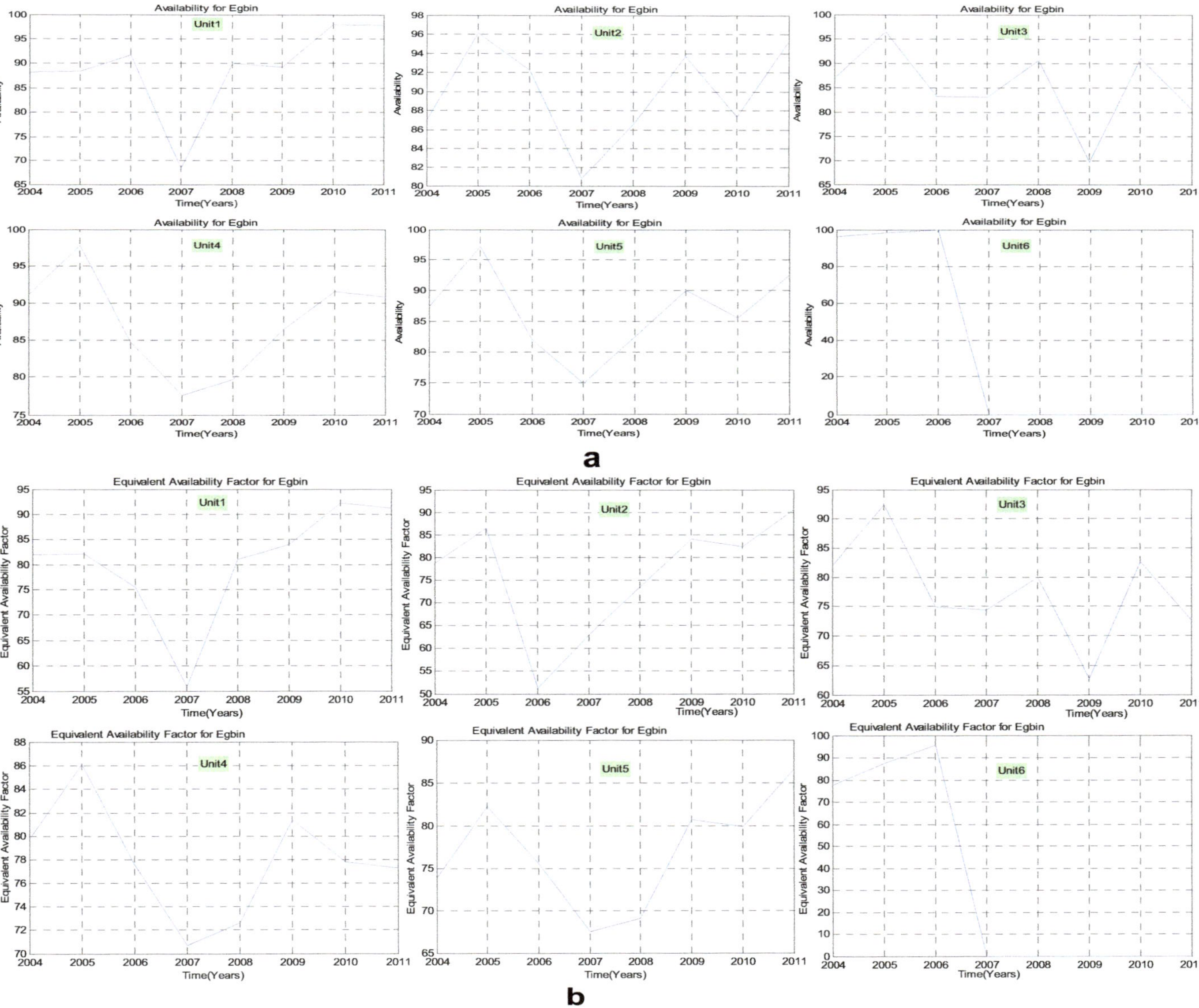

Availability for Egbin
Unit1
Availability
Time(Years)
Availability for Egbin
Unit2
Availability
Time(Years)
Availability for Egbin
Unit3
Availability
Time(Years)
Availability for Egbin
Unit4
Availability
Time(Years)
Availability for Egbin
Unit5
Availability
Time(Years)
Availability for Egbin
Unit6
Availability
Time(Years)
a
Equivalent Availability Factor for Egbin
Unit1
Equivalent Availability Factor
Time(Years)
Equivalent Availability Factor for Egbin
Unit2
Equivalent Availability Factor
Time(Years)
Equivalent Availability Factor for Egbin
Unit3
Equivalent Availability Factor
Time(Years)
Equivalent Availability Factor for Egbin
Unit4
Equivalent Availability Factor
Time(Years)
Equivalent Availability Factor for Egbin
Unit5
Equivalent Availability Factor
Time(Years)
Equivalent Availability Factor for Egbin
Unit6
Equivalent Availability Factor
Time(Years)
b

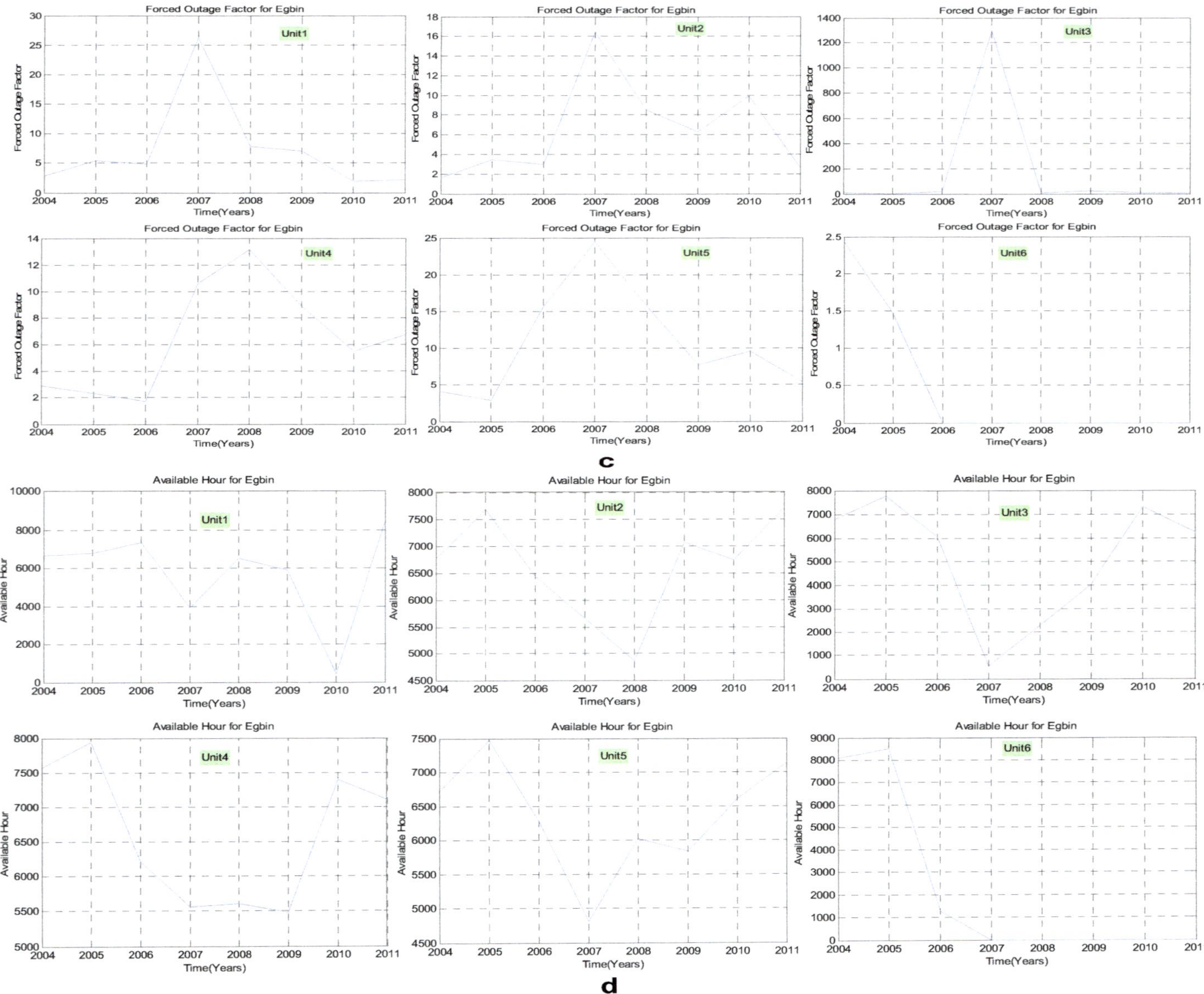
Forced Outage Factor for Egbin
Unit1
Unit2
Unit3
Unit4
Unit5
Unit6
Forced Outage Factor
Time(Years)
c
Available Hour for Egbin
Unit1
Unit2
Unit3
Unit4
Unit5
Unit6
Available Hour
Time(Years)
d

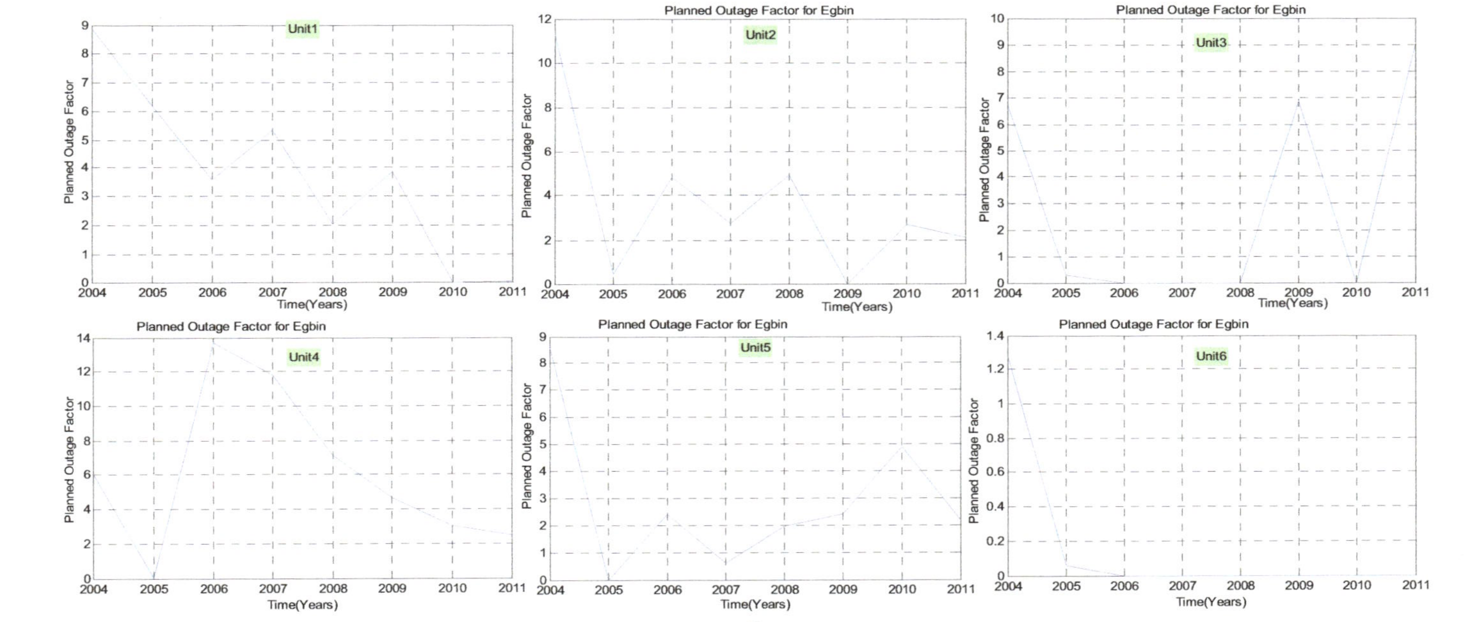

Figure 2a-e. Graphical outputs from analysis of data using MATLAB a. Egbin Availability for 2004 – 2011; b. Egbin Equivalent Availability for 2004 – 2011; c. Egbin Forced Outage Factor for 2004 – 2011; d. Egbin Available Hours for 2004 – 2011; e. Egbin Planned Outage Factor for 2004-2011.

Egbin output parameters result data op4(6)

Egbin availability from 2005-2011 =

88.2866	87.1148	87.1161	91.1398	87.4070	96.2982
88.4509	96.1323	96.5450	97.7360	97.1865	98.4844
91.6727	92.2084	83.2195	84.6210	81.9940	100.0000
68.4186	80.8300	83.1511	77.6002	74.8853	0
90.1052	86.5999	90.5897	79.6958	82.4783	0
89.1786	93.7098	69.7794	86.4976	89.9949	0
98.0677	87.3922	91.0024	91.5219	85.5928	0
97.8052	95.2711	80.5018	90.8272	92.4916	0

Egbin availability factor from 2005-2011 =

88.2866	87.1148	87.1161	91.1398	87.4070	96.2982
88.4509	96.1323	96.5450	97.7360	97.1865	98.4844
91.6727	92.2084	83.2195	84.6210	81.9940	100.0000
68.4186	80.8300	83.1511	77.6002	74.8853	0
90.1052	86.5999	90.5897	79.6958	82.4783	0
89.1786	93.7098	69.7794	86.4976	89.9949	0
98.0677	87.3922	91.0024	91.5219	85.5928	0
97.8052	95.2711	80.5018	90.8272	92.4916	0

Egbin equivalent availability factor from 2005-2011 =

81.9622	78.6865	82.1057	79.8130	73.8332	77.7170
82.0997	86.5614	92.4206	86.1182	82.2539	87.6416
75.5116	51.2137	74.9010	77.5416	75.5235	95.7546
55.7841	62.9753	74.3655	70.6976	67.5000	0
81.0247	73.4538	79.8920	72.6292	69.0699	0
83.9507	84.1697	62.7833	81.3840	80.7260	0
92.3018	82.3279	82.6638	77.8177	79.8841	0
91.1707	90.6550	72.5234	77.3147	86.5772	0

Egbin forced outage factor from 2005-2011 =

1.0e+003 *

0.0029	0.0016	0.0061	0.0029	0.0040	0.0024
0.0054	0.0034	0.0031	0.0023	0.0028	0.0015
0.0048	0.0030	0.0168	0.0017	0.0156	0
0.0263	0.0164	1.2926	0.0106	0.0245	0
0.0078	0.0085	0.0094	0.0132	0.0156	0
0.0070	0.0063	0.0233	0.0089	0.0076	0
0.0019	0.0099	0.0090	0.0055	0.0095	0
0.0022	0.0026	0.0104	0.0067	0.0053	0

Egbin service factor from 2005-2011 =

86.4337	84.2835	86.8486	88.3495	82.7559	83.6156
86.7995	95.7092	86.6106	90.6193	94.5728	84.5742
72.7252	87.8458	64.8841	63.7138	64.7047	73.5850
66.7821	80.3461	68.5262	62.1228	73.8520	0
74.1570	83.2585	67.7529	70.6668	81.7479	0
52.2791	60.4215	44.3085	51.1365	56.5098	0
78.8139	87.7802	83.3863	86.7940	78.0200	0
90.6704	80.0132	79.7840	85.5786	80.8300	0

Egbin starting reliability from 2005-2011 =

89.2857	90.3226	92.1053	100.0000	100.0000	95.0000
95.6522	100.0000	87.2340	72.7273	96.9697	90.9091
94.2857	100.0000	96.9697	95.4545	94.8718	50.0000
97.6190	97.6190	0	76.9231	100.0000	0
80.0000	110.0000	92.8571	87.1795	95.0000	0
94.5946	100.0000	100.0000	92.3077	84.8485	0
100.0000	97.8261	90.7407	95.4545	93.3333	0
96.0000	90.4762	96.8750	78.2609	93.5484	0

Egbin planned outage factor from 2005-2011 =

8.8611	11.2484	6.7397	5.9913	8.5801	1.2725
6.1984	0.4434	0.3088	0.0062	0	0.0551
3.5489	4.8167	0	13.7160	2.4219	0
5.3313	2.7339	0	11.8168	0.5915	0
2.0456	4.8722	0	7.1334	1.9395	0
3.8429	0	6.8745	4.6444	2.3946	0
0	2.7005	0	3.0074	4.9061	0
0.0222	2.1179	9.0576	2.4572	2.1592	0

Egbin capacity factor from 2005-2011 =

0.8054	0.7662	0.8211	0.7831	0.7109	0.6841
0.8088	0.8701	0.8267	0.8032	0.8168	0.7496
0.5950	0.5989	0.5751	0.5734	0.5882	0.6957
0.5616	0.6574	0.6074	0.5591	0.6714	0
0.6607	0.7191	0.5851	0.6424	0.7023	0
0.4753	0.5218	0.3827	0.4649	0.4855	0
0.7344	0.8299	0.7581	0.7496	0.7270	0
0.8449	0.7565	0.7253	0.7391	0.7532	0

Egbin forced outage rate from 2005-2011 =

3.1946	1.9051	6.6072	3.1451	4.6247	2.8233
5.8065	3.4542	3.5053	2.4310	2.8891	1.6976
6.1653	3.2756	20.5481	2.5437	19.4100	0
28.2161	16.9826	94.9657	14.5559	24.9282	100.0000
9.5715	9.2910	12.1953	15.7099	16.0097	100.0000
11.7765	9.4289	34.5078	14.7646	11.8691	100.0000
2.3930	10.1419	9.7393	5.9293	10.8557	100.0000
2.3401	3.1601	11.5718	7.2763	6.2070	100.0000

Egbin Fp from 2005-2011 =

0.9790	0.9675	0.9969	0.9694	0.9468	0.8683
0.9813	0.9956	0.8971	0.9272	0.9731	0.8588
0.7933	0.9527	0.7797	0.7529	0.7891	0.7358
0.9761	0.9940	0.8241	0.8005	0.9862	0
0.8230	0.9614	0.7479	0.8867	0.9911	0
0.5862	0.6448	0.6350	0.5912	0.6279	0
0.8037	1.0044	0.9163	0.9483	0.9115	0
0.9271	0.8398	0.9911	0.9422	0.8739	0

Egbin Ff from 2005-2011 =

0	0	0	0	0	0
0	0	0	0	0	0
0	0	0	0	0	0
0	0	0	0	0	0
0	0	0	0	0	0
0	0	0	0	0	0
0	0	0	0	0	0
0	0	0	0	0	0

Egbin EFORD from 2005-2011 =

10.0367	12.9122	7.7364	6.5738	9.8163	1.3214
7.0077	0.4613	0.3198	0.0063	0	0.0559
3.8713	5.2237	0	16.2088	2.9538	0
7.7921	3.3822	0	15.2278	0.7899	0
2.2702	5.6261	0	8.9508	2.3515	0
4.3093	0	9.8517	5.3694	2.6608	0
0	3.0901	0	3.2860	5.7319	0
0.0227	2.2230	11.2514	2.7053	2.3345	0

Egbin maintenance outage factor from 2005-2011 =

0	2.2468	0	0	1.7694	0.4584
2.6338	6.0182	6.1203	0.2954	5.1770	0
0	0	0	0	0	622.1587
13.7139	2.8507	0	0	0	0
0	4.5432	266.3607	4.5605	2.7390	0
15.2001	0.0620	10.4177	13.1107	14.2720	0
75.3592	1.3012	0	0	0	0
0.8183	3.4500	0	2.8298	5.9783	0

SUMMARY

Egbin Power Plant had most of the failures related to incessant outages occasioned by both issues within plant management control and outside plant management control. Majority of the reasons within and without plant management control as shown in the outages reasons earlier. Availability of turbine unit generator as expressed earlier is the percent of time; the turbine is available to generate power in any given period at its acceptance load. This specifies that higher percent value means high availability of the plant units, while low indicates limitations in power generation capacity.

The results obtained are a combination of the graphical output trend and the results from the output summaries and averages above. From the results, the availability of the units peaked at various times of the years. ST01 peaked at 98.07% in 2010, while ST02, 03, 04, 05 and 06 peaked at 96.13, 96.55, 97.19 and 98.48%, respectively in 2005. The overall availability of the entire units averaged at 89.00%. ST01 has the highest availability in 2010. But the year 2007 had the lowest percent values for majority of the station units. Units ST01, ST02, ST04, and ST05 had their lowest values in 2007 as 68.42, 80.83, 77.60, and 74.89, respectively. ST06 went out on Furnace explosion in 2006 but had its lowest value at 0.0% but for the period it run, it had 96.30% as it's lowest, while ST03 decreased significantly between 2008 and 2009 to as low of about 69.78% availability. Equivalent availability factor which indicates that both full forced outages and deratings which has characterized the entire units has been considered in the evaluations and also shows that availability is limited majorly by outages which is also revealed in the same trend as seen from the graphical output result above (Tables 2 and 3).

FINDINGS

There is gross inconsistency in data presentation coupled with incoherent and non-uniform presentation of operational activities, particularly in data presentations. The failure rate which is a determinant of reliability and availability is a reasonable measure for stability of generating units and indication for economical effectiveness of repairs. On the overall, the trend of availability and other indices and parameters fluctuated greatly within the period of investigation and on the average, could not reach up to the expected benchmark within the seven years span owning to reasons given above for their unavailability.

When we reconcile these results output values to the parameters and indices definitions and implications on generators (NERC/IEEE std 762), it becomes clear that some of the units' generators performed below potentials. The high values of availability and other parameters were due to the fact that full and prorated partial forced outage hours are not accounted for. However, it is likely that the time to restore a unit to full capability would average more than five hours for a single generator during demand periods. It is much more probable that the total forced outage hours would be several times higher (some previous studies suggest that the average restoration time for a gas turbine forced outage is on the order of 24 h for base loads) (Richwine, 2004).

However, equivalent availability is another index considered very effective in this regards. It is another measurement which can be tracked based on outage reporting style; it has become increasingly popular in the new power performance measurement. This is not same with the traditional time-based availability measurement expressed above (GE Power systems, 2000). Equivalent availability considers the lost capacity effects of partial equipment deratings and reports those effects as equivalent unavailable hours (GE Power systems, 2000). For example, if a unit operated for 100 h with an equipment limitation at 80% of nominal rated capacity, it would be considered to have accrued 100 h x 20% derating = 20 equivalent derated hours. For operating hours of 100 h, the traditional (time-based) availability would show as 100%; but, the equivalent availability

Table 2. The averages of overall summary of all parameters (Total) and indices for Egbin Power Station (Source: Fergus (2015) Unpublished material: The Inherent Energy Crisis in Nigeria).

Generator parameter	Unit						Station sum	Averages
	1	2	3	4	5	6		
Availability for 2004-2011	89	89.91	85.24	87.45	86.5	25.71	463.81	77.3
Availability factor for 2004-2011	89	89.91	85.24	87.45	86.5	25.69	463.79	77.3
Equivalent availability for 2004-2011	80.48	76.26	77.71	77.91	76.92	32.64	421.91	70.32
Forced outage factor for 2004-2011	0.01	0.01	0.17	0.01	0.01	0	0.2	0.03
Service factor for 2004-2011	76.08	82.46	72.76	74.87	76.62	30.22	413.02	5.04
Starting reliability for 2004-2011	93.43	98.28	82.1	87.29	94.82	29.49	485.41	80.9
Planned outage factor for 2004-2011	3.73	3.62	2.87	6.1	2.87	0.17	19.36	3.23
Capacity factor for 2004-2011	0.69	0.71	0.66	0.66	0.68	0.27	3.67	0.61
Forced outage rate for 2004-2011	8.68	7.2	24.21	8.29	12.1	63.07	123.55	20.59
Partial forced outage, Pf for 2004-2011	0.86	0.92	0.85	0.85	0.89	0.31	4.67	0.78
Full forced outage for 2004-2011	0	0	0	0	0	0	0	0
Equiv. forced outage rate Dd 2004-2011	4.41	4.11	3.64	7.29	3.33	0.17	22.97	3.83
Maintenance factor for 2004-2011	13.47	1.99	2.64	35.32	3.97	78.17	135.56	22.59

Table 3. Summary of all availability and performance parameters for Egbin Power Stations (2005-2011). (Source: Fergus E.O. (2015) Unpublished material: The Inherent Energy Crisis in Nigeria).

Power station	Availability	Availability factor	Equivalent availability factor	Forced outage factor	Service factor	Starting reliability	Planned outage factor	Capacity factor	Forced outage rate	Partial forced outage, Fp	Full Forced outage factor, Ff	EFORD	Maintenance outage factor
Average stations values	88.35	88.34	80.36	0.04	78.67	92.46	3.69	0.70	23.53	0.89	0.00	4.37	25.8

would equal 100 available hours minus the 20 equivalent derated hours for a measure of 80% (GE Power systems, 2000). This parameter could however not be used because incomplete data recording style observed generally in this Power stations.

For a good and balanced power generation system, the availability requirements should be as follows:

i. The unit generator should be = 97% which means a maximum of 11 days in a given year period of unavailability for reason of unplanned repair or maintenance etc. The important components of the unit generator should have availability of 94% minimum.

ii. The fuel supply should have the availability of 99.5% etc, but these were not the case here.

iii. The evaluation of power plant performance should be one of the most important tasks at any power station. Without its availability records, the plant staff and stakeholders cannot determine ways to improve performance of the equipment and make the plant more profit-oriented for plant owners. The causes of unavailability must be thoroughly analysed to identify the areas for generators performance improvement.

This study can be said to have provided some corresponding levels of potential and cost-effective improvements from the use of performance parameters to improve unit availability. This can be justified by using the Richwine model of electricity generation standards to justify the subject of availability using this illustration: For instance, assuming total installed power capacity in the power station within this period under review as 1320MW. On the basis that we consider the total installed capacity of 1320MW. From study findings, most of the units have derated either due to spare supply shortage or due to ageing, and hence we consider this value for illustration only.

One percent improvement in availability that can be achieved and sustained is equivalent to approximately 15.53MW of new capacity at 85% availability. To arrive at that figure we calculate the Available Capacity as the product of the capacity times the availability. Therefore a 1% improvement in Availability would result in a 13.2MW increase in Available Capacity only if that capacity were 100% available. But for a more realistic availability goal we might chose 85% (considering the average of the running units' availability) so that the 7.6 MW at 100% availability would be equal to 15.53MW at 85% availability (13.2/0.85). However, it is also apparent that not all plants and sectors have equal opportunity to achieve the same levels of cost-effective availability improvement. Hence, if the total availability improvement that can be achieved and sustained is 14%, then the total equivalent capacity represented by this availability improvement would be 217.4MW.

The assumption of 14% is made based on the nature of data available and the performance of their peers in other parts of the world, and considering the unique set of conditions in some of these generators (base loads).

It should be noted, however, that this improvement will not happen overnight, but rather will be a process that will take place over several years. The time required for the performance improvement can be minimized by taking advantage of other company's experiences to 'get down the learning curve' as quickly as possible:

i.e.: at 1% improvement in availability;

$\Rightarrow 1320\text{MW} \times \frac{1}{100} = 13.2\text{MW}$

Then if we consider a realistic availability goal of 85% of the above 13.2MW,

Then, we have: $\frac{13.2}{0.85} \cong 15.53\text{MW}$

But at 14% achievable and sustainable availability for these steam turbine-units;

Will give 15.53 × 14 = 217.4MW;

The total equivalent capacity represented by this availability will be = 217.4MW.

Some basic questions with regards to information gathering, data sourcing, collation and analysis to evaluate the inherent energy crisis have been formulated into action statements used to remedial actions to fill some of the existing gaps in the Egbin and energy sector at large.

Thus, we can conclude here that this research analysis highlights significantly the amount of potential "equivalent energy producing capability increase" that can be cost-effectively achieved by improving the availability of existing electricity generating units in Egbin power station to optimum levels.

Some basic questions with regards to information gathering, data sourcing, collation and analysis to evaluate station availability in order to ameliorate the energy crisis in Nigeria have been formulated into action statements used to remedial actions to fill some of these existing gaps in the power plant management.

MANAGING THE FUTURE

The benefits of pooling data for performance and availability monitoring system henceforth depends – in addition to the current procedures described in this paper - on the commitment of power plant operators and the energy regulators to enhance them. The underlying goal is to encourage increased production and international participation.

Key factors influencing plant performance should be identified and analysed to allow a cost benefit analysis of any activity/programme before its implementation. Strong political will is needed to handle the implementation of deregulation policies.

To analyze plant availability performance, the energy losses/outages should be scrutinised to identify the causes of unplanned or forced energy losses and to reduce the planned energy losses. Reducing planned outages increases the number of operating hours, decreases the planned energy losses and therefore, increases the energy availability factor. Reducing unplanned outages leads to a safe and reliable operation, and also reduces energy losses and increases energy availability factor (Pierre et al., 2008).

Conclusion

The inherent energy availability of power generation units in Egbin power plant in Nigeria has been investigated. Some possible causes of unavailability have been identified. Ways to overcome the causes comparable to international peers have been presented. The results of analysis through the use of software have justifiably outlined the areas of weakness in the power units. The study has touched areas of availability likely to be encountered by power plants generation managers in other power stations in Nigeria.

The study is a lead study product especially in the area of conventional power plant units' availability management that satisfies international standards as well as foundation for further researches in the field of National power availability and performances analysis in Nigeria. Generally, the facts presented alone in the study are sufficient to exhibit the importance of power availability and performance measurement in enhancing the Nigeria's energy revolution and development.

This paper challenges the widespread practice of abuse in the use of relevant parameters and indices for the determination of generator performance improvements for a healthy electricity generation, profitability and sustainability in the plant and in Nigeria.

The analysis is self-contained and gives a useful practical introduction to standard availability performance evaluations and monitoring. The indices and parameters analysis are presented in most lucid and compact manner for proper understanding especially in data arrangement and tabulations. The process and techniques applied to achieve this goal are fully articulated. Results output presentations and analysis have been covered in the most logical manner from the IEEE power plant standard availability evaluations ideology. However, to design all-encompassing tables of indices and parameters for effective availability measurement more detailed than the NERC/IEEE std 762 typically put forward requires in-depth field experience for sustainable robust results. The introduction of reasonable key performance measures, such as some Availability Value Indicators (a measure of Commercial Availability) will enable the Power station to be one of the leaders in measuring the economic value of its generators in Nigeria. Some of these new indicators have prototyped and showed success in other countries energy industries. Hence, the research provides a comprehensive strategy for other power stations to follow, and appears to be a positive step towards achieving more satisfactory integration in the industry. The evolution of "data analysis" and statistics ensures other factors/ goals are set.

RECOMMENDATIONS

1. Government through NERC should set up generating plant examining board. The board members should comprise well selected best-qualified and most respected individuals in their respective fields (Plant Engineering Design, Plant Management, Operations, Maintenance, etc.) from amongst all of the operating power plants in the country. The board shall review annually the condition of each power plant and make recommendations to executive management and owners of plants concerning actions and expenditures required to achieve performance improvement. This will help local staff to gain knowledge and also help the plant owners to allocate resources equitably.
2. The power station should align in the development of very well enhanced equipment specific Operations and Maintenance (O & M) procedures programs.
3. The power station should embrace the use of powerful software for analyses of the various performance parameters and indices. The result will be beneficial in the exchange of information and monitoring of station units performance trend allowable for improvement of performance of power generating assets in the station and to improve the quality of life to its users.
4. In alignment with other typical industry players, there is need for optimum spare parts management. Spare parts management plays a very important role in the achievement of the desired power generation availability at optimum cost. This will remove the unique problems of controlling and managing spare parts such as element of uncertainty and unavailability.
5. There should be pre-fixed meeting day for plant manager and senior executives in the power station to review all outages where each department is required to explain each outage event and to state the root cause of the problem, the immediate short term solution applied and results in addition to the long term solution that would eliminate or minimize the problem. This will enable plant managers to offer their insights and perspectives to help find the best solution.
6. Load growth should be monitored locally from the station based on subsequent demand rates and frequency. This will help regulate incidences of system collapses.
7. The new owners of the plant (Generation) must now come out with a tested and trusted blueprint in system operations that must be flexible in implementations in the Nigeria environment to guarantee availability of electricity supply to consumers.
8. The plant staff should be fully involved in decision making when a considerable decision is to be made about the management of any power station particularly in the area of maintenances. After all, "The man that wears shoes knows where it pains/ hurts". This will improve performance and availability of the plant units and make the plant profit-oriented.
9. The economics of scale should apply when sitting Power Stations. In another way, the sitting of Power

stations should not be influenced politically or affected by ethnic sentiments. This guarantees adequate gas supply or other raw materials at the long run.

10. Generally, the regulatory authority should benchmark the unit generators in the power industry. The benchmarking philosophy will help Nigeria to achieve the following if properly implemented:

i. Set realistic, achievable goals,
ii. Identify best areas for potential improvement,
iii. Give advance warning of threats,
iv. Trade knowledge and experience with peers,
iv. Quantify and manage performance risks,
v. Create increased awareness of the potential for and the value of increased plant performance

11. There is need to set up a well-equipped effective efficiency department for data collection and analysis using the applicable KPIs and standards. The results of analysis and study will help to enable us have a good planning system in the station. The data collection and monitoring should align with the industry requirement to enable all the power plants harmonize reporting standard and procedure.

12. The issue of gas shortage or low gas pressure climaxed the unavailability of the various unit generators as deduced from system collapse records as well as reasons for outages summarized. A good fuel supply policy should be put in place. This will encourage consistent supply of raw material to the power stations.

13. The "best practices" in computer database should be developed for use by all Power industry's' staff. Nigeria must as a result of urgency align with the international community in providing the various generation parameters and performance data for the operation and regulation of the power industry.

14. The plant design organizations should henceforth provide increased engineering support to the operating plants staff particularly during design upgrade projects. This is very important in Nigeria as we seek to upgrade most of the old power plants either to increase availability or dependable capacity.

Conflict of Interest

The authors have not declared any conflict of interest.

REFERENCES

André Caillé C, Al-Moneef M, Barnés de Castro F (2007). "Performance of Generating Plant Managing the Changes", Executive Summary World Energy Council.

Casazza J, Delea F (2003). "Understanding Electric Power Systems, 'An Overview of the Technology and the Marketplace'" IEEE Press Understanding Science and Technology Series. A John Wiley & Sons, Inc., Publication 2003.

Chirikutsi J (2007). "Plant Performance Measurement in Hydro Power Plants: Zimbabwe Power Company Kariba South Power station", 19th African hydro symposium November 2007.

Egbin Thermal Power Station Business Unit, Lagos state, Nigeria.

GE Power Systems (2000). "Utility Advanced Turbine System (ATS) Technology Readiness Testing Phase 3 Restructured (3R)",Predicted, Predicted Reliability, Availability, Maintainability for the GE 7H Gas Turbine. Schenectady, NY, 2000.

IEEE Power Engineering Society (2006). "IEEE Standard Definitions for Use in Reporting Electric Generating Unit Reliability, Availability, and Productivity"(Revision of IEEE Std 762-1987).

Javad B (2005). "Improvement of System Availability Using Reliability and Maintainability Analysis", December 31:1-12.

Killich M (2006). "Operational Flexibility for Steam Turbines based on Service Contracts with Diagnostics Tools", Siemens Power Generation, Germany Power Gen Europe, Cologne, 2006.

Kopman SE, Barry R, Ronald MF (1995). "Description of Electric Generating Unit Availability Improvement Programs & Analysis Techniques" The Generating Availability Data System (GADS) Application and Benefits, June, 1995.

Richwine RR (2004). Performance of Generating Plant Committee, World Energy Council. 18.

Richwine (Bob) RR (2011). "Using Reliability Data to Improve Power Plant Performance", NERC-GADS Workshop presented by Richwine Reliability Management Consultant, Richwine Consulting Group, LLC 27 Oct, 2011.

Stein S, Cohen D (2003). "Rehabilitation of Steam Power Plants, an Approach to improve the Economy of thermal Power Generation" - ALSTOM Power Generation AG Germany & Brasil, 2003.

Tell magazine Energy Survey (2009). Nigeria is World's Darkest Nation, by Fola Adekey, July 27 2009, pp.18-24.

Pierre G, Barnés de Castro F, Asger BJ, Franco de Medeiros N, Richard D, Alioune F, Jain CP, Younghoon DK,M'Mukindia M, Marie-Jose N, Abubakar S, Johannes T, Elias VG, Zhang G, Gerald D (2008). Performance of Generating Plant: Managing the Changes" World Energy Council, 2008. 17.

Mahmoud A, Tavanir I, Reza S (2000). Identification of the Relationship between Availability, Reliability and Productivity" Power Plant Performance University of Hertfordshire, UK. 16.

Gad T (2007). Reviews. Orbis Litterarum, 19:47–48. doi: 10.1111/j.1600-0730.1964.tb00347.x

Multi-objective optimization of hybrid PV/wind/diesel/battery systems for decentralized application by minimizing the levelized cost of energy and the CO_2 emissions

B. Ould Bilal[1], D. Nourou[3], C. M. F Kébé[2], V. Sambou[2], P. A. Ndiaye[2] and M. Ndongo[3]

[1]Ecole des Mines de Mauritanie (EMiM), BP : 5259 Nouakchott-Mauritanie, Sénégal.
[2]Centre International de Formation et de Recherche en Energie Solaire (C.I.F.R.E.S), ESP-UCAD, BP: 5085 Dakar Fann Sénégal.
[3]Laboratoire de Recherche Appliquée aux Energies Renouvelables de l'eau et du froid (LRAER), FST- Université des Sciences de Technologies et de Médecines (USMT), BP: 5026 Nouakchott-Mauritanie, Sénégal.

The main objective of this paper is to propose a methodology to design and optimize a stand-alone hybrid PV/wind/diesel/battery minimizing the Levelized Cost of Energy (LCE) and the CO_2 emission using a Multi-Objectives Genetic Algorithm approach. The methodology developed was applied using the solar radiation, temperature and the wind speed collected on the site of Potou located in the northwestern coast of Senegal. The LCE and the CO_2 emission were computed for each solution and the results were presented as a Pareto front between LCE and the CO_2 emission. These results show that as the LCE increases the CO_2 emission decreases. For example, the solution A (left solution on the Pareto front) presents 2.05 €/kWh and 11.89 kgCO_2/year, however the solution E (right solution on the Pareto front) shows 0.77 €/kWh and 10,839.55 kgCO_2 /year. It was also noted that the only PV/battery or Wind/ battery was not an optimal configuration for this application on the site of Potou with the use of the load profile and the specifications of the used devices. For all solutions, the PV generator was more adapted to supply the energy demand than the wind turbines.

Key words: Hybrid system, optimization, genetic algorithm, cost of energy, CO_2 emission.

INTRODUCTION

The scarcity of conventional energy resources, the rise in the fuel prices and the harmful emissions from the burning of fossil fuels has made power generation from conventional energy sources unsustainable and unviable. It is estimated that this supply demand gap will continue to rise exponentially unless it is met by some other means of power generation. Further, inaccessibility of the grid power to the remote places and the lack of rural electrification have prompted to use other sources of energy (Prabodh and Vaishalee, 2012). In the remote regions, far from the grids, electric energy is usually supplied using diesel generators. In most of cases, supplying demand energy using diesel fuel is so expensive and increases the amount of CO_2 emitted. So, renewable energy resources (e.g. solar and wind energies) have become the better alternatives for conventional energy resources. However, the use of a single renewable energy source such as wind energy or solar energy is not adequate to meet the demand for long periods due to the intermittent nature of the renewable energy high, the cost of system as well as storage subsystem (Ayong et al., 2013; Bekele and Palm, 2010; Diaf et al., 2008 ; Ekren-O and Ekren- by, 2010; Kalantar and Mousavi, 2010; Kanase-Patil et al., 2011; Saheb-Koussa et al., 2009; Zhou et al., 2010). To meet this challenge, the renewable sources such as wind and solar energy can be used in combination with the conventional energy systems making a hybrid PV/wind/diesel/battery system. These kinds of systems could allow dropping the investment, operation and the maintenance costs of systems (Colle et al., 2004).

Hongxing et al. (2009); Kyoung-Jun et al. (2013); Mukhtaruddin et al. (2015) have designed PV/wind hybrid systems coupled or now to battery bank and diesel generator by using different methods. A performance and feasibly study of hybrid renewable hybrid system coupled to batteries was done in the work of Kyoung-Jun et al. (2013); Ajay et al. (2011); Suresh-Kumar and Manoharan (2014) and Ismail et al. (2014, 2013). However, in the most case, the system was PV/diesel, PV/wind or PV/wind/batteries system. Also, in these studies, the type of devices was not taking into account in the optimization.

Several others studies on the feasibility, performance, and economic viability of hybrid power systems have been conducted using Homer (Hybrid Optimization Modeling Software) in the works of Chong et al. (2013); Bahtiyar (2012); Mir-Akbar et al. (2011); Sanjoy and Himangshu (2009); Patrick et al. (2014); Ahmad et al. (2010); Rohit and Subhes (2014); Ahmed et al. (2011); Zeinab et al. (2012); Mei and Chee (2012); Dalton et al. (2009); Belgin and Ali (2011); Eyad (2009); Alam and Manfred (2010) and Muyiwa et al. (2014). With the use of Homer, the best hybrid renewable energy configuration is which have the lower cost. However, Homer does not take into account the number of components such as variable. Also, Homer uses the monthly meteorological data variation (solar and wind speed). With the hourly solar and wind speed variation, the configuration chosen is more adapted to supply the demand.

Iterative methods of optimization have used to minimize the Annualized Cost of System (ACS) (Yang et al., 2008, 2007, 2003; Yang and Lu, 2004; Diaf et al., 2008; Shen, 2009; Kellogg et al., 1998; Koutroulis et al., 2006). These methods made possible to study the optimization and the performance of a hybrid system. In the works of Duffo-Lopes and Bernal- Agustin (2005, 2008); Senjyu et al. (2007); Ekren and Ekren (2008) and Hongxing et al. (2009), authors have studied the performance of hybrid systems using genetic algorithms minimizing the cost of the system. The methods outlined in these works did not take into account all devices of the system such as wind turbine, PV module, regulator, battery, inverter and diesel generator. Ould Bilal et al. (2010) have designed and optimized hybrid PV/wind/battery systems minimizing the ACS and the Loss of Power Supply Probability (LPSP). The authors did not take into account the diesel generator. So, the CO_2 emission was not evaluated. Since the presented problem is a Multi-objective Optimization Problem (MOP), it requires a multi-objective method for solving. This paper utilizes a Pareto-based approach which can obtain a set of optimal solutions instead of only one (Mohammad et al., 2014). The objectives to be minimized in this paper are the levelized cost of energy and the CO_2 emission.

The developed methodology was applied using the solar radiation, temperature and the wind speed collected in the site of Potou located in the northwestern coast of Senegal. The decision variables included in the optimization process are the number of PV modules, the number of wind turbines, the number of batteries, the number of regulators, the number of inverters, the number of diesel generators and the type of each device.

APPROACH AND METHODOLOGY

Hybrid solar-wind-diesel power generation system coupled to the battery bank consists of a PV module, wind turbine, diesel generator, regulator, battery bank and an inverter. A schematic diagram of the basic hybrid system is shown in Figure 1. The PV module and the wind turbine work together to meet the load demand. When the energy sources (solar and wind energy) are sufficient, the generated power, after meeting the load demand, provides energy to the battery up to its full charge. The battery supplies energy demand to help the system to cover the load requirements, when energy from renewable energy is inferior to the load demand. The load will be supplied by diesel generators when power generation by both wind turbine and PV array is insufficient and the storage is depleted. The mathematical model of the components used in this study are detailed by the Ould Bilal et al. (2012a, b).

In this paper, the Levelized Cost of consumed Energy (LCE) was considered. We do not consider the cost of the energy generating because, in the remote village, most of the energy generated was lost. For example, if the PV generator or wind turbine generator produces energy during an hour when the electrical load is zero and the batteries are fully charged, then the energy produced was lost. In addition to that, the energy is also lost in the charge and discharge processes of the batteries.

OBJECTIVE FUNCTIONS

The objectives function to be minimized is the LCE and the pollutant emission (kg of CO_2 which is the main cause of the

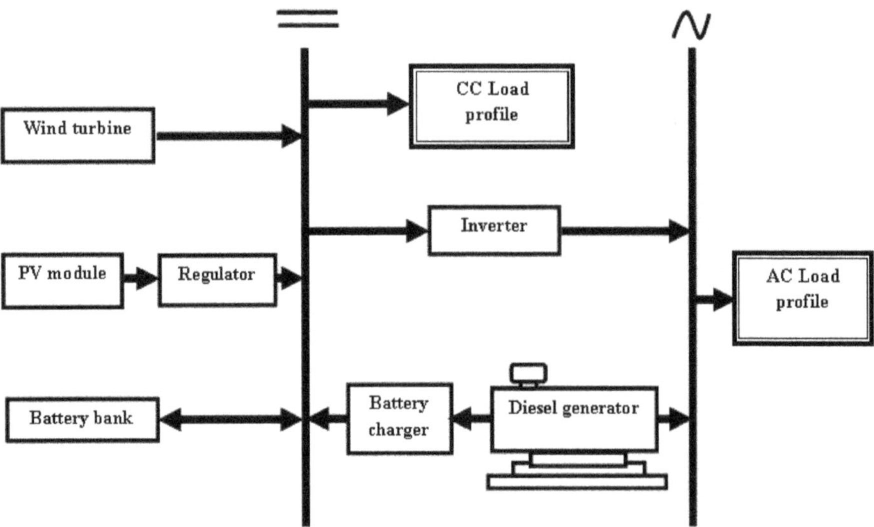

Figure 1. Bloc diagram of the hybrid solar/wind/diesel/battery system.

greenhouse effect).

Economic model based on LCE concept

The optimal combination of a hybrid solar-wind-diesel-batteries system can make the best compromise between the system pollutant emission and the total cost of energy. The economical approach, according to the concept of LCE is developed to be the best benchmark of system cost analysis in this study. According to the studied system, the LCE is composed of the capital levelized cost C_{acap}, maintenance and operation levelized cost C_{amain} and the replacement levelized cost C_{arep}.

Six devices of the hybrid system were considered: PV module, wind turbine, diesel generator, battery, regulator and inverter. The levelized cost of the kWh/year is defined as in Equation (1).

$$ACE = \frac{J(x)}{E_{annual}} \quad (1)$$

J(x) is the levelized cost of energy given by Equation (2).

$$J(x) = C_{acap}(x) + C_{amain}(x) + C_{arep}(x) \quad (2)$$

Where E_{annual} is the annual consumed energy (kWh/year), $x=[N_{pv},N_{ag},N_{dg},N_{rg},N_{bt}, N_{inv,pv},T_{pv}, T_{ag},T_{dg},T_{bt},T_{rg},T_{in}]$ is the decision vector of variables. Where T_{pv}, T_{ag}, T_{dg}, T_{bt}, T_{rg}, T_{in} are the types of PV module, wind turbine, diesel generators, batteries, solar regulators and inverters. C_{acap}, C_{amain} and C_{arap} are the levelized capital cost, levelized maintenance cost and the levelized replacement cost.

Pollutant emissions

The parameter considered in this paper to measure the pollutant emission is the (kg of CO_2). It represents the large percentage of the emission of fuel combustion (Sonntag et al., 2002). Further, CO_2 represents the main cause of the greenhouse effect. So, we evaluate the amount of the CO_2 produced by the diesel generator in the PV/wind/ diesel/battery system. The fuel consumption of the diesel generator depends on the output power. It is given by Equation (3).

$$Cons = B \cdot P_{NG} + A \cdot P_{OG} \quad (3)$$

A=0.246 l/kWh and B=0.08145 l/kWh are the coefficient of the consumption curve, defined by the user in (l/kWh) (Belfkira et al., 2011). The factor considered in this work to assess the emission of CO_2 was 3.15 $kgCO_2/l$ (Fleck and Huot, 2009).

System optimization model using multi-objectives genetic algorithm

The main objective of this work is to design and optimize hybrid PV/wind/diesel/battery systems by minimizing the LCE and the CO_2 emission. These objectives are antagonist e.g. the increase of the LCE implies the decrease of the CO_2 emission and vice versa. So, it is important to find an efficient way to solve this Multi-Objective Problem (MOP) which parameters are also independent. The Multi-Objective Genetic Algorithm, which has the important characteristics of the concept of optimal Pareto front (Coello et al., 2002) can be used to solve this problem. A Pareto front is a set of a possible solution obtained after a search process.

The Multi-Objective function used in this study was implemented by employing genetic algorithm (GA) developed by Leyland and Molyneaux (Leyland, 2002; Molyneaux, 2002). This tool was designed for the optimization of the engineering energy systems. That is generally non-linear and uses a statistical technique of grouping of the individual basis on the independent variable (creation of the families which evolves in independent manner). This method has the advantage of maintaining the diversity of the population and making coverage the algorithm towards optima even difficult to find (Sambou, 2008).

Multi-objective solution strategy

Multi-objective problem (MOP) refers to the simultaneous optimization of multiple conflicting objectives, which produces a set of solutions instead of one particular solution while some constraints should be met. In fact, most of time, we find a set of solutions, owning to the contradictory objectives. MOP can be formulated (Mohammad et al., 2013 ; Azizipanah-Abarghooee et al., 2012; Anvari Moghaddam et al., 2011) as:

$$\text{Minimize} \quad F(X) = \left(F_1(X),\ F_2(X), \ldots,\ F_{Nobj}(X)\right) \tag{4}$$

Subject to $u_i(X) < 0$, i=1, 2,..., H and $v_i(X) < 0$, i=1, 2,..., H
Where, Fi is the ith objective function, X is a determination vector that presents a solution, N_{obj} is the number of objectives. L and H are the number of the equality and the inequality constraints, respectively. In our cas $N_{obj} = 2$.

Operation system strategy

The PV generator and the wind turbine outputs are calculated according to the PV modules and the wind turbine system model by using the specifications of the PV module and the wind turbine as well as the solar radiation, the temperature and the wind speed data. The battery bank with the total nominal capacity φ_r is permitted to discharge up to a limit defined by the minimum state of charge. For a good knowledge of the real state of charge (SOC) of a battery, it is necessary to know the initial SOC, the charge or discharge time and the current. However, most storage systems are not ideal, losses occur during charging and discharging and also during storing periods (Hongxing et al., 2009). Taking these factors into account, the SOC of the battery at time t + 1 can be calculated by Equation (5).

$$SOC(t+1) = SOC(t)\cdot\left(1 - \frac{\sigma\cdot\Delta t}{24}\right) + \eta_{bt}\cdot\frac{P_{bt}(t)\cdot\Delta t}{U_{bt}\cdot\varphi_{bt}} \tag{5}$$

Where σ is the self-discharge rate which depends on the accumulated charge and the battery state of health (Guasch and Silvestre, 2003) and a proposed value of 0.2% per day is recommended, η_{bt} is the battery charging and discharging efficiency. It is difficult to measure separate charging and discharging efficiency, so manufacturers usually specify roundtrip efficiency. In this paper, the batteries charge efficiency is set equal to 80%, and the discharge efficiency is set equal to 100% (Duffos Lopes et al., 2005). U_{bt} (V) is the nominal batteries voltage, which is equal to the nominal system operating voltage and P_{bt} (t) is the power received by the battery from generators or requested by the demand. The minimum state of charge of the battery bank (SOC_{min}) can be expressed by Equation (6).

$$SOC_{min} = (1 - DOD)\cdot SOC_r \tag{6}$$

Were DOD (%) is the depth of discharge and SOC_r is the rated state of charge of the battery bank. In our case, the DOD assumed equal to 60%. So, the minimum state of charge (SOC_{min}) that the battery bank can achieve is of 40% SOC_r.
The input/output battery bank power can be computed according to the following strategy:

(i) If $PT(t) = \frac{P_{ch}(t)}{n_{ond}}$, all the produced energy is consumed by the demand. So $P_{bt}(t)=0$

(ii) If $PT(t) > \frac{P_{ch}(t)}{n_{ond}}$, then the surplus of power $P_{bt}(t) = PT(t) - \frac{P_{ch}(t)}{n_{ond}}$ is used to charge the battery. The new battery state of charge is then calculated by using the Equation (5)

(iii) if $PT(t) < \frac{P_{ch}(t)}{n_{ond}}$, then the lack of power $P_{bt}(t) = PT(t) - \frac{P_{ch}(t)}{n_{ond}}$ is provided by the battery. The new battery state of charge is calculated by using the Equation (5).

(iv) if $PT(t) < \frac{P_{ch}(t)}{n_{ond}}$ and the battery bank are depleted ($SOC=SOC_{min}$) or the energy providing from the battery bank is not enough to supply the load, then the diesel generators supply needed by the load and the surplus energy (if any), is used to charge the battery bank (Gupta et al., 2011). PT (t) is he total output power of the wind turbine and PV module generators (Ould Bilal et al., 2012a, b). The initial assumption of system configuration will be a subject to the following inequalities constraints:

$$\begin{cases} SOC_{min} \le SOC \le SOC_{max} = SOC_r \\ I_{max} \le I_{rrg} \\ P_{max} \le P_{rond} \end{cases} \tag{7}$$

Where: SOC_{min} and SOC_{max} are the minimum and the maximum of the state of charge of the battery bank, I_{max} is the maximum current delivered by the PV generated, I_{rrg} is the nominal current of the designed regulators (A), P_{max} is the maximum power of the demand and P_{rond} is the nominal power of the inverter (W).

Application on the site of Potou in Senegal

Presentation of the site

The methodology developed was applied using the solar radiation, the temperature and the wind speed collected for eight month on the site of Potou (16.27° of longitude West, 14.3° of latitude North and 21 m of altitude) located in the Northwestern coast of Senegal. This region is characterized by a wind potential adapted to small wind turbines (about 0. 2 to 10 kW) on the one side and on the other side, this region is characterized by a very sunny weather which can be used to produce energy with the use of PV module.

In this area an anemometer, a pyranometer and temperature sensor have been installed to collect the solar radiation, temperature and the wind speed. A data acquisition system was used to record the parameters every one second. Then, the data are averaged over 10 min intervals and was recorded in the memory of the datalogger. The used data were averaged over each one hour. Results presenting the real distribution and the theoretical distribution of Weibull of the mean wind speed are shown in Figure 2. Figure 3 shows the hourly radiation for a typical day on the site of Potou. The wind power density and the solar energy are 95 W/m² and 3.90 kWh/m²/d respectively.

Load profile

The daily load profile is represented by a sequence of powers and it is considered as constant over a time-step of 1 h. The used load profile (Figure 4) denotes the consumption of a typical isolated town which fluctuation during the day corresponds to the operation of public and domestic equipments (refrigerators, television, radio, domestic mill, welding machines, sewing machine and other equipment). The peak of demand observed at night corresponds to the use of the domestic equipment (refrigerators, lighting, television, radio) and some commercial equipment (refrigerator and lighting in

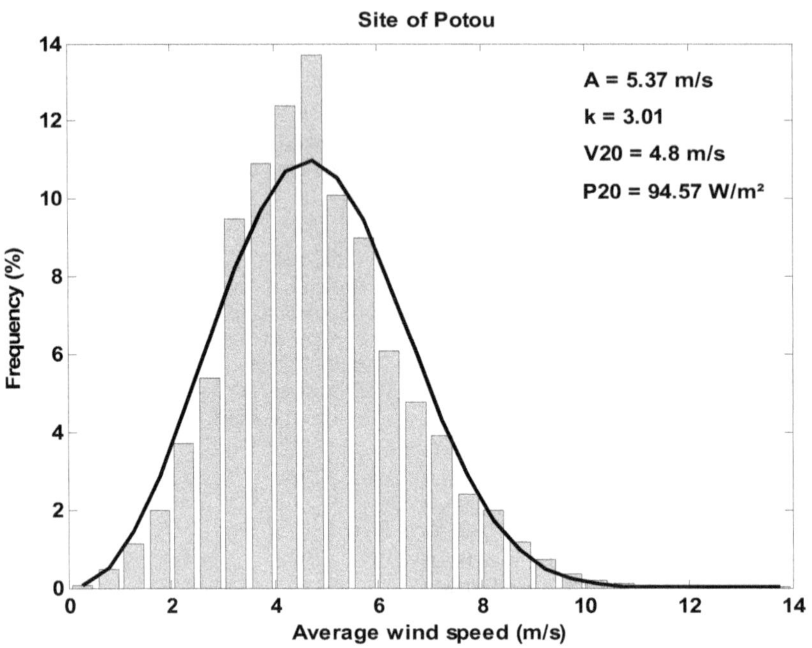

Figure 2. Real distribution and Weibull distribution on the site of Potou.

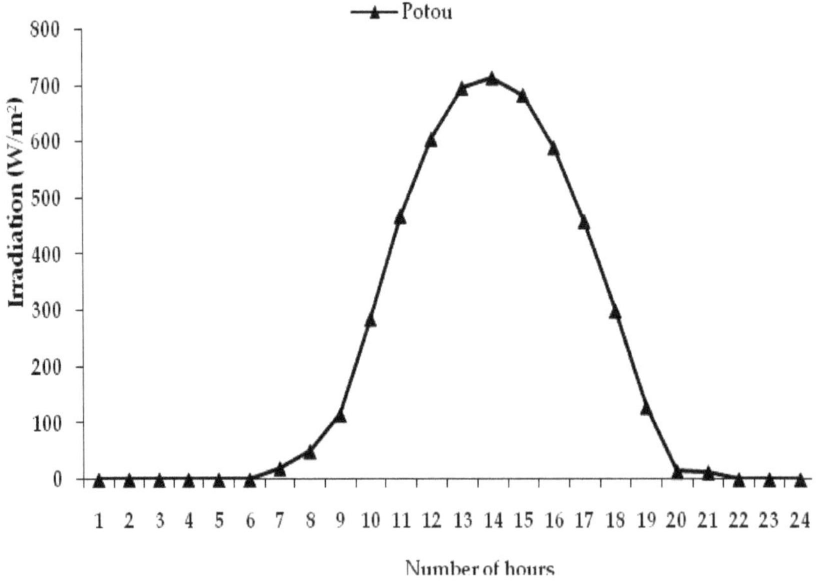

Figure 3. Solar Irradiation profile on the site of Potou.

the shops). The light and the television are the main element used overnight in the remote village. The total consumption energy of the used load profile is 94 kWh/d.

Components characteristics

The specifications of the components used to design and optimize hybrid PV/wind/diesel/battery are presented in Table 1. Five types of wind turbines and four types of: PV modules, batteries, regulators, inverters and diesel generators were used respectively.

RESULTS AND DISCUSSION

The optimization of PV/wind/diesel generator/battery hybrid systems by minimizing the LCE and the CO_2

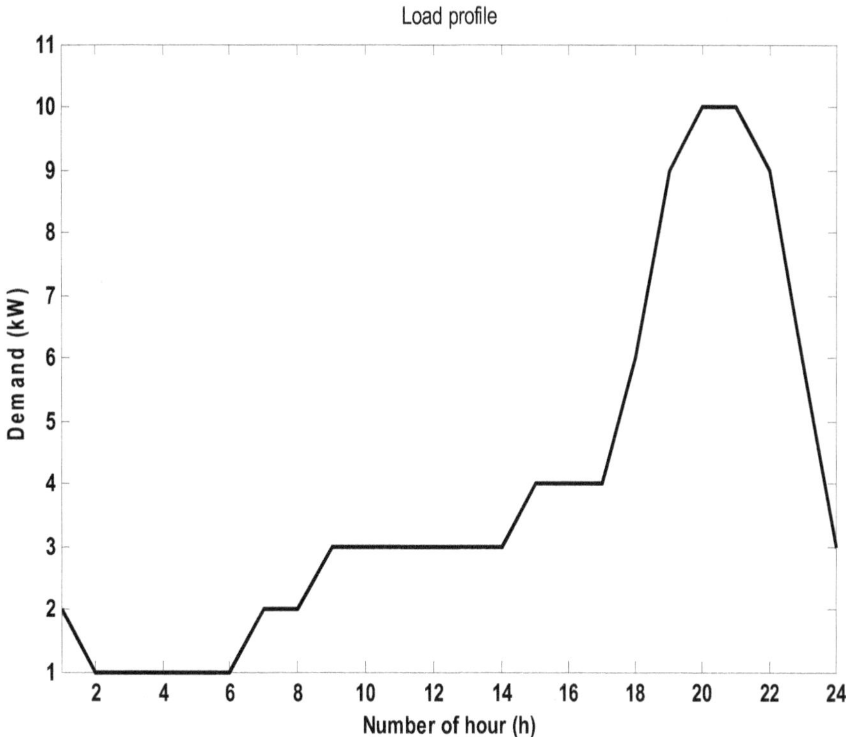

Figure 4. load profile of the demand energy.

emission were carried out by using multi-objectives genetic algorithm approach. Obtained results have appeared as an optimal Pareto front. Every solution of the best Pareto front was formed by a combination of hybrid systems and control strategy, with a different LCE and the pollutant emission.

Figure 5 shows the optimal Pareto front between the LCE and the CO_2 emission. It was noted that the increasing of LCE implies the decreasing of the CO_2 emission. To illustrate the results given by Figure 5, five solutions (A, B, C, D and E) on different position of the Pareto front curve were indicated.

Table 2 depicts the size and the output energy of these five optimal solutions. It can be noted that the optimal type of the PV module, wind generator, battery bank, regulator, inverter and diesel generator was respectively the N°4, N°5, N°3, N°1, N°3, N°1 and N°3. It is, also, possible to note that the LCE decreases by 27 and 40% while passing from the solution A to solutions B and C respectively. That is because of the diminution of the components numbers of the system, specially the PV modules and the wind turbines in the systems. In the contrast, the diesel generator was more solicited, thus, the operation hours of diesel generator increases. For example the operations hours of the diesel generators pass from 6 h to 31 and to 119 h when the solution passes to B and C from A. So, the CO_2 emission increases by 89.40 $kgCO_2$/year and by 750.19 $kgCO_2$/year respectively when the solutions pass to B and to C from A. From Table 2, it can be seen as the PV modules and wind turbines decrease, the size of diesel generator increases. It can be also noted that the output energy from the wind turbine was lower (with fraction of 46% for the solution-A, 11% for the solution-B and 0% for the solution-C, D and E) compared to the output energy from PV generator. That can be explained by the higher potential of the solar and its variations which are more suitable to the variation of the load profile. The output energy from the diesel generators was lower for the solution A, B and C compared to solutions D and E. The highest fraction of the output energy observed for these three solutions was 16% observed for the solution C. The corresponding hours of operation was 119 h. However, the output energy from diesel generators was higher for the latest solutions (D and E) which are solutions without renewable energies (fraction of renewable energy was 0%). While, the amount of the CO_2 emitted was higher (4,870.37 and 10,839.55 $kgCO_2$/year respectively).

It can be noted from Table 2 that the battery bank was more solicited for the solution C (PV/diesel system). Figure 6 (solution-C) shows the distribution of the state of charge of this solution. It can be seen that the battery bank discharges up to 60 for 35% of the time. The lowest SOC was 43 observed for the solution-B, but this state of charge was observed for only 0.06% of time. The lowest average state of charge (SOC) was 75% observed for the

Table 1. Specifications of the components.

Specifications of the wind turbine						
Type of wind turbines	Cut-in wind speed Vci (m/s)	Rate wind speed Vr (m/s)	Cut-off wind speed Vco (m/s)	Rated power Pr (W)	Output voltage (V)	Cost (€)
1	2	9	12	500	48	3051
2	3.5	11	13	600	48	1995
3	3.5	12	12	1500	48	2995
4	2.5	14	25	5600	48	8870

Specifications of the PV module						
Type of of PV module	Rate voltage (V)	Nominal peak power k (W)	Current of short-circuit ¶(A)	Voltage of open circuit (V)	Fill factor	Cost (Euro)
1	12	75	4.70	21.50	0.74	590
2	12	80	5.31	21.30	0.71	540
3	12	100	6.46	20.00	0.77	559
4	12	150	8.40	21.60	0.74	900

Specifications of the batteries			
Type of batteries	Nominal capacity (Ah)	Nominal voltage (V)	Cost (Euro)
1	80	12	195
2	100	12	215
3	200	12	416
4	720	2	2059

Specifications of the regulators			
Type of the regulators	Nominal current (A)	Nominal voltage (V)	Cost (Euro)
1	30	48	230
2	40	48	250
3	45	48	289
4	60	48	295

Specifications of the inverters			
Type of the inverters	Nominal power (W)	Nominal voltage (V)	Cost (Euro)
1	3500	48	2799
2	2400	48	2165
3	4500	48	4185
4	5000	48	5350

Specifications of the diesel generators		
Type of diesel generator	Nominal output power	Cost (Euro)
1	3050	668
2	4000	862
3	4600	879
4	4860	895

solution-C and the highest (89%) was observed for the solution A. Figure 6 (solution-A), shows that the battery bank remain at the SOC 100% for 42% of time. So, the battery bank is less solicited. That allows the increasing of the lifetime of batteries, thus the diminution of the replacement cost of batteries.

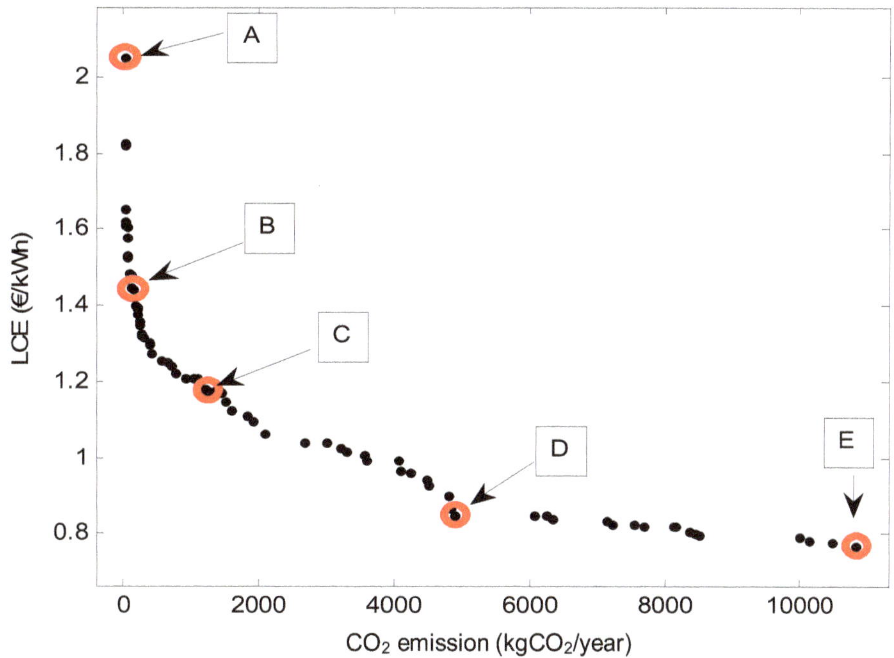

Figure 5. Optimal Pareto front of hybrid PV/wind/diesel/battery system

Table 2. Five solutions of the optimal pareto front.

Solution	Solution A	Solution B	Solution C	Solution D	Solution E
Number of PV modules	28	40	28	0	0
Number of Wind turbine	5	1	0	0	0
Number Batteries	60	60	64	48	24
Number of Regulators	2	3	2	0	0
Number of Inverters	5	5	5	5	5
Number Diesel generators	1	2	5	10	10
Type of Wind turbines	5	5	--	--	--
Type of PV modules	4	4	4	4	4
Type of Batteries	3	3	3	3	3
Type of Regulators	1	1	1	--	--
Type of Inverters	1	1	1	1	1
Type of diesel generators	3	3	2	3	3
Annual electrical energy delivered by PV generator (kWh/year)	26994.00	38563.00	26994.00	0.00	0.00
Annual electrical energy delivered by wind turbine (kWh/year)	23843.00	4768.60	0.00	0.00	0.00
Annual electrical energy delivered by diesel generator (kWh/year)	14.93	114.38	689.93	24102.70	25116.10
Annual operating hours of diesel (h)	6	31	119	4357	4589
Annual excess of energy (kWh/year)	28386.934	20980.38	5217.93	1636.70	2650.10
SCO_{min} (%)	56	43	55	61	60
Mean of SCO (%)	89	78	75	80	77
Annualized cost system of energy (€/kWh)	2.05	1.48	1.22	0.85	0.77
CO2 emission (kg CO2/year)	11.89	99.29	762.08	4870.37	10839.55

In order to highlight the hourly behavior of the obtained optimal configuration, the solution B (hybrid PV/wind/diesel/battery) has been used. A simulation was conducted on a period from 1st January to 19th February

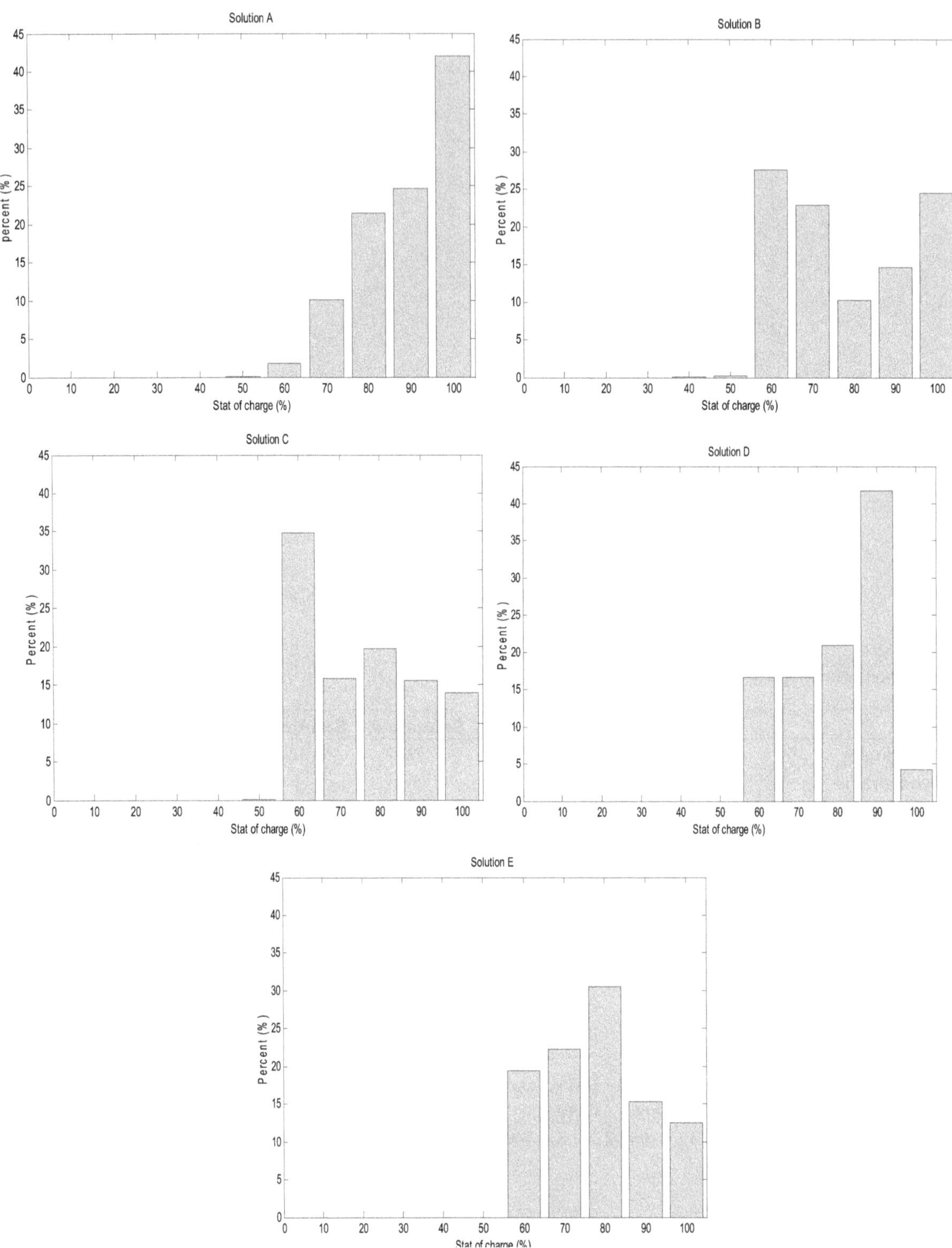

Figure 6. State of charge of the five solutions (A, B, C, D and E) of the Pareto front.

(12:00 hours) and is reported in Figure 7. Figure 7a, b, c and d show the output power from PV generator, wind turbine, the diesel generator and the output/input battery bank power. Figure 8 gives the state of charge of the battery bank for the indicated period.

According to the strategy denoted above and to the

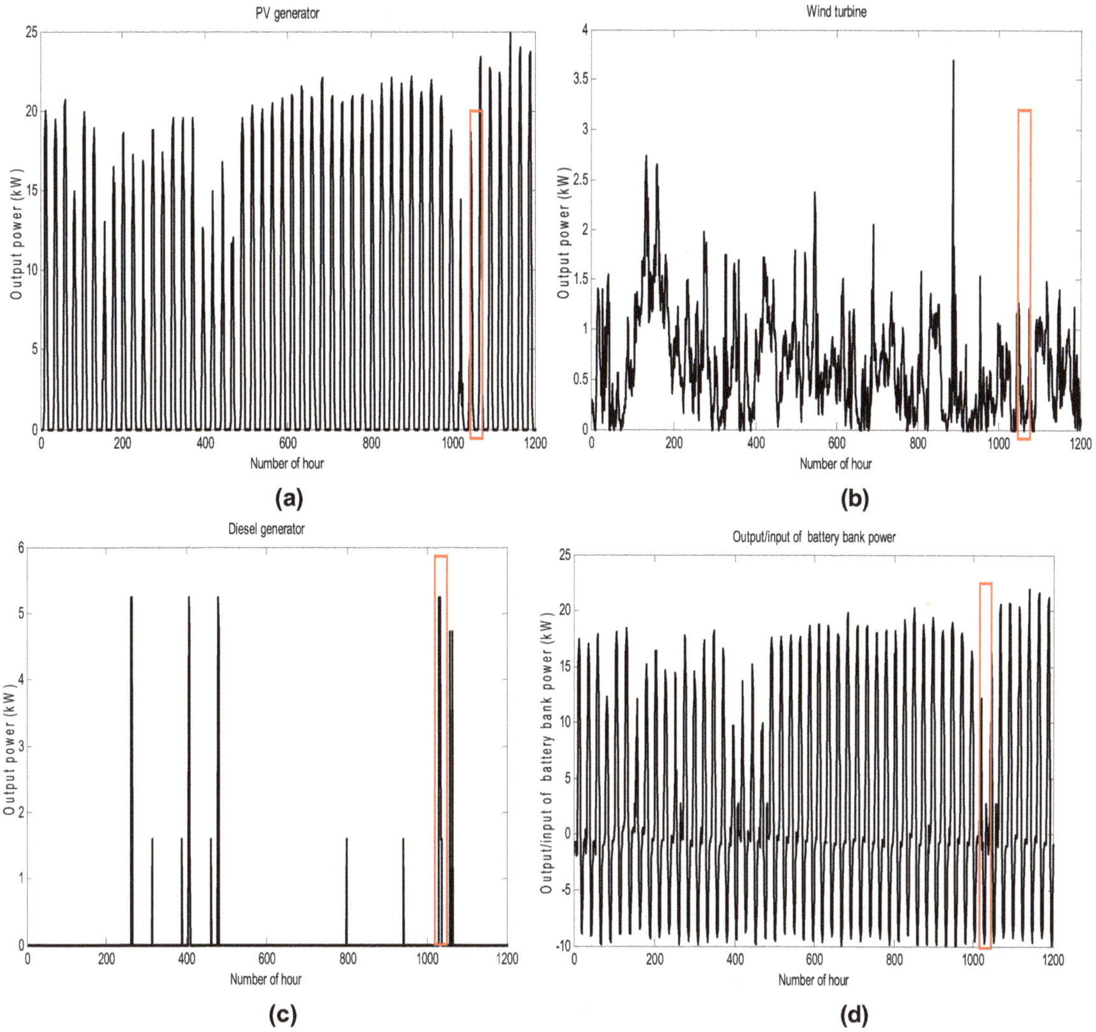

(a)

(b)

(c)

(d)

Figure 7. Behavior of the PV generator, Wind turbine, Diesel generator and the battery bank (solution B).

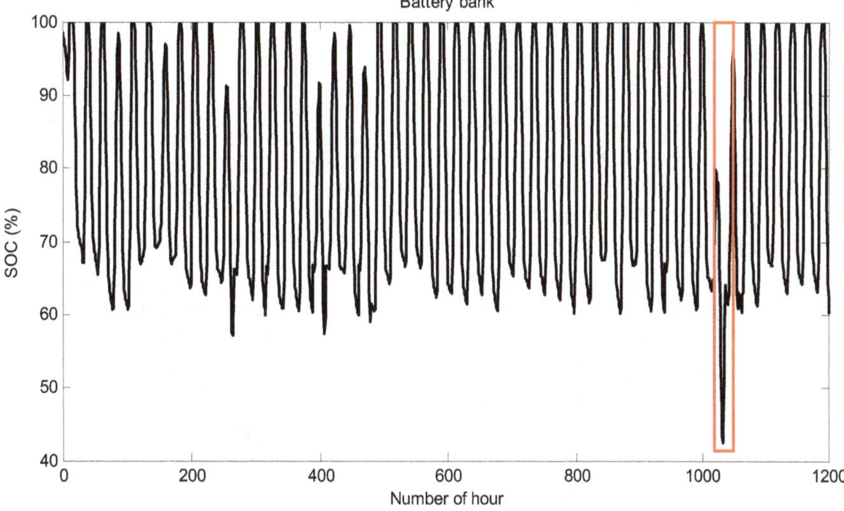

Figure 8. State of charge of the battery bank (solution B).

operation under the model constraints given by expression 7, it can be verified that, when the renewable sources power is greater than the power demand, the surplus power is stored in the battery bank, then P_{bt} >0 (Figure 7a, b and d). When the renewable energy is smaller, the lack of energy is provided by the battery bank and or by the diesel generator (Figure 7c), then the P_{bt} <0. Moreover, when the output energy from the PV generator and wind turbine is greater than the demand energy and the battery bank is fully charged then, the surplus of the energy produced can be used for the water pumping, water desalination, or to supply other demand of energy according to the needs of the village where the system is installed. From Figure 8, it can be seen that, the SOC remain between SOC_{max} (100%) and the SOC_{min} (40%). The minimum SOC achieved is 43% and observed for the 12th February at 23 h. In this hour of February day, it was noted 0 kW output energy from wind turbine and from PV generator. So the battery bank was deeply discharged and the diesel generator is operated to supply the load in the on hand and to charge the batteries in the other hand. The average of the SOC during this period of operations (1st January to 19th February) was 75% and the hour number of diesel generator time operation was 19 h.

Conclusion

The methodology for optimal sizing of multi-objective hybrid PV/wind/diesel/battery bank systems minimizing the LCE and the CO_2 emission by using a Genetic Algorithm approach was developed in this paper. The obtained results were depicted on the optimal Pareto front. From the results, we can outline the following points:

(i) The increasing of the LCE implies the decreasing of the CO_2 emission.
(ii) The LCE decreases by 27 and 40 % while passing from the solution A to the solutions B and C. In the contrast the CO_2 emission increases by 89.40 kgCO_2/year and 750.19 kgCO_2/year respectively when the solutions pass to B and C from A.
(iii) The PV generator was more solicited than the wind turbine generator for the hybrid PV/wind/diesel/batteries in the site of Potou. For the solutions, the highest fraction of the wind turbine was 46% observed for the solution A.
(iv) The only PV/battery or Wind/ battery were not an optimal configuration for this application on the site of Potou with the use of the load profile and the indicated specifications of the devices. It would be interested to perform modeling, incorporating the objectives of availability and reliability constraints of components to achieve a more accurate assessment of the cost of ownership system.

Conflict of Interest

The authors have not declared any conflict of interest.

ACKNOWLEDGEMENTS

Authors wish to thank the Programme pour la promotion des énergies renouvelables, de l'électrification rurale et de l'approvisionnement durable en combustibles domestiques (PERACOD) for the data provided in 2007-2008 and the African Union for the financial support.

REFERENCES

Ahmad R, Kazem M, Hossein K (2010). Modeling of a Hybrid Power System for Economic Analysis and Environmental Impact in HOMER. IEEE. 978-1-4244-6760-0.

Ahmed MAH, Priscilla NJ and Mohd S (2011). Optimal configuration assessment of renewable energy in Malaysia. Renewable Energy. 36:881-888.

Ajay KB, Gupta RA, Rajesh K (2011). Optimization of Hybrid PV/wind Energy System using Meta Particle Swarm Optimization (MPSO). IEEE. 978-1-4244-7882-8.

Alam HM, Manfred D (2010). Hybrid systems for decentralized power generation in Bangladesh. Energy for Sustainable Development. 14:48–55.

Anvari Moghaddam A, Seifi A, Niknam T, Alizadeh Pahlavani MR (2011). Multiobjective operation management of a renewable MG (micro-grid) with backup micro-turbine/fuel cell/battery hybrid power source. Energy 36:6490-6507.

Ayong H, Rudi K, Managam R, Yohannes MS, Junaidi M (2013). Techno-economic analysis of photovoltaic/wind hybrid system for onchore/ remote area in Indonesia. Energy 59:652-657.

Azizipanah-Abarghooee R, Niknam T, Roosta A, Malekpour AR, Zare M (2012). Probabilistic multiobjective wind-thermal economic emission dispatch based on point estimated method. Energy 37:322-335.

Bahtiyar D (2012). Determination of the optimum hybrid renewable power generating systems for Kavakli campus of Kirklareli University, Turkey. Renewable Sustainable Energy Rev. 16:6183–6190.

Bekele G, Palm G (2010). Feasibility study for a standalone solar–wind-based hybrid energy system for application in Ethiopia. Appl. Energy. 87(2):487–495.

Belfkira R, Zhang L, Barakat G (2011). Optimal sizing study of hybrid wind/PV/diesel power generation unit. Solar Energy 85:100-110.

Belgin ET, Ali YT (2011). Economic analysis of standalone and grid connected hybrid energy systems. Renewable Energy 36:1931-1943.

Chong L, Xinfeng G, Yuan Z, Chang X, Yan R, Chenguang S, Chunxia Y (2013). Techno-economic feasibility study of autonomus hybrid wind-PV-battery power system for a household in Urumqi, China. Energy 55:263-272.

Coello CA, Veldhuizen DAV, Lamont GB (2002). Evolutionary algorithms for solving multi-objective problems. New York: Kluwer Academic/Plenum Publishers.

Colle S, Abreu SL, Ruther R (2004). Economic evaluation and optimisation of hybrid diesel/photovoltaic systems integrated to electricity grid. Solar Energy 76:295–299.

Dalton GJ, Lockington DA and Baldock TE (2009). Feasibility analysis of renewable energy supply options for a grid-connected large hotel. Renewable Energy 34:955–964.

Diaf S, Notton G, Belhamel M, Haddadi M, Louche A (2008). Design and techno-economical optimization for hybrid PV/wind system under various meteorological conditions. Appl. Energy 85(10):968–987.

Duffo-Lopes R, Bernal- Agustin JL (2008). Multio-bjective design of wind-diesel-hydrogen-battery systems. Renewable Energy 33:2559-2572.

Duffo-Lopes R, Bernal- Agustin JL (2005). Design and control strategies of PV-Diesel systems using genetic algorithm. Solar Energy 79:33-46.

Eyad SH (2009). Techno-economic analysis of autonomous hybrid photovoltaic-diesel-battery system. Energy for Sustainable Development. 13:143–150.

Ekren O, Ekren BY (2010). Size optimization of a PV/wind hybrid energy conversion system with battery storage using simulated annealing. Appl. Energy 87(2):592–598.

Ekren O, Ekren BY (2008). Size optimization of a PV/wind hybrid energy conversion system with battery storage using response surface methodology. Appl. Energy 85:1086-1101.

Fleck B, Huot M (2009). Comparative life-cycle assessment of a small wind turbine for residential off-grid use. Renewable Energy 34:2688-2696.

Guasch D, Silvestre S (2003). Dynamic battery model for photovoltaic applications. Prog Photovoltaics: Res. Appl. 11:193–206.

Gupta A, Saini RP, Sharma MP (2011). Modelling of hybrid energy system d Part I: Problem formulation and model development. Renewable Energy 36:459-465.

Hongxing Y, Wei Z, Chengshi L (2009). Optimale design and techno-economic analysis of a hybrid solar-wind power generation system. Appl. Energy 86:163-169.

Ismail MS, Moghavvemi M, Mahlia TMI (2014). Genetic algorithm based optimization on modeling and design of hybrid renewable energy systems. Energy Convers.Manage. 85:120–130.

Ismail MS, Moghavvemi M, Mahlia TMI (2013). Techno-economic analysis of an optimized photovoltaic and diesel generator hybrid power system for remote houses in a tropical climate. Energy Conversion and Management. 69:163–173.

Kanase-Patil AB, Saini RP, Sharma MP (2011). Sizing of integrated renewable energy system based on load profilesand reliability index for the state of Uttarakhand in India. Renewable Energy *36*:2809-2821.

Kalantar M, Mousavi GSM (2010). Dynamic behavior of a stand-alone hybrid power generation system of wind turbine, micro turbine, solar array and battery storage. Appl. Energy 87(10):3051–3064.

Kellogg WD, Nehrir NH, Venkataramanan G, Gerez V (1998). Generation unite sinzing and cost analysis for stand-alone wind, photovoltaïque and hybrid wind/PV systems. IEEE transaction on Energy Convers. 13(1):70-75.

Koutroulis E, Kolokotsa D, Potirakis A, Kalaitzakis K (2006). Methodology for optimal sizing of stand-alone photovoltaic/wind generator systems using genetic algorithms. Solar Energy 80:1072-1088.

Kyoung-Jun L, Dongsul S, Dong-Wook Y, Han-Kyu C, Hee-Je Kim (2013). Hybrid photovoltaic/diesel greenship operating in standalone and grid-connected mode – Expeimental Investigation. Energy. 49:475-483.

Leyland G (2002). Multi-objective optimization applied to industrial energy problems.188p. These EPFL, n°: 2572 Lausanne.

Mei SN, Chee WT (2012). Assessment of economic viability for PV/wind/diesel hybrid energy system in southern Peninsular Malaysia. Renewable Sustainable Energy Rev. 16:634–647.

Mir-Akbar H, Hugh C, Christopher S (2011). A feasibility study of hybrid wind power systems for remote communities. Energy Policy 39:877–886.

Mohammad RN, Ali AV, Rasoul AA, Mahshid J (2014). Enhanced gravitational search algorithm for multi-objective distribution feeder reconfiguration considering reliability, loss and operational cost. IET Generation, Transmission & Distribution. 1:55–69.

Mohammad RN, Rasoul AA, Behrouz ZMS, Kayvan G (2013). A novel approach to multi-objective optimal power flow by a new hybrid optimization algorithm considering generator constraints and multi-fuel type. Energy 49:119-136.

Molyneaux A (2002). A practical evolutionary method for the multi-objective optimization of complex integrated energy systems including vehicul drive-trains. These EPFL, n°: 2636, Lausanne. 194 pp.

Mukhtaruddin RNSR, Rahman HA, Hassan MY, Jamian JJ (2015). Optimal hybrid renewable energy design in autonomous system usingIterative-Pareto-Fuzzy technique. Electrical Power Energy Systems. 64:242–249.

Muyiwa SA, Samuel SP and Olanrewaju MO (2014). Assessment of decentralized hybrid PV solar-diesel power system for applications in Northern part of Nigeria. Energy Sustainable Dev. 19:72–82.

Ould Bilal B, Ndiaye PA, Kebe CMF, Sambou V, Ndongo M (2012a). Methodology to Size an Optimal Standalone Hybrid Solar-Wind-Battery System using Genetic Algorithm. Int. J. Phys. Sci. 7(18):2647-2655.

Ould Bilal B, Sambou V, Kebe CMF, Ndiaye PA and Ndongo M (2012b). Methodology to Size an Optimal Stand-Alone PV/wind/diesel/battery System Minimizing the Levelized cost of Energy and the CO2 Emissions. Energy Procedia 14:1636-1647.

Ould Bilal B, Sambou V, Ndiaye PA, Kebe CMF, Ndongo M (2010). Optimal design of a hybrid Solar-Wind-Battery System using the minimization of the annualized cost system and the minimization of the loss of power supply probability (LPSP). Renewable Energy 35:2388-2390.

Patrick MM, Sennoga T and Ines SM (2014). Analysis of the cost of reliabkle electricity: A new method for analyzing grid connected solar, diesel and hybrid distributed electricity systems considering an unreliable electric grid with exemples in Uganda. Energy 66:523-534.

Prabodh B, Vaishalee D (2012). hybrid renewable energy systems for power generation in stand-alone application: A review. Renewable Sustainable Energy Rev. 16:2926-2939.

Rohit S, Subhes CB (2014). Off-grid electricity generation with renewable energy technologies in India: An application of HOMER. Renewable Energ. 62:388-398.

Saheb-Koussa D, Haddadi M, Belhamel M (2009). Economic and technical study of a hybrid system (wind–photovoltaic–diesel) for rural electrification in Algeria. Appl. Energy 86:1024-1030.

Sambou V (2008). Transferts thermiques instationnaires: vres une optimisation de parois de bâtiments. Thèse, PHASE, Université Paul Sabatier. 195 pp.

Sanjoy K, Himangshu RG (2009). A wind–PV-battery hybrid power system at Sitakunda in Bangladesh. Energy Policy. 37:3659–3664.

Senjyu T, Hayashi D, Yona A, Urasaki N and Funabashi T (2007). Optimal configuration of power generating systems in island with renewable energy. Renewable Energy 32:1917-1933.

Shen WX (2009). Optimally sizing of solar array and batteriy in a standalone photovoltaic system in Malaysia. Renewable Energy 34:348-352.

Sonntag RE, Borgnakke C, Wylen GJV (2002). Fundamentals of thermodynamics, 6th ed. New York: Wiley.

Suresh-Kumar U, Manoharan PS (2014). Economic analysis of hybrid power systems (PV/diesel) in different climatic zones of Tamil Nadu. Energy Convers. Manage. 80:469–476.

Yang HX. Burnett L, Weather JL (2003). Data and probability analysis of hybrid photovoltaic–wind power generation systems in Hong Kong. Renewable Energy 28:1813–1824.

Yang HX, Lu L (2004). Study on typical meteorological years and their effect on building energy and renewable energy simulations. ASHRAE Transactions 110(2):424–431.

Yang HX, Lu L, Zhou WA (2007). Novel optimization sizing model for hybrid solar–wind power generation system. Solar Energy 81(1):76–84.

Yang HX, Zhou W, Lu L, Fang Z (2008). Optimal sizing method for stand-alone hybrid solar-wind system with LPSP technology by using genetic algorithm. Solar Energy 82:354-367.

Zeinab AME, Muhammad FMZ, Kamaruzzaman S, Abass AA (2012). Design and performance of photovoltaic power system as a renewable energy source for residential in Khartoum. Int. J. Phys. Sci. 7(25):4036-4042.

Zhou W, Lou C, Li Z, Lu L, Yang H (2010). Current status of research on optimum sizing of stand-alone hybrid solar–wind power generation systems. Appl. Energy 87:380–389.

Permissions

All chapters in this book were first published in IJPS, by Academic Journals; hereby published with permission under the Creative Commons Attribution License or equivalent. Every chapter published in this book has been scrutinized by our experts. Their significance has been extensively debated. The topics covered herein carry significant findings which will fuel the growth of the discipline. They may even be implemented as practical applications or may be referred to as a beginning point for another development.

The contributors of this book come from diverse backgrounds, making this book a truly international effort. This book will bring forth new frontiers with its revolutionizing research information and detailed analysis of the nascent developments around the world.

We would like to thank all the contributing authors for lending their expertise to make the book truly unique. They have played a crucial role in the development of this book. Without their invaluable contributions this book wouldn't have been possible. They have made vital efforts to compile up to date information on the varied aspects of this subject to make this book a valuable addition to the collection of many professionals and students.

This book was conceptualized with the vision of imparting up-to-date information and advanced data in this field. To ensure the same, a matchless editorial board was set up. Every individual on the board went through rigorous rounds of assessment to prove their worth. After which they invested a large part of their time researching and compiling the most relevant data for our readers.

The editorial board has been involved in producing this book since its inception. They have spent rigorous hours researching and exploring the diverse topics which have resulted in the successful publishing of this book. They have passed on their knowledge of decades through this book. To expedite this challenging task, the publisher supported the team at every step. A small team of assistant editors was also appointed to further simplify the editing procedure and attain best results for the readers.

Apart from the editorial board, the designing team has also invested a significant amount of their time in understanding the subject and creating the most relevant covers. They scrutinized every image to scout for the most suitable representation of the subject and create an appropriate cover for the book.

The publishing team has been an ardent support to the editorial, designing and production team. Their endless efforts to recruit the best for this project, has resulted in the accomplishment of this book. They are a veteran in the field of academics and their pool of knowledge is as vast as their experience in printing. Their expertise and guidance has proved useful at every step. Their uncompromising quality standards have made this book an exceptional effort. Their encouragement from time to time has been an inspiration for everyone.

The publisher and the editorial board hope that this book will prove to be a valuable piece of knowledge for researchers, students, practitioners and scholars across the globe.

List of Contributors

V. A. Ezekoye
Department of Physics and Astronomy, University of Nigeria, Nsukka, Nigeria

Blanca E. Carvajal-Gámez
National Polytechnic Institute, Professional Unit Engineering and Advanced Technologies, Av. IPN 2580, Electronica, Barrio La Laguna Ticoman, 07740, Mexico D. F., Mexico

Francisco J. Gallegos-Funes
National Polytechnic Institute, Professional Unit Engineering and Advanced Technologies, Av. IPN 2580, Electronica, Barrio La Laguna Ticoman, 07740, Mexico D. F., Mexico

Alberto J. Rosales-Silva
National Polytechnic Institute, Professional Unit Engineering and Advanced Technologies, Av. IPN 2580, Electronica, Barrio La Laguna Ticoman, 07740, Mexico D. F., Mexico

José L. López-Bonilla
National Polytechnic Institute, Professional Unit Engineering and Advanced Technologies, Av. IPN 2580, Electronica, Barrio La Laguna Ticoman, 07740, Mexico D. F., Mexico

A. C. Ofomatah
National Centre for Energy Research and Development, University of Nigeria, Nsukka, Enugu State, Nigeria

C. O. B. Okoye
Department of Pure and Industrial Chemistry, University of Nigeria, Nsukka. Enugu State, Nigeria

F. C. Odo
National Centre for Energy Research and Development, University of Nigeria, Nsukka, Nigeria Department of Physics and Astronomy, University of Nigeria, Nsukka, Nigeria

D. O. Ugbor
Department of Physics and Astronomy, University of Nigeria, Nsukka, Nigeria

P. E. Ugwuoke
National Centre for Energy Research and Development, University of Nigeria, Nsukka, Nigeria

Z. Ahmad
Faculty of Mechanical Engineering, Universiti Teknologi Malaysia, 81300, Johor Bahru, Malaysia

J. Campbell
School of Engineering, Cranfield University, Cranfield, Bedfordshire MK43 0AL, United Kingdom

G. Ramesh
Department of Ceramic Technology, A.C Tech. Campus, Anna University, Chennai, India

R. V. Mangalaraja
Department of Materials Engineering, University of Concepcion, Concepcion, Chile

S. Ananthakumar
Materials and Minerals Division, National Institute for Interdisciplinary Science and Technology (NIIST), Trivandrum, India

P. Manohar
Department of Ceramic Technology, A.C Tech. Campus, Anna University, Chennai, India

D. Sardari
Faculty of Engineering, Science and Research Branch, Islamic Azad University, P. O. Box 14515-775, Tehran, Iran

M. Kurudirek
Department of Physics, Faculty of Science, Ataturk University, 25240, Erzurum, Turkey

Tamrat Aragaw
Department of Biology, College of Natural and Computational Sciences, Haramaya University, P. O. Box 138, Haramaya, Ethiopia

Mebeaselassie Andargie
Department of Biology, College of Natural and Computational Sciences, Haramaya University, P. O. Box 138, Haramaya, Ethiopia

Amare Gessesse
Department of Biology, College of Natural and Computational Sciences, Addis Ababa University, P. O. Box 1176, Addis Ababa, Ethiopia

A. P. Aizebeokhai
Department of Physics, College of Science and Technology, Covenant University, P. M. B. 1023, Ota, Ogun State, Nigeria

O. A. Oyebanjo
Department of Physics, College of Science and Technology, Covenant University, P. M. B. 1023, Ota, Ogun State, Nigeria

Department of Physics/Telecommunication, Tai Solarin University of Education, Ijagun, Ogun State, Nigeria

Hourieh Abna
Department of Architecture, University of Guilan, Rasht, Iran

Mohammad Iranmanesh
Department of Architecture, University of ShahidBahonar, Kerman, Iran

Iman Khajehrezaei
Department of Architecture, University of Guilan, Rasht, Iran

Mohammad Aghaei
Department of Mechanical Engineering, Azad University, Sirjan, Iran

Salahaddin A. Ahmed
Department of Physics, Faculty of Science and Science Education, School of Science, University of Sulaimani, Iraq

Garba S. Adamu
Department of Mathematics, Waziri Umaru Federal Polytechnic, Birnin Kebbi, Nigeria

A. Danbaba
Department of Statistics, Usmanu Danfodiyo University, Sokoto, Nigeria

Mohand KESSAL
Laboratoire Génie Physique des Hydrocarbures, Faculté des hydrocarbures et de la chimie, Université M'Hamed Bougara de Boumerdés-35000- Algérie

Rachid BOUCETTA
Laboratoire Génie Physique des Hydrocarbures, Faculté des Sciences, Université M'Hamed Bougara de Boumerdés-35000-Algérie

Mohammed ZAMOUM
Laboratoire Génie Physique des Hydrocarbures, Faculté des Sciences, Université M'Hamed Bougara de Boumerdés-35000-Algérie

Mourad TIKOBAINI
Laboratoire Génie Physique des Hydrocarbures, Faculté des hydrocarbures et de la chimie, Université M'Hamed Bougara de Boumerdés-35000- Algérie

E. L. Efurumibe
Physics Department, College of Natural and Physical Sciences, Michael Okpara University of Agriculture, Umudike, Abia State, Nigeria

A. D. Asiegbu
Physics Department, College of Natural and Physical Sciences, Michael Okpara University of Agriculture, Umudike, Abia State, Nigeria

A. S. Shalaby
Applied Physics Department, Faculty of Applied Science, Taibah University, Saudi Arabia
Physics Department, Faculty of Science, Beni Suef University, Egypt

E. C. Okoroigwe
Department of Mechanical Engineering, University of Nigeria, Nsukka, Nigeria
National Centre for Energy Research and Development, University of Nigeria, Nsukka, Nigeria

M. N Eke
Department of Mechanical Engineering, University of Nigeria, Nsukka, Nigeria

H. U. Ugwu
Department of Mechanical Engineering, Michael Okpara University of Agriculture, Umudike, Abia State, Nigeria

Mansour Hosseini-Firouz
Department of Engineering, Ardabil Branch, Islamic Azad University, Ardabil, Iran

Noradin Ghadimi
Department of Engineering, Ardabil Branch, Islamic Azad University, Ardabil, Iran

F. E. Ogieva
Department of Electrical Electronic Engineering, Faculty of Engineering, University of Benin, Benin City, Edo State, Nigeria

A. S. Ike
Department of Electrical Electronic Engineering, Faculty of Engineering, University of Benin, Benin City, Edo State, Nigeria

C. A. Anyaeji
Department of Electrical Electronic Engineering, Faculty of Engineering, University of Benin, Benin City, Edo State, Nigeria

B. Ould Bilal
Ecole des Mines de Mauritanie (EMiM), BP : 5259 Nouakchott-Mauritanie, Sénéga

D. Nourou
Laboratoire de Recherche Appliquée aux Energies Renouvelables de l'eau et du froid (LRAER), FST-Université des Sciences de Technologies et de Médecines (USMT), BP: 5026 Nouakchott-Mauritanie, Sénégal

C. M. F Kébé
Centre International de Formation et de Recherche en Energie Solaire (C.I.F.R.E.S), ESP-UCAD, BP: 5085 Dakar Fann Sénégal

V. Sambou
Centre International de Formation et de Recherche en Energie Solaire (C.I.F.R.E.S), ESP-UCAD, BP: 5085 Dakar Fann Sénégal

P. A. Ndiaye
Centre International de Formation et de Recherche en Energie Solaire (C.I.F.R.E.S), ESP-UCAD, BP: 5085 Dakar Fann Sénégal

M. Ndongo
Laboratoire de Recherche Appliquée aux Energies Renouvelables de l'eau et du froid (LRAER), FST-Université des Sciences de Technologies et de Médecines (USMT), BP: 5026 Nouakchott-Mauritanie, Sénéga

www.ingramcontent.com/pod-product-compliance
Lightning Source LLC
Chambersburg PA
CBHW050455200326
41458CB00014B/5193
* 9 7 8 1 6 8 2 8 6 0 2 7 4 *